An Introduction to Solid Mechanics

S. K. Roy Chowdhury

Prasanta Sahoo

CAMBRIDGE
UNIVERSITY PRESS

CAMBRIDGE
UNIVERSITY PRESS

Shaftesbury Road, Cambridge CB2 8EA, United Kingdom

One Liberty Plaza, 20th Floor, New York, NY 10006, USA

477 Williamstown Road, Port Melbourne, VIC 3207, Australia

314–321, 3rd Floor, Plot 3, Splendor Forum, Jasola District Centre, New Delhi – 110025, India

103 Penang Road, #05–06/07, Visioncrest Commercial, Singapore 238467

Cambridge University Press is part of Cambridge University Press & Assessment, a department of the University of Cambridge.

We share the University's mission to contribute to society through the pursuit of education, learning and research at the highest international levels of excellence.

www.cambridge.org
Information on this title: www.cambridge.org/9781009479202

First published 2024

Printed in India by Shree Maitrey Printech Pvt. Ltd., Noida

A catalogue record for this publication is available from the British Library

ISBN 978-1-009-47920-2 Paperback

To my wife, Swagata.
—S. K. Roy Chowdhury

To my wife, Sarmila, and son, Arja.
—Prasanta Sahoo

An Introduction to Solid Mechanics

Solid mechanics, compared to mechanics of materials or strength of materials, is generally considered to be a higher-level course, usually taught to senior undergraduate students. This book is primarily aimed at this group of students and intended to present the basics of solid mechanics in a simple and concise manner. Following the AICTE model curriculum, it attempts to bridge the gap between complex formulations in the theory of elasticity and elementary strength of materials in a simplified manner. Each chapter begins with the learning objectives and ends with a pool of solved problems on each topic to illustrate the text material. Because of the simpler approach adopted in solving difficult problems, the book will be useful for all student groups who wish to learn the basics of solid mechanics without much difficulty.

S. K. Roy Chowdhury received BTech (Hons) in Mechanical Engineering from IIT Kharagpur, MSc in Mechanical Engineering from the University of Leeds, UK, and PhD from the University of Birmingham, UK. He conducted post-doctoral research at Lancaster University and the Universities of Birmingham and Leeds for more than six years. A former Professor of Mechanical Engineering at IIT Kharagpur, he has more than 45 years of teaching and research experience in India and the UK and is a renowned expert in the field of applied mechanics and tribology. He contributed to research in tribology and allied fields for five decades with numerous publications in international journals and conference proceedings. His present research interest includes triboemission of infrared and X-rays. He has supervised many PhD theses and also coordinated and participated in several funded research projects.

Prasanta Sahoo received Bachelor of Mechanical Engineering (Hons) from Jadavpur University and MTech and PhD from IIT Kharagpur. He is a Professor of Mechanical Engineering at Jadavpur University. He has more than 30 years of teaching and research experience in machine design and tribology and has supervised over 30 PhD theses so far. He has authored two books, edited three volumes, written over 30 book chapters, and published more than 500 articles in peer-reviewed journals and conference proceedings. He has been ranked among the top 2% global scientists in Stanford/Elsevier listing and among the top 1% global peer reviewers in Publons/Web of Science. He is a recipient of the prestigious INSA Teacher Award, 2021.

Contents

Preface

Solid mechanics, compared to mechanics of materials or strength of materials, is generally considered to be a higher level course. It is usually offered in higher semester to senior students. There are many textbooks available on solid mechanics, but they generally include a large part of theory of elasticity with in depth mathematical formulations. The usual prerequisites are one or two semester course on elementary strength of materials and a thorough mathematical background, including scalar, vector, and tensor field theory and cartesian and curvilinear index notation. The difference in levels between these books and elementary texts on strength of materials is generally formidable. However, in our experience of teaching this course for many years at premier institutes like IIT Kharagpur and Jadavpur University, despite its complexity, senior students generally cope well with the course using the readily available textbooks.

However, there is a vast student population pursuing mechanical, civil, or allied engineering disciplines across the country in colleges where AICTE curriculum is followed. Through several years of interaction with this group of students, we have found that there is no suitable textbook that suits their requirements. The book is primarily aimed at this group of students, attempting to bridge the gap between complex formulations in the theory of elasticity and elementary strength of materials in a simplified manner for better understanding. Index notations have been avoided, and the mathematical derivations are restricted to second-order differential equations, their solution methodologies, and only a few special functions, such as stress function and Laplacian operators.

The text follows more or less the AICTE guidelines and consists of twelve chapters. The first five chapters introduce the engineering aspects of solid mechanics and establish the basic theorems of elasticity, governing equations, and their solution methodologies. The next four chapters discuss thick cylinders, rotating disks, torsion of members with both circular and noncircular cross-sections, and stress concentration in some depth using the elasticity approaches. Thermoelasticity is an important issue in the design of high-speed machinery and many other engineering applications. This is dealt with in some detail in the tenth chapter. Problems on contact between curved bodies in two-dimensional and three-dimensional situations can be challenging, and they have wide applications in mechanical engineering such as in bearing and gear technology. Detailed discussions on this topic are provided in the eleventh chapter. In many engineering problems, the energy method offers a simpler solution, and this is discussed in the final chapter.

The book is intended to present the basics of solid mechanics in a simple and concise manner for beginners, and it includes a large number of solved problems on each topic to illustrate the text material. Because of the simpler approach adopted in solving complex problems, we believe that the book will be useful for all student groups who wish to learn basic solid mechanics without much difficulty.

We wish to thank all our former and present postgraduate and research students who have helped us at different stages in the preparation of the manuscript. Finally, we thank our publisher, Cambridge University Press, especially their editorial and production teams, for their care and attention in publishing the book.

S. K. Roy Chowdhury
Prasanta Sahoo

1

Introduction to Solid Mechanics

Learning Objectives

After careful study of this chapter, students should be able to do the following:

LO1: Identify the difference between engineering mechanics and the theory of elasticity approach.

LO2: Explain yielding and brittle fracture.

LO3: Describe the stress–strain behavior of common engineering materials.

LO4: Compare hardness, ductility, malleability, toughness, and creep.

LO5: Explain different hardness measurement techniques.

1.1 INTRODUCTION [LO1]

Mechanics is one of the oldest physical sciences, dating back to the times of Aristotle and Archimedes. The subject deals with force, displacement, and motion. The concepts of mechanics have been used to solve many mechanical and structural engineering problems through the ages. Because of its intriguing nature, many great scientists including Sir Isaac Newton and Albert Einstein delved into it for solving intricate problems in their own fields.

Engineering mechanics and mechanics of materials developed over centuries with a few experiment-based postulates and assumptions, particularly to solve engineering problems in designing machines and structural parts. Problems are many and varied. However, in most cases, the requirement is to ensure sufficient strength, stiffness, and stability of the components, and eventually those of the whole machine or structure. In order to do this, we first analyze the forces and stresses at different points in a member, and then select materials of known strength and deformation behavior, to withstand the stress distribution with tolerable deformation and stability limits. The methodology has now developed to the extent of coding that takes into account the whole field stress, strain, deformation behaviors, and material characteristics to predict the probability of failure of a component at the weakest point. Inputs from the theory of elasticity and plasticity, mathematical and computational techniques, material science, and many other branches of science are needed to develop such sophisticated coding.

The theory of elasticity too developed but as an applied mathematics topic, and engineers took very little notice of it until recently, when critical analyses of components in high-speed machinery, vehicles, aerospace technology, and many other applications became necessary. The types of problems considered in both the elementary strength of material and the theory of elasticity are similar, but the approaches are different. The strength of the materials approach is generally simple. Here the emphasis is on finding practical solutions to a problem with simplifying assumptions. On the other hand, in the theory of elasticity, the approach is to obtain a general solution that often requires in-depth knowledge of mathematics and physics. *Solid Mechanics* is probably a middle ground between these two extreme approaches. Initially the theory of elasticity approach is adopted. It helps understand the problem to a greater depth. Then essential equations and procedures of their general solution are taken up methodically. With this clear understanding of physical phenomena, practical problems and their solutions are sought.

The book is intended to present initial learners with the basics of solid mechanics in a simple and concise manner. In order to understand the stress, strain, and displacement distributions in a solid medium subjected to external forces, with the usual assumptions of linear and small displacement theory, and to follow them up to establish mathematical models that allow solutions of elasticity problems, knowledge of several areas of elementary mathematics is needed. In Chapter 2, Cartesian tensors and matrix transformation, which are of particular interest in elasticity, are discussed. This allows derivation of small strain tensors and compatibility equations in a concise manner. Stress tensor and Cauchy relations are derived in Chapter 3. This leads to discussions on principal stresses and their directions. Equilibrium equations are important in solving elasticity equations. Symmetry equations are then discussed. In Chapter 4, linear elastic material behavior and the generalized Hooke's law are discussed. Boundary value problems, along with the concept of uniqueness and superposition, are also discussed in this chapter. Chapter 5 looks at two-dimensional (2D) problems in the form of plane stress and plane strain problems in Cartesian coordinates. Chapter 5 also introduces the governing equations of elasticity in cylindrical polar coordinates. Axi-symmetry problems are also discussed here. Chapters 6, 7, 8, and 9 discuss thick cylinder, rotating discs, torsion of members with both circular and noncircular cross-sections, and stress concentration in machine members, respectively. All these problems are considered in some depth with elasticity approach. Thermoelasticity is an important topic in mechanical, civil, aeronautical, and mining engineering. This is dealt with in some detail in Chapter 10. Problems on contact between solids even in 2D are challenging and this has wide applications in engineering such as in bearing and gear technology. This is discussed in Chapter 11. In many engineering problems, the energy method gives a simpler solution. This is discussed in Chapter 12.

1.2 MATERIAL PROPERTIES [LO2–5]

Solid mechanics essentially deals with deformable bodies. In fact all solids are deformable to different extents; for example, rubber is highly deformable while concrete or stone deforms negligibly. It is necessary here to discuss some of the important mechanical properties of common engineering materials such as elasticity, plasticity, creep, and fatigue. The most useful application of the study of solid mechanics is probably in the design of machine parts or structures, and there the failure modes and criteria are most important. All the failure criteria used in machine design are based on a single measured parameter, viz. the yield point of the materials. This is obtained in the tension test carried out in an apparatus commonly known as the universal testing machine (UTM). Failure in machine design is commonly classified as

(a) Yielding (b) Brittle fracture

When the stress developed within a machine part exceeds the yield strength of the material, the component may elongate to the extent that it is no longer useful for the purpose it is designed for. This applies to ductile materials, whereas brittle materials may fracture mostly with no warning beyond the yield strength. This shows that some knowledge of material properties is a prerequisite for a better understanding of solid mechanics. First, we consider some details of a tension test. The widely used UTM for these tests is generally hydraulically operated and essentially consists of a table, an upper crosshead and a movable crosshead. A tensile specimen is held between the upper and movable crossheads while a compression test specimen is held between the table and the movable crosshead with the help of jaws fitted to the crossheads. Load and displacement are measured by load and displacement transducers. Modern machines have sophisticated control and data acquisition system for precise control of loading and strain rate. The final output from the system is a load–displacement or more commonly a stress–strain diagram for a material. Standard specimens are used for uniformity and comparative study of material behavior.

A schematic diagram of a typical UTM is shown in Figure 1.1(a). Here a standard specimen (shown in Figure 1.1(b)) is held between two jaws in one fixed and other movable crossheads and the load is applied hydraulically. Here the stress is measured as force/area and the strain as change in length/original length over the gauge length of the specimen shown in Figure 1.1(b). The stress–strain diagram shown in Figure 1.1(c) is linear up to the elastic limit, and essentially it is similar to the load–displacement curve. A few important issues near the yield point must be noted. Although E is taken as the elastic limit, if load is withdrawn at point P the specimen will regain its original shape and size, indicating that point P is the proportional limit and this occurs at around 0.1% strain. Plastic yielding starts at E beyond which yield point Y is noted. This is an unstable region and in a carefully controlled experiment we may note one upper yield point Y_1 and a lower yield point Y_2. Many small peaks may also be found but they are of little significance. Beyond Y_2 the curve rises nonlinearly until it reaches a peak at U, the ultimate tensile stress. For an elastic-perfectly plastic material the curve is expected to be parallel to the strain axis beyond point Y_2. However, for some materials, rise in stress–strain curve is often attributed to local work-hardening. Beyond U specimen starts necking and it fractures soon after this. The fall in the curve is due to a decrease in apparent stress, calculated based on the initial cross-sectional area, during necking. The true stress–strain diagram is based on the instantaneous cross-sectional area and therefore it rises with a sharp bend at the point U. Many useful mechanical properties of engineering materials may be studied with careful experimentation in a UTM.

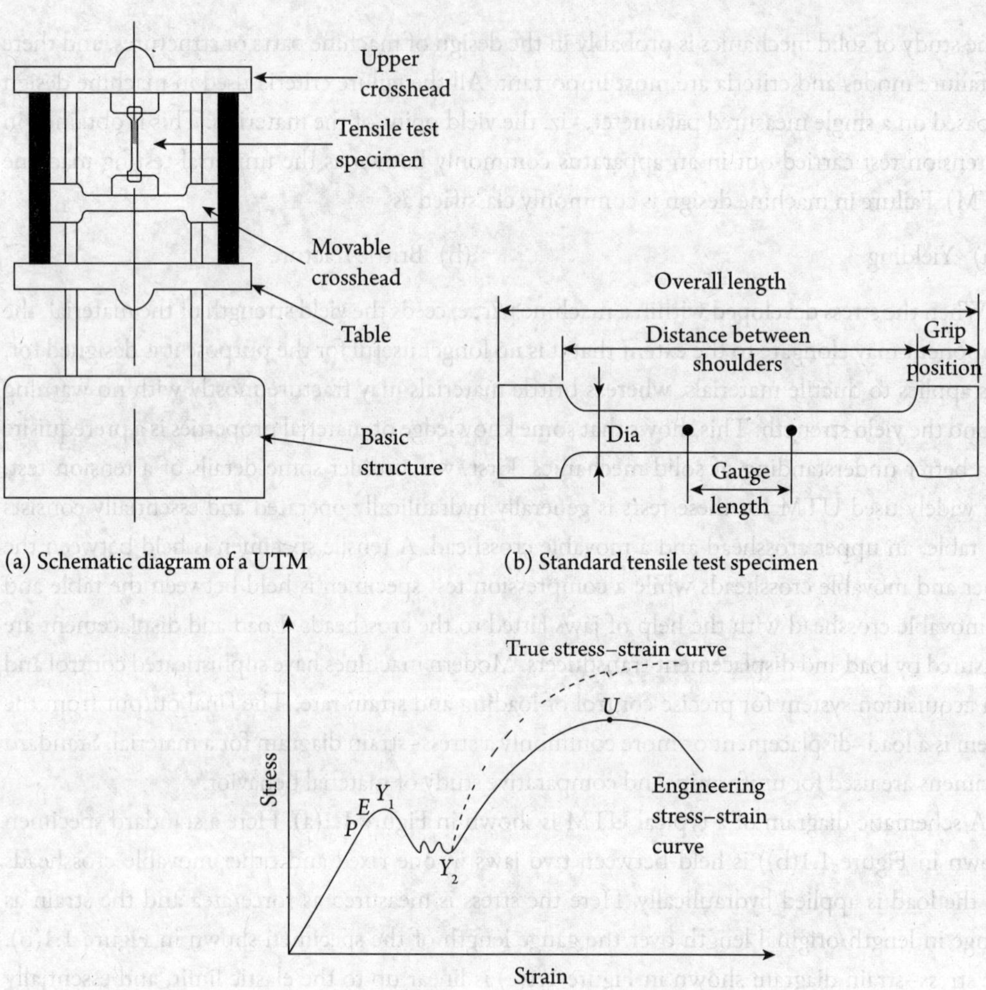

(a) Schematic diagram of a UTM (b) Standard tensile test specimen

(c) Typical stress–strain diagram for low carbon mild steels

Figure 1.1 Details of a tensile test

We shall now discuss other relevant mechanical properties. In the study of mechanics of deformable solids, elasticity and plasticity are the two important properties. The stress–strain diagram shown in Figure 1.1 is applicable for low-carbon mild steel only, which may be taken as a ductile material. The nature of the curve varies for different types of material, some of which are shown in Figure 1.2.

For many brittle and ductile materials, no knee in the stress–strain curve may be seen to denote the yield point. Only the ultimate tensile stress U is seen where the curve flattens and begins to fall. Just as linear elastic materials are given by $\sigma = E\varepsilon$ and nonlinear elastic materials are given by $\sigma = E\varepsilon^n$ where $n \neq 1$. For elastic-plastic materials, the stress–strain curve becomes parallel to the strain axis after an initial linear rise, indicating a plastic zone with increase in strain at a constant yield stress.

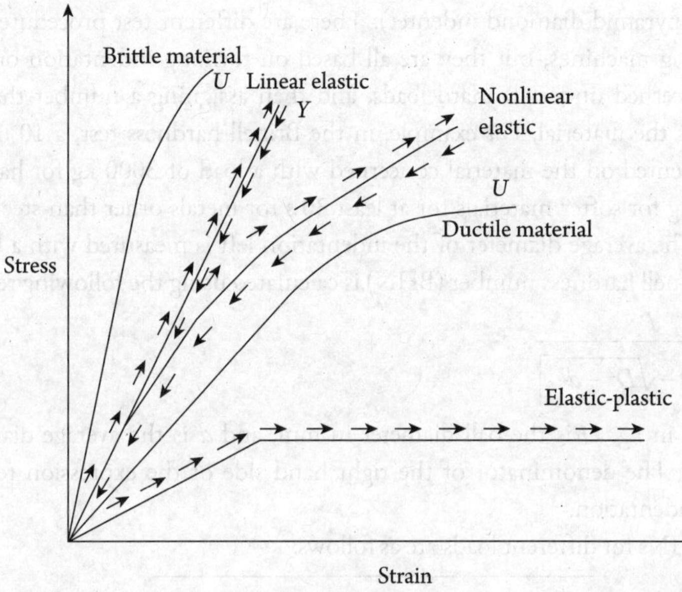

Figure 1.2 Typical stress–strain diagram of some common engineering materials

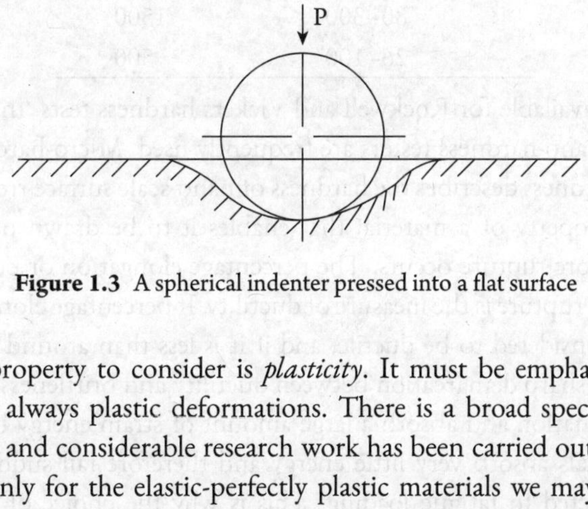

Figure 1.3 A spherical indenter pressed into a flat surface

Next important property to consider is *plasticity*. It must be emphasized that all inelastic deformations are not always plastic deformations. There is a broad spectrum of elastic-plastic deformation behavior and considerable research work has been carried out over the past century on this aspect. It is only for the elastic-perfectly plastic materials we may expect full plasticity beyond the yield point. A typical example of plastic flow is the indentation test, where a spherical ball or an indenter of any other standard shape is pressed on a semi-infinite solid and plastic flow in the solid occurs at the tip of the indenter as shown in Figure 1.3. Initiation of plasticity is often given by Mises–Hencky criterion as

$$(\sigma_1 - \sigma_2)^2 + (\sigma_2 - \sigma_3)^2 + (\sigma_1 - \sigma_3)^2 = 2\sigma_y^2, \tag{1.2.1}$$

where σ_1, σ_2, σ_3, and σ_y are the three principal stresses and the tensile yield stress, respectively.

Hardness is another mechanical property that is related to plasticity. This gives a measure of resistance to permanent indentation, deformation, or penetration. Different hardness-measuring machines are available, such as Brinell (steel ball of 10 mm diameter), Rockwell (cone shaped diamond), Vickers (pyramid diamond indenter). There are different test procedures for each of these hardness testing machines, but they are all based on forming indentation on the surface of the material concerned under standard loads, and then assigning a number that represents the flow pressure of the material. For example, in the Brinell hardness test, a 10 mm diameter hardened ball is indented on the material concerned with a load of 3000 kg for hard materials, or 1500 kg or 500 kg for softer materials for at least 30 s for metals other than steel, and 10–15 s for steel or iron. The average diameter of the indentation left is measured with a low-powered microscope. The Brinell hardness number (BHN) is calculated using the following relation.

$$BHN = \frac{P}{\frac{\pi D}{2}\left[D - \sqrt{D^2 - d^2}\right]}, \tag{1.2.2}$$

where P is the load in kg, D is the ball diameter in mm, and d is the average diameter of the indentation in mm. The denominator of the right hand side of the expression represents the surface area of the indentation.

Approximate BHNs for different loads are as follows:

Brinell Hardness Number	*Load (kg)*
160–600	3000
80–300	1500
26–100	500

Similar charts are available for Rockwell and Vickers hardness tests. In the present industrial scenario, micro- and nano-hardness testers are frequently used. Micro-hardness refers to the skin hardness and nano-hardness describes the hardness of nano-scale surface treatments.

Ductility is the property of a material that enables it to be drawn out or elongated to an appreciable extent before rupture occurs. The percentage elongation or percentage reduction in area that occurs before rupture is the measure of ductility. If percentage elongation exceeds around 15% the material is considered to be ductile, and if it is less than around 5% it is considered to be brittle. There is no sharp demarcation between ductility and brittleness. Ductile materials fail gradually by neck formation and absorb a large amount of strain energy before fracture occurs, whereas brittle materials absorb very little energy and therefore fail suddenly. Brittle materials fail faster when subjected to fatigue loading. This is why the choice of materials in design is important. Some examples of ductile materials are mild steel, aluminum, copper, and tin, etc., whereas cast iron, ceramics, glass, and concrete are typical examples of brittle materials.

Malleability is a special case of ductility where it can be rolled into thin sheets. Lead, soft steel, wrought iron, and copper are some typical malleable materials.

Other relevant properties of materials are resilience and toughness. *Resilience* is the property of the material that enables it to resist shock and impact by absorbing strain energy without permanent deformation. Resilience is measured in terms of strain energy (SE) per unit volume given by

$$SE = \frac{1}{2}\sigma\epsilon. \tag{1.2.3}$$

For linear load–deflection plots, the area under the curve represents the strain energy.

Toughness is the ability of a material to absorb energy during both elastic and inelastic deformation. The measure of toughness is the area within the stress–strain diagram including the inelastically deformed zone as shown in Figure 1.5. The difference between resilience and toughness may be noted by comparing Figure 1.4 and Figure 1.5.

Creep is a time-dependent permanent deformation under a constant stress level that may be below the yield stress. Creep is affected by temperature to a large extent. The temperature at which creep begins depends on the material and alloy composition along with the stress level it is subjected to. The end of the useful life of many high-temperature components in industry is usually limited by failure due to creep or sometimes called stress-rupture. If a component is subjected to a constant stress, there will be an immediate deformation or strain, and beyond this point if the stress continues to act for a longer time, strain gradually rises and this rise in strain is strongly dependent on temperature. Creep is a subject of long-standing research, but briefly, we can indicate its behavior as shown in Figure 1.6.

The initial strain due to a constant stress is given by OA, and beyond this, strain is both time- and temperature-dependent.

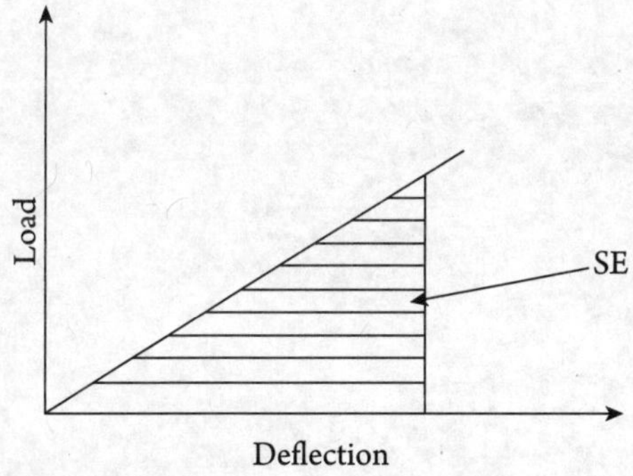

Figure 1.4 Linear load–deflection plot

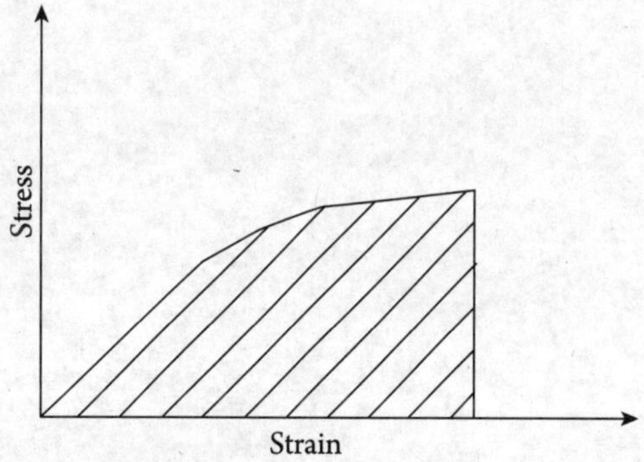

Figure 1.5 Stress–strain diagram to indicate toughness

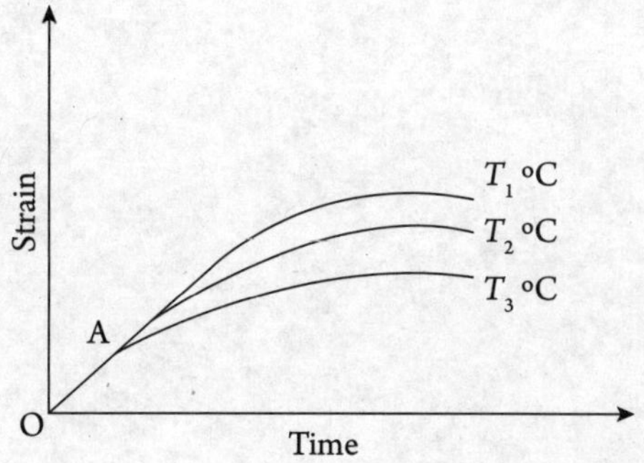

Figure 1.6 Temperature-dependent strain

Exercises

1. What is the scope of solid mechanics and how does it differ from the Theory of Elasticity and the Strength of Materials?
2. In the stress–strain diagram for low-carbon steels in a typical tensile test with a standard tensile test specimen why does the curve droop beyond the ultimate stress? Where does the typical engineering stress–strain curve differ from the true stress–strain curve for low-carbon mild steel?
3. How do you distinguish between yielding and brittle fracture in a machine part failure?
4. Write the stress–strain relation for nonlinear elastic materials.
5. How do you distinguish between ductility and brittleness in engineering materials?
6. How do you distinguish between resilience and toughness in engineering materials?

2

Principles of Strain Measures

Learning Objectives

After careful study of this chapter, students should be able to do the following:

LO1: Define scalar, vector, and tensor.

LO2: Describe strain tensor.

LO3: Describe normal and shear strain in an arbitrary direction.

LO4: Define principal strain and principal axes.

LO5: Describe strain invariants.

LO6: Recognize rotation.

LO7: State compatibility equations.

LO8: Understand the experimental method for strain measurement.

2.1 MATHEMATICAL PRELIMINARIES [LO1]

In any scientific or engineering field of study, knowledge of some mathematical techniques and methods are essential. Solid mechanics is no exception. To develop proper formulation methods and solution techniques for elasticity problems, it is necessary to have an appropriate mathematical background. In this chapter, we shall discuss Cartesian tensors, which have a special significance in the discussion of stress, strain, and displacement fields, and their manipulation. Other mathematical details will be discussed as and when they are required in solving different problems.

Tensors may be defined in a number of ways. One simple definition is that a tensor is a physical quantity that is governed by certain transformation laws when the coordinate system is changed. A tensor is invariant under any change of coordinate system, but its components along the coordinate axes change with the changed coordinate system. Tensors of order zero are called *scalars*. Common examples of scalars are temperature, density, Young's modulus, or Poisson's ratio. They have a single magnitude at each point in space, and they are invariant with coordinate transformations. A typical example of scalars is often taken as temperature T at a point in space with coordinates (x, y, z) represented as $T(x, y, z)$. Temperature at the same point does not change if we choose a different coordinate system (x', y', z') represented as $T'(x', y', z')$ and we may say

$$T = T'. \tag{2.1.1}$$

Tensors of first order are vectors, and we know that a vector has a magnitude and a direction. A typical example of a vector is a velocity vector $\boldsymbol{V}$. It is sometimes taken as a convention to represent a vector by a bold letter. Consider the velocity vector $\boldsymbol{V}$ in (x, y, z) coordinate system. In index notation, the coordinate axes may be represented as (x_1, x_2, x_3) and velocity components may be expressed as a column or a row matrix as $\begin{bmatrix} V_1 \\ V_2 \\ V_3 \end{bmatrix}$ or $[V_1, V_2, V_3]$.

In Figure 2.1 the vector $\boldsymbol{V}$ is shown with reference to two coordinate axes x_i and x_i' and velocity components are

$$\boldsymbol{V} = [V_1, V_2, V_3] \text{ or } \boldsymbol{V} = \begin{bmatrix} V_1 \\ V_2 \\ V_3 \end{bmatrix}, \tag{2.1.2}$$

If we now consider the transformation matrix, then the above equation may be written as

$$V_i' = \sum_{j=1}^{3} a_{ij} V_j, \text{ where } i = 1 \text{ to } 3. \tag{2.1.3}$$

Here a_{ij} are the elements of transformation matrix and the above equation may be written as, for example,

$$\begin{bmatrix} V_1' \\ V_2' \\ V_3' \end{bmatrix} = \begin{bmatrix} a_{11} & a_{12} & a_{13} \\ a_{21} & a_{22} & a_{23} \\ a_{31} & a_{32} & a_{33} \end{bmatrix} \begin{bmatrix} V_1 \\ V_2 \\ V_3 \end{bmatrix}. \tag{2.1.4}$$

In index notation, equation (2.1.4) may be written in a concise form as

$$V_i' = a_{ij} V_j. \tag{2.1.5}$$

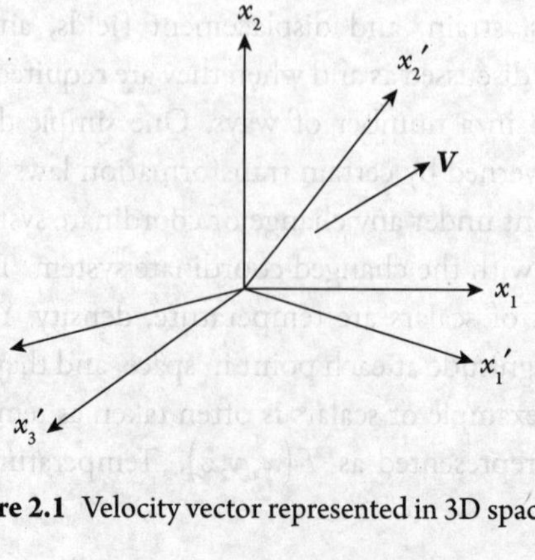

Figure 2.1 Velocity vector represented in 3D space

Here the index i that is not repeated is known as free index and the repeated index as dummy index.

A tensor of second order can be written using matrix notation and it is very important in the discussion of stress and strain. Since stress is a tensor of second order, we may represent this as σ_{ij} and it is convenient to write the transformation law as

$$\sigma'_{ij} = a_{ip} a_{jq} \sigma_{pq}. \tag{2.1.6}$$

Here σ'_{ij} are the components of the stress tensor with respect to the rotated coordinates x'_i, and σ_{pq} are the components of the stress tensor with respect to original coordinate system x_i, as shown in Figure 2.1. We may write a similar transformation law for the strain tensor. It may be noted that the product of two first order tensors is a second order tensor. Consider two first order tensors B_i and C_i in x_i coordinates. Their components in x'_i coordinates may be given in line with equation (2.1.5). We may therefore write $B'_k = a_{ki} B_i$ and $C'_l = a_{lj} C_j$ and the product of these two tensors may be given by $D_{ij} = B_i C_j$. Product $B_i C_j$ is known as outer product whereas $B_i C_i$ is the inner product of B_i and C_i. The transformation of D_{ij} may now be given by $D'_{kl} = a_{ki} a_{lj} D_{ij}$.

Similar transformation law applies for higher order tensors; for example, for a typical third-order tensor, the law would be

$$D'_{ijk} = a_{ip} a_{jq} a_{kr} D_{pqr}. \tag{2.1.7}$$

2.2 DISPLACEMENT AND STRAIN ANALYSIS [LO2]

An elastic body undergoes deformation when it is subjected to external forces. At any point P with coordinates (x, y, z) within the body, displacements in the x, y, and z directions are generally denoted by u, v, and w respectively (Figure 2.2). If the displacements vary from point to point, the body develops strain. There are normally two types of strains: normal strain denoted by $\in$ and shear strain denoted by γ.

Consider first a small element AB of length δx (Figure 2.3). If the element is displaced along x-coordinate to a new position $A'B'$ then $AA' = u$ and $BB' = u + \frac{\partial u}{\partial x}\delta x$. Normal strain in the x-direction, $\in_x$ may then be given as

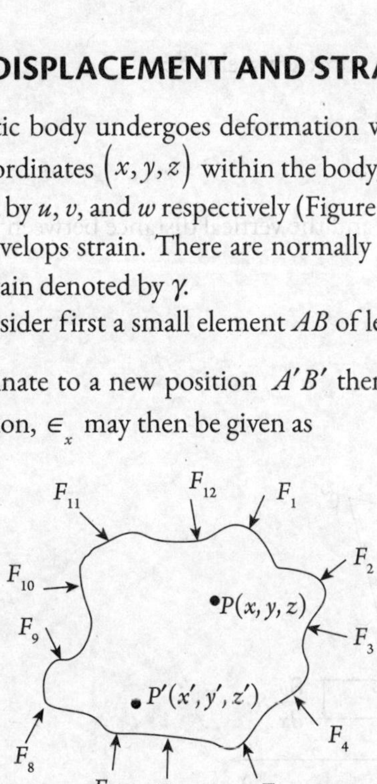

Figure 2.2 Varying displacements from point to point within an arbitrary body subjected to external forces

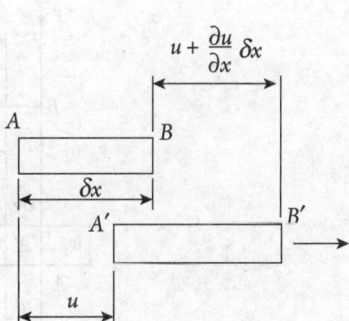

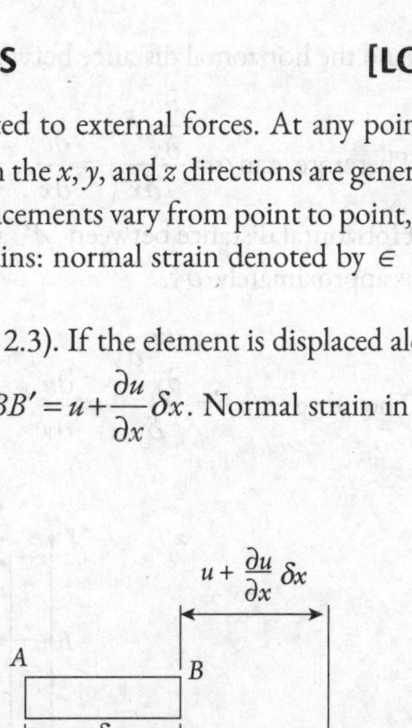

Figure 2.3 Displacement of a small element in the x-direction

$$\epsilon_x = \frac{u + \frac{\partial u}{\partial x}\delta x - u}{\delta x} = \frac{\partial u}{\partial x}.$$

In similar arguments, normal strains in the y and z directions may be given as

$$\epsilon_y = \frac{\partial v}{\partial y}$$

$$\epsilon_z = \frac{\partial w}{\partial z}.$$

The three normal strains are therefore

$$\epsilon_x = \frac{\partial u}{\partial x}, \ \epsilon_y = \frac{\partial v}{\partial y}, \text{ and } \epsilon_z = \frac{\partial w}{\partial z}. \tag{2.2.1}$$

To find shear strain, we consider elements in xy, yz, and xz planes in sequence. Consider first an element $ABCD$ in xy plane as shown in Figure 2.4. Let the element be displaced to $A'B'C'D'$. Point A is displaced to point A' with displacements u in the x direction, v in the y direction. It may be noted in Figure 2.4, the element $ABCD$ deforms as it displaces due to shearing. Line AD rotates by an angle α as it displaces to new position $A'D'$. Similarly, line AB rotates by an angle β as it displaces to new position $A'B'$.

Vertical distance between A' and D' is $\frac{\partial v}{\partial x}\delta x$.

And the horizontal distance between A' and D' is approximately δx.

Therefore, $\tan\alpha = \dfrac{\frac{\partial v}{\partial x}\delta x}{\delta x} = \dfrac{\partial v}{\partial x}$.

Horizontal distance between A' and B' is $\frac{\partial u}{\partial y}\delta y$ and the vertical distance between A' and B' is approximately δy.

Therefore, $\tan\beta = \dfrac{\frac{\partial u}{\partial y}\delta y}{\delta y} = \dfrac{\partial u}{\partial y}$.

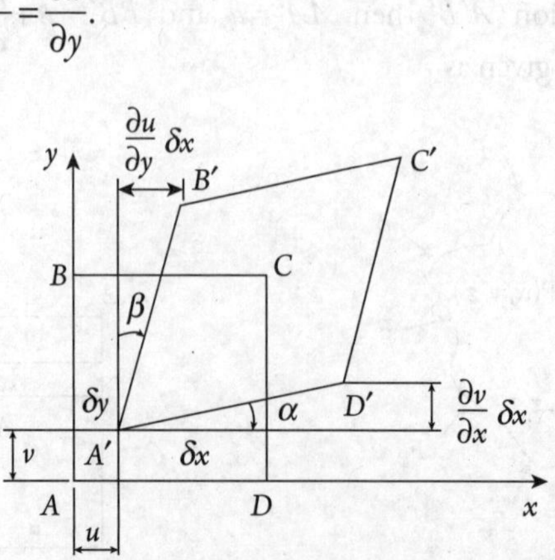

Figure 2.4 Deformation and displacement due to shearing

Shear strain in *xy* plane is $\gamma_{xy} = \alpha + \beta = \frac{\partial v}{\partial x} + \frac{\partial u}{\partial y}$.

In similar arguments, $\gamma_{yz} = \frac{\partial w}{\partial y} + \frac{\partial v}{\partial z}$ and $\gamma_{zx} = \frac{\partial u}{\partial z} + \frac{\partial w}{\partial x}$.

Three shear strains are therefore,

$$\gamma_{xy} = \frac{\partial v}{\partial x} + \frac{\partial u}{\partial y}, \quad \gamma_{yz} = \frac{\partial w}{\partial y} + \frac{\partial v}{\partial z}, \quad \text{and} \quad \gamma_{zx} = \frac{\partial w}{\partial x} + \frac{\partial u}{\partial z}. \tag{2.2.2}$$

Therefore, we may state that displacements at a point x_i can be completely given by three components, *u*, *v*, *w* (Figure 2.5) and state of strain at a point is given by the strain tensor

$$[\in] = \begin{bmatrix} \in_x & \frac{1}{2}\gamma_{xy} & \frac{1}{2}\gamma_{xz} \\ \frac{1}{2}\gamma_{yx} & \in_y & \frac{1}{2}\gamma_{yz} \\ \frac{1}{2}\gamma_{zx} & \frac{1}{2}\gamma_{zy} & \in_z \end{bmatrix}.$$

The shear strains as defined above are engineering shear strain γ_{xy}, γ_{yz}, and γ_{zx}. They are equal to twice the tensorial shear strains $\in_{xy}$, $\in_{yz}$, and $\in_{zx}$. This means

$$\gamma_{xy} = 2\in_{xy}, \gamma_{yz} = 2\in_{yz}, \text{ and } \gamma_{zx} = 2\in_{zx}. \tag{2.2.3}$$

And the strain tensor can be given in terms of strain matrix as

$$[\in] = \begin{bmatrix} \in_x & \in_{xy} & \in_{xz} \\ \in_{yx} & \in_y & \in_{yz} \\ \in_{zx} & \in_{zy} & \in_z \end{bmatrix} = \begin{bmatrix} \in_x & \frac{1}{2}\gamma_{xy} & \frac{1}{2}\gamma_{xz} \\ \frac{1}{2}\gamma_{yx} & \in_y & \frac{1}{2}\gamma_{yz} \\ \frac{1}{2}\gamma_{zx} & \frac{1}{2}\gamma_{zy} & \in_z \end{bmatrix}. \tag{2.2.4}$$

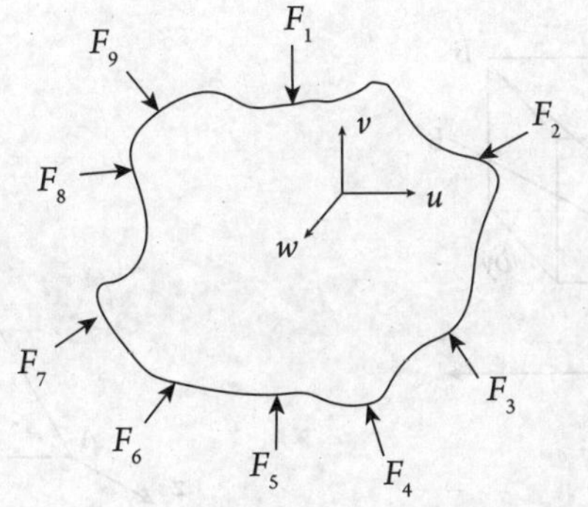

Figure 2.5 Components of displacement at a point

2.3 SMALL STRAIN IN AN ARBITRARY DIRECTION [LO3]

To proceed further with strain analysis, it is necessary to find strain in an arbitrary direction in a Cartesian coordinate system. Consider an element with sides δx, δy, and δz, as shown in Figure 2.6. Since strain deals with the change in length, let us consider the diagonal *AB*. If the coordinates of point *A* are (x, y, z) and those of *B* are $(x+\delta x,\ y+\delta y,\ z+\delta z)$, and $A'B'$ is the deformed state of line *AB*, such that the displacement components of point A' in three coordinate directions are (u, v, w) and displacement components of B' relative to A' are

$$\begin{aligned}\delta u &= \frac{\partial u}{\partial x}\delta x+\frac{\partial u}{\partial y}\delta y+\frac{\partial u}{\partial z}\delta z\\ \delta v &= \frac{\partial v}{\partial x}\delta x+\frac{\partial v}{\partial y}\delta y+\frac{\partial v}{\partial z}\delta z\\ \delta w &= \frac{\partial w}{\partial x}\delta x+\frac{\partial w}{\partial y}\delta y+\frac{\partial w}{\partial z}\delta z.\end{aligned} \tag{2.3.1}$$

Let us now denote $AB=r$; therefore $A'B'=r+\delta r$. We may thus write the increment δr in terms of direction cosines (d.c.) *l*, *m*, and *n*.

$$\delta r = l\delta u+m\delta v+n\delta w. \tag{2.3.2}$$

At this stage, it is necessary to discuss the d.c. The d.c. of a vector is defined as the cosines of the angles it makes with coordinate axes. It is convenient to explain this with reference to a tetrahedron and the normal vector to the oblique plane. The oblique plane *ABC* in the tetrahedron shown in Figure 2.7 has a normal vector *PN*. The vector makes angles α, β, and γ with the coordinate axes, and we may write the d.c. *l*, *m*, and *n* as

$$l=\cos\alpha,\ m=\cos\beta,\ n=\cos\gamma. \tag{2.3.3}$$

It is convenient to carry on further analysis in terms of d.c. Now referring to Figure 2.6, the increment in *AB* may be given as in equation (2.3.2), where

$$l=\frac{\delta x}{r},\ m=\frac{\delta y}{r},\ n=\frac{\delta z}{r}. \tag{2.3.4}$$

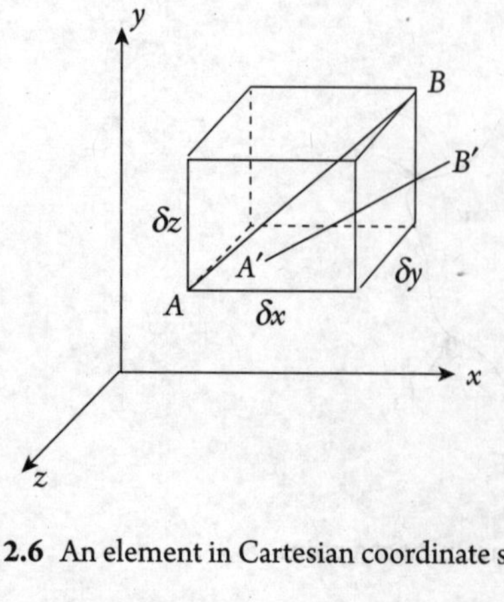

Figure 2.6 An element in Cartesian coordinate system

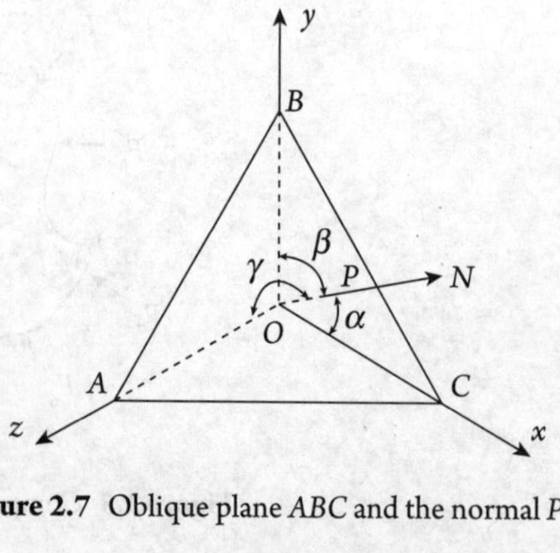

Figure 2.7 Oblique plane *ABC* and the normal *PN*

This also means $l = \dfrac{\Delta ABO}{\Delta ABC},\ m = \dfrac{\Delta AOC}{\Delta ABC},\ n = \dfrac{\Delta BOC}{\Delta ABC}.$

From Figure 2.6, r is the original length and δr the change in length of AB, as shown in equation (2.3.2).

The strain in an arbitrary direction is given by $\epsilon_n = \dfrac{\delta r}{r} = \dfrac{l}{r}\delta u + \dfrac{m}{r}\delta v + \dfrac{n}{r}\delta w.$

Substituting values of δu, δv, and δw from equation (2.3.1) and simplifying using equation (2.3.4), we have the normal strain in an arbitrary direction as

$$\epsilon_n = \epsilon_x l^2 + \epsilon_y m^2 + \epsilon_z n^2 + \gamma_{xy} lm + \gamma_{yz} mn + \gamma_{zx} nl. \tag{2.3.5}$$

We now proceed to find an expression for shear strain in an arbitrary direction. The shear strain is essentially given by the change in the angle between two lines that were originally at the right angle. Consider lines AB and AC initially at right angle, and their deformed state $A'B'$ and $A'C'$ as shown in Figure 2.8. Let the d.c.'s of AB, AC, $A'B'$, and $A'C'$ be $\{l_1, m_1, n_1\}, \{l_2, m_2, n_2\}, \{l_3, m_3, n_3\}$, and $\{l_4, m_4, n_4\}$.

Therefore, $l_1 l_2 + m_1 m_2 + n_1 n_2 = 0$, since the included angle between AB and AC is $\dfrac{\pi}{2}$.

and, $l_3 l_4 + m_3 m_4 + n_3 n_4 = \cos\theta$, since the included angle between $A'B'$ and $A'C'$ is θ.

The shear strain in an arbitrary direction γ_n is then given by

$$\gamma_n = (90° - \theta) \approx \sin(90° - \theta) = \cos\theta = l_3 l_4 + m_3 m_4 + n_3 n_4. \tag{2.3.6}$$

Now, l_3, the cosine of the angle between $A'B'$ and the x-axis is given by

$$l_3 = \frac{\delta x + \delta u}{A'B'}.$$

If we assume $A'B' = AB \approx r$, then we have

$$l_3 = \left(\frac{\delta x}{r} + \frac{\delta u}{r}\right) = \left(\frac{\delta x}{r} + \frac{\partial u}{\partial x}\frac{\delta x}{r} + \frac{\partial u}{\partial y}\frac{\delta y}{r} + \frac{\partial u}{\partial z}\frac{\delta z}{r}\right) = \left\{ l_1\left(1 + \frac{\partial u}{\partial x}\right) + m_1\frac{\partial u}{\partial y} + n_1\frac{\partial u}{\partial z} \right\}.$$

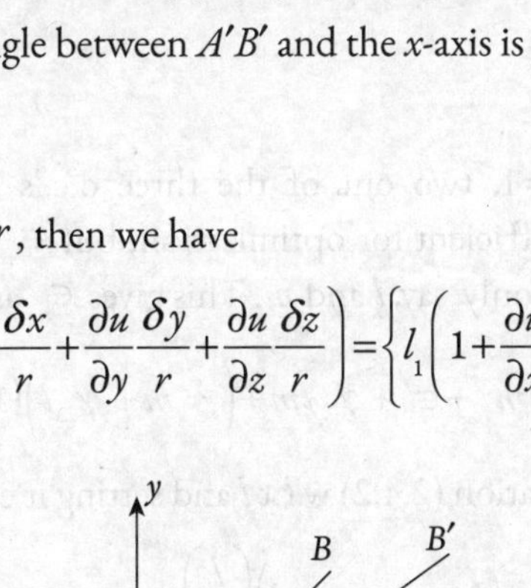

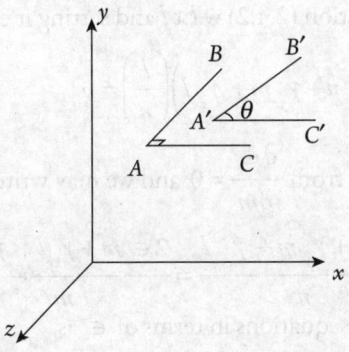

Figure 2.8 Deformed state of lines *AB* and *AC*

In similar lines, m_3, n_3, l_4, m_4, and n_4 may be expressed, and substituting those expressions in equation (2.3.6), we may write γ_n as

$$\gamma_n = 2\left(l_1 l_2 \in_x + m_1 m_2 \in_y + n_1 n_2 \in_z\right) + \left(l_1 m_2 + m_1 l_2\right)\gamma_{xy} + \left(m_1 n_2 + n_1 m_2\right)\gamma_{yz}$$

$$+\left(n_1 l_2 + l_1 n_2\right)\gamma_{zx}. \qquad (2.3.7)$$

It should be noted that in deriving equation (2.3.7) we neglected the square terms for small strains and also used the condition, $l_1 l_2 + m_1 m_2 + n_1 n_2 = 0$.

To summarize, we may say that the normal and shear strain in an arbitrary direction can be given in terms of six strain components $\in_x, \in_y, \in_z, \gamma_{xy}, \gamma_{yz}$, and γ_{zx} as shown in equations (2.3.5) and (2.3.7).

2.4 PRINCIPAL STRAINS AND PRINCIPAL AXES [LO4]

Similar to the state of strain at a point, it is always possible to find three orthogonal lines that remain orthogonal even after straining within a material, indicating that there are no shear strains associated with these directions. These are therefore *principal axes of strains*. It will be shown that these are also the directions in which the direct strains have stationary values, known as *principal strains*. To find these strains, we proceed as follows:

Normal strain in equation (2.3.5) is d.c.-dependent, and therefore there must be extreme values of these strains at some orientation given by

$$\frac{\partial \in_n}{\partial l} = \frac{\partial \in_n}{\partial m} = 0. \qquad (2.4.1)$$

Considering $l^2 + m^2 + n^2 = 1$, two out of the three d.c.'s are independent variables, and therefore equation (2.4.1) is sufficient for optimization, but $\in_n$ in equation (2.3.5) needs to be expressed in terms of two d.c.'s only, say, l and m. This gives $\in_n$ as

$$\in_n = \left(\in_x - \in_z\right)l^2 + \left(\in_y - \in_z\right)m^2 + \in_z + \gamma_{xy}\, lm + \left(\gamma_{yz}\, m + \gamma_{zx}\, l\right)\left(1 - m^2 - l^2\right)^{\frac{1}{2}}. \qquad (2.4.2)$$

Partially deriving $\in_n$ in equation (2.4.2) w.r.t l and setting it equal to zero, we get

$$\left(2\in_x l + \gamma_{xy} m + \gamma_{zx} n\right) - \left(2\in_z n + \gamma_{yz} m + \gamma_{zx} l\right)\left(\frac{l}{n}\right) = 0.$$

Similar results may be found from $\dfrac{\partial \in_n}{\partial m} = 0$ and we may write

$$\frac{2\in_x l + \gamma_{xy} m + \gamma_{zx} n}{l} = \frac{2\in_z n + \gamma_{yz} m + \gamma_{zx} l}{n} = \frac{2\in_y m + \gamma_{xy} l + \gamma_{yz} n}{m} = 2\in\ (say).$$

We may therefore write three equations in terms of $\in$ as

$$\in_x = \in - \frac{\gamma_{xy} m + \gamma_{zx} n}{2l}$$

$$\in_y = \in - \frac{\gamma_{yx} l + \gamma_{yz} n}{2m}$$

$$\in_z = \in - \frac{\gamma_{zx} l + \gamma_{zy} m}{2n}. \tag{2.4.3}$$

It is interesting to note that substituting these expressions in equation (2.3.5) $\in_n$ turns out to be $\in$ as shown below:

$$\in_n = \in_x l^2 + \in_y m^2 + \in_z n^2 + \gamma_{xy} lm + \gamma_{yz} mn + \gamma_{zx} nl = \in \left(l^2 + m^2 + n^2\right) = \in.$$

This means that the stationary value of the normal strain in an arbitrary direction, as obtained from the condition $\frac{\partial \in_n}{\partial l} = \frac{\partial \in_n}{\partial m} = 0$, is $\in$.

$\in$ therefore represents the principal strain.

Equation (2.4.3) may now be written in the matrix form as

$$\begin{bmatrix} \in_x - \in & \frac{1}{2}\gamma_{xy} & \frac{1}{2}\gamma_{xz} \\ \frac{1}{2}\gamma_{yx} & \in_y - \in & \frac{1}{2}\gamma_{yz} \\ \frac{1}{2}\gamma_{zx} & \frac{1}{2}\gamma_{zy} & \in_z - \in \end{bmatrix} \begin{bmatrix} l \\ m \\ n \end{bmatrix} = \begin{bmatrix} 0 \\ 0 \\ 0 \end{bmatrix}. \tag{2.4.4}$$

A nontrivial solution to the above equation is

$$\begin{vmatrix} \in_x - \in & \frac{1}{2}\gamma_{xy} & \frac{1}{2}\gamma_{xz} \\ \frac{1}{2}\gamma_{yx} & \in_y - \in & \frac{1}{2}\gamma_{yz} \\ \frac{1}{2}\gamma_{zx} & \frac{1}{2}\gamma_{zy} & \in_z - \in \end{vmatrix} = 0. \tag{2.4.5}$$

Here $\in$ represents the eigenvalue of the problem and the principal strain. On expansion, equation (2.4.5) yields the following cubic equation of $\in$.

$$\in^3 - \left(\in_x + \in_y + \in_z\right)\in^2 + \left(\in_x \in_y + \in_y \in_z + \in_z \in_x - \frac{1}{4}\gamma_{xy}^2 - \frac{1}{4}\gamma_{yz}^2 - \frac{1}{4}\gamma_{zx}^2\right)\in$$
$$-\left(\in_x \in_y \in_z + \frac{1}{4}\gamma_{xy}\gamma_{yz}\gamma_{zx} - \frac{1}{4}\in_x \gamma_{yz}^2 - \frac{1}{4}\in_y \gamma_{zx}^2 - \frac{1}{4}\in_z \gamma_{xy}^2\right) = 0. \tag{2.4.6}$$

Three roots of this cubic equation give the three principal strains. If the three principal strains are denoted as $\in_1 > \in_2 > \in_3$, then we may also write

$$\left(\in - \in_1\right)\left(\in - \in_2\right)\left(\in - \in_3\right) = 0,$$

which gives the cubic equation in terms of principal strains as

$$\in^3 - \left(\in_1 + \in_2 + \in_3\right)\in^2 + \left(\in_1\in_2 + \in_2\in_3 + \in_1\in_3\right)\in - \in_1\in_2\in_3 = 0 . \tag{2.4.7}$$

Therefore, the roots of equations (2.4.6) and (2.4.7) yield the same results.

Eigenvectors give the principal axis directions. Eigenvalue problems for both strain and stress tensors are similar. The problems and their solutions procedure will be treated in depth in Chapter 3. However, we discuss here some essential properties of strain tensors. Strain tensors can be split into spherical and deviatorial tensors as follows:

$$\begin{bmatrix} \in_x & \frac{1}{2}\gamma_{xy} & \frac{1}{2}\gamma_{xz} \\ \frac{1}{2}\gamma_{yx} & \in_y & \frac{1}{2}\gamma_{yz} \\ \frac{1}{2}\gamma_{zx} & \frac{1}{2}\gamma_{zy} & \in_z \end{bmatrix} = \begin{bmatrix} \in_m & 0 & 0 \\ 0 & \in_m & 0 \\ 0 & 0 & \in_m \end{bmatrix} + \begin{bmatrix} \in_x - \in_m & \frac{1}{2}\gamma_{xy} & \frac{1}{2}\gamma_{xz} \\ \frac{1}{2}\gamma_{yx} & \in_y - \in_m & \frac{1}{2}\gamma_{yz} \\ \frac{1}{2}\gamma_{zx} & \frac{1}{2}\gamma_{zy} & \in_z - \in_m \end{bmatrix}. \tag{2.4.8}$$

The first component of the RHS consisting of only mean strain, given by $\in_m = \dfrac{\in_x + \in_y + \in_z}{3}$, is the spherical component and this is due to volume change, whereas the second component is the deviatorial component and this is due to distortion. Similar stress components will be found in the next section, and they are widely used in failure theories of machine parts.

Now we consider an interesting aspect of normal and shear strains in arbitrary directions given in equations (2.3.5) and (2.3.7). It is convenient to express these strains in terms of principal strains for simplicity. To demonstrate this, we first consider that the coordinate axes coincide with the principal axes. This means $\in_x = \in_1$, $\in_y = \in_2$, $\in_z = \in_3$, and $\gamma_{xy} = \gamma_{yz} = \gamma_{zx} = 0$. Since there are no shear strains in principal planes, equations (2.3.5) and (2.3.7) yield the following.

$$\in_n = \in_x l^2 + \in_y m^2 + \in_z n^2 + \gamma_{xy} lm + \gamma_{yz} mn + \gamma_{zx} nl = \in_1 l^2 + \in_2 m^2 + \in_3 n^2 \tag{2.4.9}$$

$$\gamma_{n1} = 2\left(l_1 l_2 \in_x + m_1 m_2 \in_y + n_1 n_2 \in_z\right) + \left(l_1 m_2 + m_1 l_2\right)\gamma_{xy} + \left(m_1 n_2 + n_1 m_2\right)\gamma_{yz}$$
$$+ \left(n_1 l_2 + l_1 n_2\right)\gamma_{zx} = 2\left(l_1 l_2 \in_1 + m_1 m_2 \in_2 + n_1 n_2 \in_3\right). \tag{2.4.10}$$

Equation (2.4.10) essentially gives the shear strain between an arbitrary direction "n", defined by (l_1, m_1, n_1) and an orthogonal direction "2", defined by (l_2, m_2, n_2). We call this γ_{n1}.

Now to find the resultant shear strains associated with an arbitrary direction "n" defined by (l_1, m_1, n_1), we need to find first the shear strain between "n" and another orthogonal "5" defined by (l_5, m_5, n_5), which is orthogonal to 2 and 5 as shown in Figure 2.9, and this is given by

$$\gamma_{n2} = 2\left(l_1 l_5 \in_1 + m_1 m_5 \in_2 + n_1 n_5 \in_3\right).$$

And the resultant is

$$\gamma_n^2 = \gamma_{n1}^2 + \gamma_{n2}^2 . \tag{2.4.11}$$

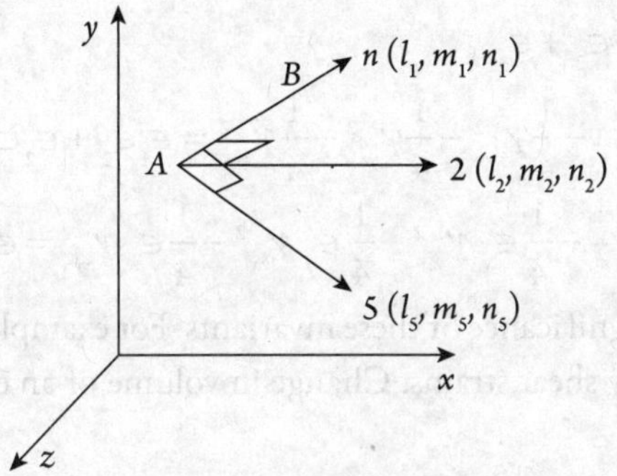

Figure 2.9 Orthogonal directions

This gives $\gamma_n^2 = 4[\in_1^2 l_1^2 l_2^2 + \in_1^2 l_1^2 l_5^2 + \in_2^2 m_1^2 m_2^2 + \in_2^2 m_1^2 m_5^2 + \in_3^2 n_1^2 n_2^2 + \in_3^2 n_1^2 n_5^2 +$
$2(l_1 l_2 m_1 m_2 \in_1 \in_2 + l_1 l_2 n_1 n_2 \in_1 \in_3 + m_1 m_2 n_1 n_2 \in_2 \in_3 + l_1 l_5 m_1 m_5 \in_1 \in_2 + l_1 l_5 n_1 n_5 \in_1 \in_3$
$+ m_1 m_5 n_1 n_5 \in_2 \in_3)]$.

Rearranging this and letting $l_2 l_5 + m_2 m_5 + n_2 n_5 = 0$, since they are mutually orthogonal, we have

$$\gamma_n^2 = 4[(\in_1 - \in_2)^2 l^2 m^2 + (\in_2 - \in_3)^2 m^2 n^2 + (\in_3 - \in_1)^2 l^2 n^2].$$

Following the method for finding the principal strain, we may find the maximum shear strain.

We let $\dfrac{\partial \gamma_n^2}{\partial l} = \dfrac{\partial \gamma_n^2}{\partial m} = 0$, since only two d.c.'s are independent.

This gives us 12 planes of maximum shear strains. The planes of maximum shear strains and maximum strains, and the corresponding normal strains from equations (2.4.11) and (2.4.9) are given as follows:

$$l = 0;\ m = \pm\frac{1}{\sqrt{2}};\ n = \pm\frac{1}{\sqrt{2}};\ \gamma_{max} = \pm(\in_2 - \in_3);\ \in_n = \frac{\in_2 + \in_3}{2}$$

$$l = \pm\frac{1}{\sqrt{2}};\ m = 0;\ n = \pm\frac{1}{\sqrt{2}};\ \gamma_{max} = \pm(\in_1 - \in_3);\ \in_n = \frac{\in_1 + \in_3}{2}$$

$$l = \pm\frac{1}{\sqrt{2}};\ m = \pm\frac{1}{\sqrt{2}};\ n = 0;\ \gamma_{max} = \pm(\in_1 - \in_2);\ \in_n = \frac{\in_1 + \in_2}{2}. \quad (2.4.12)$$

2.5 STRAIN INVARIANTS [LO5]

Another important property of a strain tensor is the strain invariant. Equations (2.4.6) and (2.4.7) yield the same principal strains irrespective of the coordinate axes used. Therefore, the coefficients must remain invariant. The three strain invariants are thus given as:

$$J_1 = \epsilon_x + \epsilon_y + \epsilon_z = \epsilon_1 + \epsilon_2 + \epsilon_3.$$

$$J_2 = \epsilon_x \epsilon_y + \epsilon_y \epsilon_z + \epsilon_z \epsilon_x - \frac{1}{4}\gamma_{xy}^2 - \frac{1}{4}\gamma_{yz}^2 - \frac{1}{4}\gamma_{zx}^2 = \epsilon_1\epsilon_2 + \epsilon_2\epsilon_3 + \epsilon_1\epsilon_3. \tag{2.5.1}$$

$$J_3 = \epsilon_x \epsilon_y \epsilon_z + \frac{1}{4}\gamma_{xy}\gamma_{yz}\gamma_{zx} - \frac{1}{4}\epsilon_x\gamma_{yz}^2 - \frac{1}{4}\epsilon_y\gamma_{zx}^2 - \frac{1}{4}\epsilon_z\gamma_{xy}^2 = \epsilon_1\epsilon_2\epsilon_3.$$

There is some physical significance of these invariants. For example, volume changes are caused by the normal strains, not by shear strains. Change in volume of an element of sides δx, δy, and δz is given by

$$\Delta V = (\delta x + \epsilon_x \delta x)(\delta y + \epsilon_y \delta y)(\delta z + \epsilon_z \delta z) - \delta x \delta y \delta z$$

and we may write dilatation as

$$\frac{\Delta V}{V} = \frac{(\delta x + \epsilon_x \delta x)(\delta y + \epsilon_y \delta y)(\delta z + \epsilon_z \delta z) - \delta x \delta y \delta z}{\delta x \delta y \delta z}$$

$$= \epsilon_x + \epsilon_y + \epsilon_z \quad \text{(Neglecting the product of strains)}$$

$$= J_1 = \epsilon_1 + \epsilon_2 + \epsilon_3. \tag{2.5.2}$$

Dilatation is thus given by the first strain invariant. Many such examples may be cited and more detailed discussions will follow in the stress invariants section.

2.6 ROTATIONS [LO6]

There is another variable related to strained element known as rotation. Consider the strained element in the *xy* plane as shown in Figure 2.4. There shear strain is given by $\gamma_{xy} = \alpha + \beta$ whereas the average rotation of the element about the *z*-axis is given by

$$\omega_{xy} = \frac{\alpha - \beta}{2} = \frac{1}{2}\left[\frac{\partial u}{\partial y}\delta y / \delta y - \frac{\partial v}{\partial x}\delta x / \delta x\right] = \frac{1}{2}\left(\frac{\partial u}{\partial y} - \frac{\partial v}{\partial x}\right).$$

We may therefore write the rotations about *x*, *y*, and *z* axes respectively as

$$\omega_{yz} = \frac{1}{2}\left(\frac{\partial v}{\partial z} - \frac{\partial w}{\partial y}\right), \ \omega_{zx} = \frac{1}{2}\left(\frac{\partial w}{\partial x} - \frac{\partial u}{\partial z}\right), \ \omega_{xy} = \frac{1}{2}\left(\frac{\partial u}{\partial y} - \frac{\partial v}{\partial x}\right). \tag{2.6.1}$$

2.7 COMPATIBILITY CONDITIONS [LO7]

As we consider the variation of strains from point to point, physical constraints are that as the material deforms, it must remain continuous, i.e., no holes or overlaps are created. Mathematically, this means that the displacements are continuous and differentiable function of position. Since the six strain components are just functions of three displacements, strain components are not independent. There must be some relations connecting these strain components. These relations are generally known as compatibility relations. Let us consider first the three strains in the *xy* plane

$$\epsilon_x = \frac{\partial u}{\partial x},\ \epsilon_y = \frac{\partial v}{\partial y},\ \gamma_{xy} = \frac{\partial u}{\partial y} + \frac{\partial v}{\partial x}.$$

Second derivatives of these strain components may be written as

$$\frac{\partial^2 \epsilon_x}{\partial y^2} = \frac{\partial^3 u}{\partial x \partial y^2};\ \frac{\partial^2 \epsilon_y}{\partial x^2} = \frac{\partial^3 v}{\partial y \partial x^2};\ \frac{\partial^2 \gamma_{xy}}{\partial x \partial y} = \frac{\partial^3 u}{\partial x \partial y^2} + \frac{\partial^3 v}{\partial y \partial x^2}. \quad (2.7.1)$$

This gives

$$\frac{\partial^2 \gamma_{xy}}{\partial x \partial y} = \frac{\partial^2 \epsilon_x}{\partial y^2} + \frac{\partial^2 \epsilon_y}{\partial x^2}.$$

Similarly, two more such equations in *yz* and *zx* planes can be written. All the three equations are given as follows:

$$\frac{\partial^2 \gamma_{xy}}{\partial x \partial y} = \frac{\partial^2 \epsilon_x}{\partial y^2} + \frac{\partial^2 \epsilon_y}{\partial x^2}.$$

$$\frac{\partial^2 \gamma_{yz}}{\partial y \partial z} = \frac{\partial^2 \epsilon_y}{\partial z^2} + \frac{\partial^2 \epsilon_z}{\partial y^2}.$$

$$\frac{\partial^2 \gamma_{zx}}{\partial z \partial x} = \frac{\partial^2 \epsilon_z}{\partial x^2} + \frac{\partial^2 \epsilon_x}{\partial z^2}. \quad (2.7.2)$$

Note that the equations are independent of displacement terms.

Three more relations may be obtained by eliminating the displacements from the shear strains as shown below:

$$\gamma_{xy} = \frac{\partial u}{\partial y} + \frac{\partial v}{\partial x},\ \gamma_{yz} = \frac{\partial v}{\partial z} + \frac{\partial w}{\partial y},\ \gamma_{zx} = \frac{\partial w}{\partial x} + \frac{\partial u}{\partial z}.$$

We may now differentiate the shear strains as follows:

$$\frac{\partial \gamma_{xy}}{\partial z} = \frac{\partial^2 u}{\partial y \partial z} + \frac{\partial^2 v}{\partial x \partial z}$$

$$\frac{\partial \gamma_{yz}}{\partial x} = \frac{\partial^2 v}{\partial z \partial x} + \frac{\partial^2 w}{\partial y \partial x} \quad (2.7.3)$$

$$\frac{\partial \gamma_{zx}}{\partial y} = \frac{\partial^2 w}{\partial x \partial y} + \frac{\partial^2 u}{\partial z \partial y}.$$

We may now manipulate equation (2.7.3) to come up with an equation eliminating the displacements. In order to do this, we first add the left-hand side of the first and third equations in (2.7.3), and subtract the left-hand side of the second equation. This gives

$$\frac{\partial \gamma_{xy}}{\partial z} - \frac{\partial \gamma_{yz}}{\partial x} + \frac{\partial \gamma_{zx}}{\partial y} = 2\frac{\partial^2 u}{\partial y \partial z}.$$

Now we differentiate both sides with respect to x and that gives us

$$\frac{\partial}{\partial x}\left(\frac{\partial \gamma_{xy}}{\partial z}-\frac{\partial \gamma_{yz}}{\partial x}+\frac{\partial \gamma_{zx}}{\partial y}\right)=2\frac{\partial^3 u}{\partial x \partial y \partial z}=2\frac{\partial^2 \in_x}{\partial y \partial z}.$$

Note that the resultant equation is independent of any displacement term.

Similarly, we may write two more equations, and all three equations are given as follows:

$$\begin{aligned}
\frac{\partial}{\partial x}\left(\frac{\partial \gamma_{xy}}{\partial z}-\frac{\partial \gamma_{yz}}{\partial x}+\frac{\partial \gamma_{zx}}{\partial y}\right)&=2\frac{\partial^2 \in_x}{\partial y \partial z}\\
\frac{\partial}{\partial y}\left(\frac{\partial \gamma_{yz}}{\partial x}-\frac{\partial \gamma_{zx}}{\partial y}+\frac{\partial \gamma_{xy}}{\partial z}\right)&=2\frac{\partial^2 \in_y}{\partial x \partial z}\\
\frac{\partial}{\partial z}\left(\frac{\partial \gamma_{zx}}{\partial y}-\frac{\partial \gamma_{xy}}{\partial z}+\frac{\partial \gamma_{yz}}{\partial x}\right)&=2\frac{\partial^2 \in_z}{\partial x \partial y}.
\end{aligned} \tag{2.7.4}$$

Six equations in (2.7.2) and (2.7.4) are known as the compatibility equations.

In problems where strains vary significantly in three directions, all six equations in (2.7.2) and (2.7.4) must be satisfied. For two-dimensional (2D) cases, generally five equations are automatically satisfied, leaving only (say)

$$\frac{\partial^2 \in_x}{\partial y^2}+\frac{\partial^2 \in_y}{\partial x^2}=\frac{\partial^2 \gamma_{xy}}{\partial x \partial y}. \tag{2.7.5}$$

2.8 EXPERIMENTAL STRAIN MEASUREMENT [LO8]

In the study of strain analysis, it is important to have some basic understanding of strain measurement. There are many different methods of strain measurements. Among the methods, some are whole-field and some are point-to-point.

(a) *Whole-field methods*

These techniques help considerably in the understanding of the strain field graphically or pictorially. These include:

- Photo-elastic methods
- Birefringent coating methods
- Brittle coating methods
- Grid techniques
- Moire method

(b) *Point-to-point methods*

- Mechanical
- Optical
- Acoustic strain gauge
- Electrical strain gauge

Among these techniques, the electrical strain gauge method is simple and useful in strain analysis at a point. Here we shall limit our discussion to experimental strain measurement using the electrical strain gauge technique only for its simplicity and wide usage. Here the strain measurement is based on the change in electrical resistance in an electrical strain gauge. Clearly this gives axial strain, and therefore a group of strain gauges are fixed at a point in some specified orientation to compute both normal and shear strains. These are known as strain rosettes. There may be three or four element rosettes.

1. *Three-element rosettes*
   (a) Rectangular type
   (b) Delta type
2. *Four-element rosettes*
   (a) Rectangular type
   (b) Tee-delta rosette

These are shown in Figure 2.10.

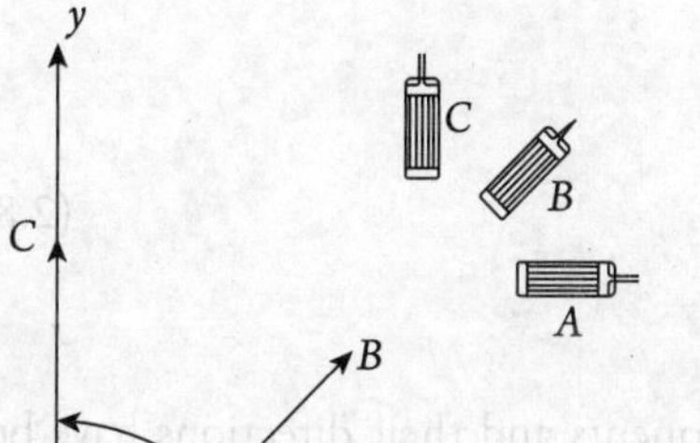

(a) Three-element rectangular rosette

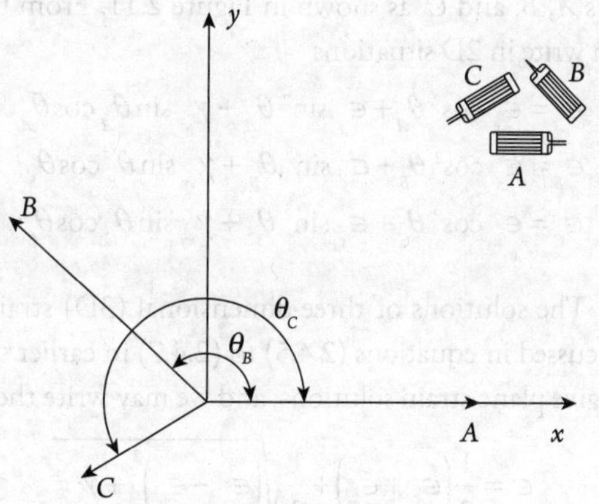

(b) Three-element delta rosette

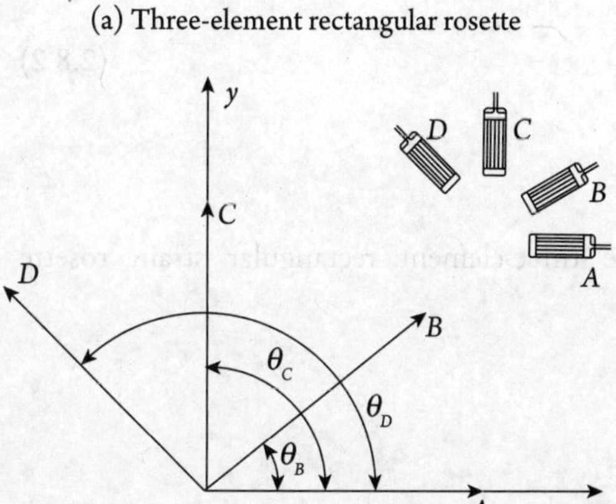

(c) Four-element rectangular rosette

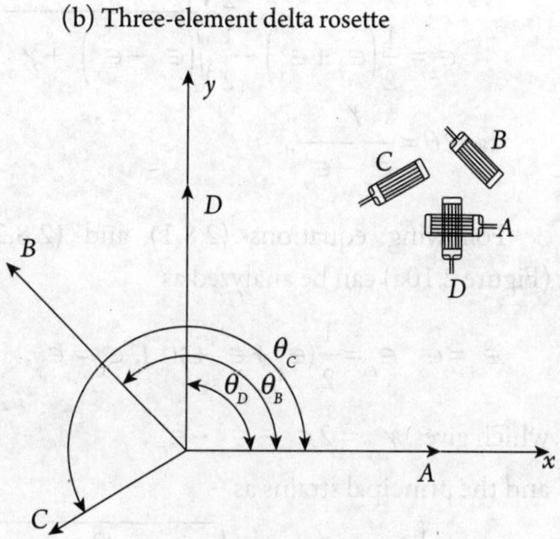

(d) Tee-delta rosette

Figure 2.10 Strain rosettes

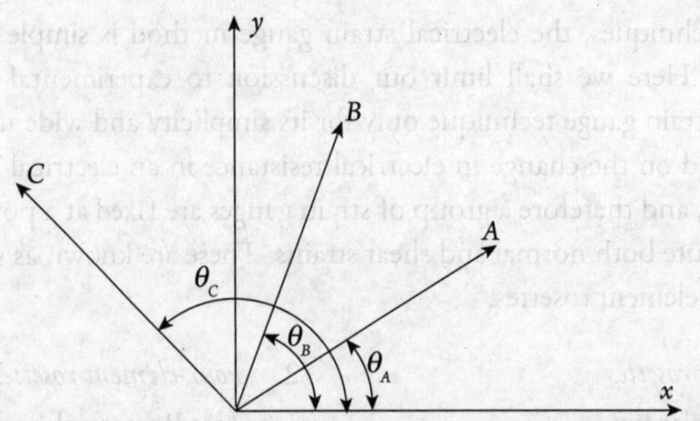

Figure 2.11 Three strain gauges placed at arbitrary angles relative to *x* and *y* axes

In order to determine the principal strains, three or more measurements for normal strains at a point in different directions are necessary. Let us now consider the strain gauges aligned along axes *A*, *B*, and *C*, as shown in Figure 2.11. From the strain transformation equations (2.3.5), we can write in 2D situations

$$\begin{aligned} \epsilon_A &= \epsilon_x \cos^2\theta_A + \epsilon_y \sin^2\theta_A + \gamma_{xy} \sin\theta_A \cos\theta_A \\ \epsilon_B &= \epsilon_x \cos^2\theta_B + \epsilon_y \sin^2\theta_B + \gamma_{xy} \sin\theta_B \cos\theta_B \\ \epsilon_C &= \epsilon_x \cos^2\theta_C + \epsilon_y \sin^2\theta_C + \gamma_{xy} \sin\theta_C \cos\theta_C. \end{aligned} \tag{2.8.1}$$

The solutions of three-dimensional (3D) strain components and their directions have been discussed in equations (2.4.5) to (2.4.7) in earlier sections. These can be reduced to 2D problems to give plane strain solutions, and we may write the principal strains in 2D situations as

$$\begin{aligned} \epsilon_1 &= \frac{1}{2}\left(\epsilon_x + \epsilon_y\right) + \frac{1}{2}\sqrt{\left(\epsilon_x - \epsilon_y\right)^2 + \gamma_{xy}^{\ 2}} \\ \epsilon_2 &= \frac{1}{2}\left(\epsilon_x + \epsilon_y\right) - \frac{1}{2}\sqrt{\left(\epsilon_x - \epsilon_y\right)^2 + \gamma_{xy}^{\ 2}} \\ \tan 2\theta &= \frac{\gamma_{xy}}{\epsilon_x - \epsilon_y}. \end{aligned} \tag{2.8.2}$$

Following equations (2.8.1) and (2.8.2), the three-element rectangular strain rosette (Figure 2.10a) can be analyzed as

$$\epsilon_A = \epsilon_x,\ \epsilon_B = \frac{1}{2}(\epsilon_x + \epsilon_y + \gamma_{xy}),\ \epsilon_C = \epsilon_y,$$

which gives $\gamma_{xy} = 2\epsilon_B - \epsilon_A - \epsilon_C$,

and the principal strains as

$$\epsilon_{1,2} = \frac{1}{2}\left(\epsilon_A + \epsilon_C\right) \pm \frac{1}{2}\sqrt{\left(\epsilon_A - \epsilon_C\right)^2 + \left(2\epsilon_B - \epsilon_A - \epsilon_C\right)^2}.$$

Similar treatment may be adopted for other cases. In four-element rosettes, one extra strain gauge is available, and this is not strictly needed, but this is used for checking purposes. For example, consider a four-element rectangular rosette as shown in Figure 2.10 (c), the angles between the gauges are 0°, 45°, 90°, and 135° for the gauges *A*, *B*, *C*, and *D*. The configuration ensures a quick check of strain invariant $J_1 = \epsilon_x + \epsilon_y = \epsilon_A + \epsilon_C = \epsilon_B + \epsilon_D$. This gives an independent check on the accuracy of strain evaluation.

Example 2.1

The displacement field in an elastic medium is given by

$$\left.\begin{aligned} u &= 15x^2 y + 3z \\ v &= 5y^2 + 3xz \\ w &= 3z^2 + yz + 5 \end{aligned}\right\} \times 10^{-3} \text{ mm.}$$

Find the strain tensor at a point whose coordinates are (1, 2, 0.5). Find also the principal strains and principal axes.

Solution:

The strain components at the point are

$$\epsilon_x = \frac{\partial u}{\partial x} = 30xy \times 10^{-3} = 60 \times 10^{-3}$$

$$\epsilon_y = \frac{\partial v}{\partial y} = 10y \times 10^{-3} = 20 \times 10^{-3}$$

$$\epsilon_z = \frac{\partial w}{\partial z} = (6z + y) \times 10^{-3} = 5 \times 10^{-3}$$

$$\gamma_{xy} = \frac{\partial u}{\partial y} + \frac{\partial v}{\partial x} = (15x^2 + 3z) \times 10^{-3} = 16.5 \times 10^{-3}$$

$$\gamma_{yz} = \frac{\partial v}{\partial z} + \frac{\partial w}{\partial y} = (3x + z) \times 10^{-3} = 3.5 \times 10^{-3}$$

$$\gamma_{zx} = \frac{\partial w}{\partial x} + \frac{\partial u}{\partial z} = 3 \times 10^{-3}.$$

Therefore, the strain tensor may be given as

$$\epsilon_{ij} = \begin{bmatrix} 60 & 8.25 & 1.5 \\ 8.25 & 20 & 1.75 \\ 1.5 & 1.75 & 5 \end{bmatrix} \times 10^{-3}.$$

Now to find the principal strains we need to solve the characteristic equation. Following equations (2.4.7) and (2.5.1), this is given as

$$\epsilon^3 - J_1 \epsilon^2 + J_2 \epsilon - J_3 = 0,$$

where J_1, J_2, and J_3 are strain invariants and the roots of the cubic equation, i.e., the eigenvalues give the principal strains. Here

$$J_1 = \epsilon_x + \epsilon_y + \epsilon_z = (60+20+5)\times 10^{-3} = 85\times 10^{-3}$$

$$J_2 = \epsilon_x\epsilon_y + \epsilon_y\epsilon_z + \epsilon_z\epsilon_x - \frac{1}{4}\left(\gamma_{xy}^2 + \gamma_{yz}^2 + \gamma_{zx}^2\right)$$

$$= \left[60\times 20 + 20\times 5 + 60\times 5 - \frac{1}{4}\left(16.5^2 + 3.5^2 + 3^2\right)\right]\times 10^{-6} = 1526.625\times 10^{-6}$$

$$J_3 = \epsilon_x\epsilon_y\epsilon_z + \frac{1}{4}\gamma_{xy}\gamma_{yz}\gamma_{zx} - \frac{1}{4}\left(\epsilon_x\gamma_{yz}^2 + \epsilon_y\gamma_{zx}^2 + \epsilon_z\gamma_{yx}^2\right)$$

$$= 60\times 20\times 5 + \frac{1}{4}\left[(16.5\times 3.5\times 3) - \left(60\times 3.5^2 + 20\times 3^2 + 5\times 16.5^2\right)\right]\times 10^{-9}$$

$$= 5474.25\times 10^{-9}.$$

Substituting J_1, J_2, and J_3 in the characteristic equation and solving for ϵ we get three principal stains as

$$\epsilon_1 = 0.0617$$

$$\epsilon_2 = 0.0185$$

$$\epsilon_3 = 0.0048.$$

Next, we find the eigenvectors that represent the principal axes. There are many different methods of finding the eigenvectors of a characteristic equation. One easy method is given below:

First, we find the eigenvectors (say) l_1, m_1, and n_1 corresponding to ϵ_1 and to do that we may write the following three equations using the strain tensor calculated above and $\epsilon_1 = 0.0617$.

$$\begin{cases} (0.06-0.0617)l_1 + 0.00825m_1 + 0.0015n_1 = 0 \\ 0.00825l_1 + (0.02-0.0617)m_1 + 0.00175n_1 = 0 \\ 0.0015l_1 + 0.00175m_1 + (0.005-0.0617)n_1 = 0. \end{cases}$$

Since $l_1^2 + m_1^2 + n_1^2 = 1$, there are only two independent equations in the above set. Let us therefore set $n_1 = 1$ and solve the first two equations for l_1 and m_1. This gives

$$l_1 = 27.229 \quad m_1 = 5.429 \quad n_1 = 1.$$

Normalizing these by dividing by $\left(l_1^2 + m_1^2 + n_1^2\right)^{1/2} = 27.783$, we get one set of eigenvectors as

$$l_1 = 0.98 \quad m_1 = 0.195 \quad n_1 = 0.036,$$

Similarly, the other two sets of eigenvectors are

$$l_2 = -0.1976 \qquad m_2 = 0.9759 \qquad n_2 = 0.0941$$

$$l_3 = -0.0107 \qquad m_3 = -0.1085 \qquad n_3 = 0.994$$

Example 2.2

The state of strain at a point in a solid body is given by

$$\epsilon_{ij} = \begin{bmatrix} 0.01 & \frac{1}{2}(0.03) & \frac{1}{2}(0.05) \\ \frac{1}{2}(0.03) & 0.02 & \frac{1}{2}(0.04) \\ \frac{1}{2}(0.05) & \frac{1}{2}(0.04) & 0.03 \end{bmatrix}.$$

Find the normal strain in a direction that is equally inclined to the three coordinate axes. Find also the normal strain in the direction that makes angles 60°, 75°, and 30° with *x*, *y*, and *z* coordinate axes respectively.

Solution:

Since $l_1^2 + m_1^2 + n_1^2 = 1$ in the first case, $l = m = n = \dfrac{1}{\sqrt{3}}$.

Following equation (2.3.5), we write the strain in an arbitrary direction as

$$\epsilon_n = \epsilon_x l^2 + \epsilon_y m^2 + \epsilon_z n^2 + \gamma_{xy} lm + \gamma_{yz} mn + \gamma_{zx} nl.$$

Here

$$\epsilon_x = 0.01,\ \epsilon_y = 0.02,\ \epsilon_z = 0.03$$

$$\gamma_{xy} = 0.03,\ \gamma_{yz} = 0.04,\ \gamma_{zx} = 0.05.$$

Substituting the values in equation (2.3.5) reproduced above, we have $\epsilon_n = 0.06$.

In the second case $l = \cos 60° = 0.5$, $m = \cos 75° = 0.259$ and $n = \cos 30° = 0.866$.

Again, substituting these values in equation (2.3.5), we have

$$\epsilon_n = 0.06085.$$

Example 2.3

The components of the strain field in a body are given by $\epsilon_x = 0.03xy, \epsilon_y = 0.01y$, $\epsilon_z = 0.006z + 0.001y$, $\gamma_{xy} = 0.015x^2 + 0.03z$, $\gamma_{yz} = 0.03x + 0.01z$, and $\gamma_{xz} = 0.01x$.

Check if it is a compatible strain field.

Solution:

For the strain field to be compatible, it is necessary that the strain field satisfies the six compatibility equations given below:

$$\frac{\partial^2 \epsilon_x}{\partial y^2} + \frac{\partial^2 \epsilon_y}{\partial x^2} = \frac{\partial^2 \gamma_{xy}}{\partial x \partial y}$$

$$\frac{\partial^2 \epsilon_y}{\partial z^2} + \frac{\partial^2 \epsilon_z}{\partial y^2} = \frac{\partial^2 \gamma_{yz}}{\partial y \partial z}$$

$$\frac{\partial^2 \in_z}{\partial x^2}+\frac{\partial^2 \in_x}{\partial z^2}=\frac{\partial^2 \gamma_{xz}}{\partial z \partial x}$$

$$\frac{\partial}{\partial x}\left(\frac{\partial \gamma_{xy}}{\partial z}-\frac{\partial \gamma_{yz}}{\partial x}+\frac{\partial \gamma_{zx}}{\partial y}\right)=2\frac{\partial^2 \in_x}{\partial y \partial z}$$

$$\frac{\partial}{\partial y}\left(\frac{\partial \gamma_{yz}}{\partial x}-\frac{\partial \gamma_{zx}}{\partial y}+\frac{\partial \gamma_{xy}}{\partial z}\right)=2\frac{\partial^2 \in_y}{\partial x \partial z}$$

$$\frac{\partial}{\partial z}\left(\frac{\partial \gamma_{zx}}{\partial y}-\frac{\partial \gamma_{xy}}{\partial z}+\frac{\partial \gamma_{yz}}{\partial x}\right)=2\frac{\partial^2 \in_z}{\partial x \partial y}.$$

It can be easily shown that the strain field satisfies all the compatibility equations; therefore, strain field is compatible.

Example 2.4

The displacement field for a body is given by

$$\left.\begin{aligned} u&=6x^2+4y^2+z \\ v&=-3x+2y^2 \\ w&=2x+6y^2-6z^2 \end{aligned}\right\}\times 10^{-3}\text{ mm.}$$

Determine the components of strain tensor at a point (2, –1, 3).

Solution:

The strain tensor is given by

$$\in_{ij}=\begin{bmatrix} \in_x & \frac{1}{2}\gamma_{xy} & \frac{1}{2}\gamma_{xz} \\ \frac{1}{2}\gamma_{yx} & \in_y & \frac{1}{2}\gamma_{yz} \\ \frac{1}{2}\gamma_{zx} & \frac{1}{2}\gamma_{zy} & \in_z \end{bmatrix},$$

where $\in_x=\frac{\partial u}{\partial x}, \in_y=\frac{\partial v}{\partial y}, \in_z=\frac{\partial w}{\partial z}$,

$$\gamma_{xy}=\frac{\partial v}{\partial x}+\frac{\partial u}{\partial y},\ \gamma_{yz}=\frac{\partial w}{\partial y}+\frac{\partial v}{\partial z},\ \gamma_{zx}=\frac{\partial w}{\partial x}+\frac{\partial u}{\partial z}.$$

Substituting $u(x, y, z)$, $v(x, y)$, and $w(x, y, z)$ in the above relations, we obtain:

$$\in_x=12x\times 10^{-3},\ \in_y=4y\times 10^{-3},\ \in_z=-12z\times 10^{-3},$$

$$\gamma_{xy}=(8y-3)\times 10^{-3},\ \gamma_{yz}=12y\times 10^{-3},\ \gamma_{zx}=3\times 10^{-3}.$$

At point $(2, -1, 3)$,

$\in_x = 24\times10^{-3}, \in_y = -4\times10^{-3}, \in_z = -36\times10^{-3}$,

$\gamma_{xy} = -11\times10^{-3}, \gamma_{yz} = -12\times10^{-3}, \gamma_{zx} = 3\times10^{-3}$.

Hence, the strain tensor is given as

$$\in_{ij} = \begin{bmatrix} 24 & -5.5 & 1.5 \\ -5.5 & -4 & -6 \\ 1.5 & -6 & -36 \end{bmatrix}\times10^{-3}. \qquad \text{...(Ans.)}$$

Example 2.5

The strain tensor at a point in a body is given by

$$\in_{ij} = \begin{bmatrix} 7.5 & 2 & 2.5 \\ 2 & 2 & 3 \\ 2.5 & 3 & 10 \end{bmatrix}\times10^{-3}.$$

Find the principal strains and principal axes.

Solution:

$$\in_{ij} = \begin{bmatrix} 7.5 & 2 & 2.5 \\ 2 & 2 & 3 \\ 2.5 & 3 & 10 \end{bmatrix}\times10^{-3} = \begin{bmatrix} \in_x & \frac{1}{2}\gamma_{xy} & \frac{1}{2}\gamma_{xz} \\ \frac{1}{2}\gamma_{yx} & \in_y & \frac{1}{2}\gamma_{yz} \\ \frac{1}{2}\gamma_{zx} & \frac{1}{2}\gamma_{zy} & \in_z \end{bmatrix}.$$

Hence, $\in_x = 7.5\times10^{-3}$, $\in_y = 2\times10^{-3}$, $\in_z = 10\times10^{-3}$,

$\gamma_{xy} = 4\times10^{-3}$, $\gamma_{yz} = 6\times10^{-3}$, $\gamma_{zx} = 5\times10^{-3}$.

(a) *Computing Principal Strain*

Strain invariants are given as

$$J_1 = \in_x + \in_y + \in_z = 19.5\times10^{-3}$$

$$J_2 = \in_x\in_y + \in_y\in_z + \in_z\in_x - \frac{1}{4}\left(\gamma_{xy}^2 + \gamma_{yz}^2 + \gamma_{zx}^2\right) = 90.75\times10^{-6}$$

$$J_3 = \det|\in| = 60\times10^{-9}.$$

Characteristic equation $\in^3 - J_1\in^2 + J_2\in - J_3 = 0$.

Solving,
$$\in_1 = 12.753\times10^{-3}$$
$$\in_2 = 5.957\times10^{-3} \qquad \text{...(Ans.)}$$
$$\in_3 = 0.7897\times10^{-3}.$$

(b) *Computing Principal Strain Directions*

$$\left(\epsilon_x - \epsilon_1\right) l_1 + \frac{1}{2}\gamma_{xy} m_1 + \frac{1}{2}\gamma_{xz} n_1 = 0$$

For $\epsilon_1 = 12.753\times10^{-3}$; $$\frac{1}{2}\gamma_{xy} l_1 + \left(\epsilon_y - \epsilon_1\right) m_1 + \frac{1}{2}\gamma_{yz} n_1 = 0$$

$$l_1^2 + m_1^2 + n_1^2 = 1.$$

Solving, $l_1 = 0.503$, $m_1 = 0.318$, $n_1 = 0.803$. ...(Ans.)

$$\left(\epsilon_x - \epsilon_2\right) l_2 + \frac{1}{2}\gamma_{xy} m_2 + \frac{1}{2}\gamma_{xz} n_2 = 0$$

For $\epsilon_2 = 5.957\times10^{-3}$; $$\frac{1}{2}\gamma_{xy} l_2 + \left(\epsilon_y - \epsilon_2\right) m_2 + \frac{1}{2}\gamma_{yz} n_2 = 0$$

$$l_2^2 + m_2^2 + n_2^2 = 1.$$

Solving, $l_2 = -0.844$, $m_2 = -0.0197$, $n_2 = 0.536$. ...(Ans.)

$$\left(\epsilon_x - \epsilon_3\right) l_3 + \frac{1}{2}\gamma_{xy} m_3 + \frac{1}{2}\gamma_{xz} n_3 = 0$$

For $\epsilon_3 = 0.7897\times10^{-3}$; $$\frac{1}{2}\gamma_{xy} l_3 + \left(\epsilon_y - \epsilon_1\right) m_3 + \frac{1}{2}\gamma_{yz} n_3 = 0$$

$$l_3^2 + m_3^2 + n_3^2 = 1.$$

Solving, $l_3 = 0.186$, $m_3 = -0.948$, $n_3 = 0.258$. ...(Ans.)

Example 2.6

The strain tensor at a point in a body is given by

$$\epsilon_{ij} = \begin{bmatrix} 2 & 0 & 4 \\ 0 & 1 & 2 \\ 4 & 2 & 2 \end{bmatrix} \times 10^{-3}.$$

Find the invariants of the tensor and split the tensor into hydrostatic (spherical) and deviatoric tensors.

Solution:

(a) Strain invariants are calculated as

$$J_1 = \epsilon_x + \epsilon_y + \epsilon_z = 5\times10^{-3}$$

$$J_2 = \epsilon_x\epsilon_y + \epsilon_y\epsilon_z + \epsilon_z\epsilon_x - \frac{1}{4}\left(\gamma_{xy}^2 + \gamma_{yz}^2 + \gamma_{zx}^2\right) = -12\times10^{-6}$$

$$J_3 = \det|\epsilon| = -20\times10^{-9}.$$ ...(Ans.)

(b) Hydrostatic and deviatoric strains:

$$\epsilon_{mean} = \frac{\epsilon_x + \epsilon_y + \epsilon_z}{3} = 1.667\times10^{-3} = \epsilon_m.$$

Therefore, we have hydrostatic strain tensor:

$$[\in]_{hydrostatic} = \begin{bmatrix} 1.667 & 0 & 0 \\ 0 & 1.667 & 0 \\ 0 & 0 & 1.667 \end{bmatrix} \times 10^{-3}$$ ...(Ans.)

Deviatoric strain tensor: $$[\in]_{deviatoric} = \begin{bmatrix} \in_x - \in_m & \frac{1}{2}\gamma_{xy} & \frac{1}{2}\gamma_{xz} \\ \frac{1}{2}\gamma_{yx} & \in_y - \in_m & \frac{1}{2}\gamma_{yz} \\ \frac{1}{2}\gamma_{zx} & \frac{1}{2}\gamma_{zy} & \in_z - \in_m \end{bmatrix}$$

$$= \begin{bmatrix} 0.333 & 0 & 4 \\ 0 & -0.667 & 2 \\ 4 & 2 & 0.333 \end{bmatrix} \times 10^{-3}.$$ ...(Ans.)

Example 2.7

The components of the strain tensor at a point in a body are as follows:

$$\in_x = a_1x^2 + a_2xy + a_3y^2;\ \gamma_{xy} = a_4xy$$
$$\in_y = b_1x^2 + b_2xy + b_3y^2;\ \gamma_{yz} = b_4yz$$
$$\in_z = c_1x^2 + c_2xy + c_3y^2;\ \gamma_{xz} = c_4xz.$$

Find the conditions among the constants for the strain field to be compatible.

Solution:

Substituting the strain functions in the six compatibility equations:

$$\frac{\partial^2 \in_x}{\partial y^2} + \frac{\partial^2 \in_y}{\partial x^2} = \frac{\partial^2 \gamma_{xy}}{\partial x \partial y};\ \Rightarrow\ 2(a_3 + b_1) = a_4.$$

$$\frac{\partial^2 \in_y}{\partial z^2} + \frac{\partial^2 \in_z}{\partial y^2} = \frac{\partial^2 \gamma_{yz}}{\partial y \partial z};\ \Rightarrow\ 2c_3 = b_4.$$

$$\frac{\partial^2 \in_z}{\partial x^2} + \frac{\partial^2 \in_x}{\partial z^2} = \frac{\partial^2 \gamma_{xz}}{\partial z \partial x};\ \Rightarrow\ 2c_1 = c_4.$$

$$\frac{\partial}{\partial x}\left(\frac{\partial \gamma_{xy}}{\partial z} - \frac{\partial \gamma_{yz}}{\partial x} + \frac{\partial \gamma_{zx}}{\partial y}\right) = 2\frac{\partial^2 \in_x}{\partial y \partial z};\ \Rightarrow \text{no relation found.}$$

$$\frac{\partial}{\partial y}\left(\frac{\partial \gamma_{yz}}{\partial x} - \frac{\partial \gamma_{zx}}{\partial y} + \frac{\partial \gamma_{xy}}{\partial z}\right) = 2\frac{\partial^2 \in_y}{\partial x \partial z};\ \Rightarrow \text{no relation found.}$$

$$\frac{\partial}{\partial z}\left(\frac{\partial \gamma_{zx}}{\partial y}-\frac{\partial \gamma_{xy}}{\partial z}+\frac{\partial \gamma_{yz}}{\partial x}\right)=2\frac{\partial^2 \epsilon_z}{\partial x \partial y}; \Rightarrow c_2=0.$$

Thus, the constants are related as: $\begin{cases} 2(a_3+b_1)=a_4 \\ 2c_3=b_4 \\ 2c_1=c_4 \\ c_2=0 \end{cases}$ ...(Ans.)

Example 2.8

Strain measurements by a three-element rectangular strain rosette at a point in a body give: $\epsilon_A=100\times10^{-6}$, $\epsilon_B=50\times10^{-6}$, and $\epsilon_C=50\times10^{-6}$. Find the principal strains and their directions at the point.

Solution:

For a three-element strain rosette,

$$\epsilon_A=100\times10^{-6},\ \epsilon_B=50\times10^{-6},\ \epsilon_C=50\times10^{-6}$$

(a) Principal strains:

$$\epsilon_{1,2}=\frac{1}{2}(\epsilon_A+\epsilon_B)\pm\frac{1}{2}\sqrt{(\epsilon_A-\epsilon_C)^2+(2\epsilon_B-\epsilon_A-\epsilon_C)^2}$$

$$=\left(75\pm25\sqrt{2}\right)\times10^{-6} = 110.35\times10^{-6} \text{ and } 39.64\times10^{-6}.$$ ...(Ans.)

(b) Direction of principal strain:

$$\epsilon_A=\epsilon_x=100\times10^{-6}$$

$$\epsilon_B=\frac{1}{2}(\epsilon_x+\epsilon_y+\gamma_{xy})=50\times10^{-6}$$

$$\epsilon_C=\epsilon_y=50\times10^{-6}$$

$$\therefore\ \gamma_{xy}=2\epsilon_B-\epsilon_A-\epsilon_C=-50\times10^{-6}$$

$$\tan 2\theta=\frac{\gamma_{xy}}{(\epsilon_x-\epsilon_y)}=-1$$

Hence, $\theta_1=-22.5°$, $\theta_2=90°+\theta_1=67.5°$ (principal strains are orthogonal). ...(Ans.)

Exercises

1. Define tensors and identify the order of tensor of the following physical quantities:
   (a) Temperature
   (b) Velocity
   (c) Strain

   State also their transformation laws.

2. If the displacement components at a point in a body in the *x, y, z* directions are *u, v, w* respectively in Cartesian coordinate system, find the expressions for the normal and shear strains from the first principle.

   If the coordinates of the points *A, B, C,* and *D* in the rectangular element, shown in Figure 2.4, are (0, 0). (0, 80), (100, 80), and (100, 0) respectively in its undeformed position, and the coordinates of the points *A′, B′, C′,* and *D′* in its deformed position are approximately (1, 1), (2.5, 81), (103.5, 83), and (102, 3) respectively in arbitrary units, find the strain components ϵ_x, ϵ_y, and γ_{xy}. Find also rotation ω_{xy} in *xy* plane.

   (0.01, 0.0187, 0.0385, –0.000525)

3. Show that the normal strain ϵ_n in an arbitrary direction is given by

   $$\epsilon_n = \epsilon_x l^2 + \epsilon_y m^2 + \epsilon_z n^2 + \gamma_{xy} lm + \gamma_{yz} mn + \gamma_{zx} nl,$$

   where $\epsilon_x, \epsilon_y, \epsilon_z, \gamma_{xy}, \gamma_{yz}$, and γ_{zx} are the normal and shear strains in Cartesian coordinate system and the d.c. of the normal vector are l, m, and n.

4. The state of strain at a point in a material is defined by the following components:

   $$\epsilon_x = 1.5\times10^{-3}, \epsilon_y = 0.6\times10^{-3}, \epsilon_{xz} = -0.5\times10^{-3}$$

   $$\gamma_{xy} = -0.9\times10^{-3}, \gamma_{yz} = 0.4\times10^{-3}, \gamma_{zx} = 0.3\times10^{-3}.$$

   Find the normal strain in the directions:
   (i) Equally inclined to the three coordinate axes,
   (ii) Making angles 30°, 60°, and –30° to the axes *x, y,* and *z* respectively.

   ($0.467 \times 10^{-3}, 0.91 \times 10^{-3}$)

5. Decompose the strain tensor at a point in a body in hydrostatic and deviatoric components and discuss the properties of the constituent terms.

6 Write the strain tensor in the case of a 2D strain analysis, where ϵ_z, γ_{yz}, and γ_{xz} may be assumed to be negligible and the strains are independent of *z*-direction, i.e., $\frac{\partial}{\partial z} = 0$. Write also the strain invariants and the characteristic equation in terms of strain invariants.

   Find also the principal stresses from the characteristic equation.

7. What are the basic considerations in the development of compatibility equations. Show that for a 2D case as stated in problem 6, the equation reduces to

   $$\frac{\partial^2 \epsilon_x}{\partial y^2} + \frac{\partial^2 \epsilon_y}{\partial x^2} = \frac{\partial^2 \gamma_{xy}}{\partial x \partial y}.$$

8. The state of strain at a point in a body is given by

$$\epsilon_x = 1.5\times10^{-3}, \epsilon_y = 0.6\times10^{-3}, \epsilon_z = 0$$

$$\gamma_{xy} = -0.5\times10^{-3}, \gamma_{yz} = 0, \gamma_{zx} = 0.$$

Find

(a) Normal and shear strains on the plane whose normal is at 30° to *x*-axis and perpendicular to *z*-axis.

(b) The strain invariants.

(c) Principal strains.

(d) Maximum shear strains at the point.

[(a) 1.06×10^{-3}, 1.95×10^{-3} (b) 2.1×10^{-3}, 0.837×10^{-3}, 0 (c) 1.565×10^{-3}, 0.535×10^{-3}, 0 (d) 1.03×10^{-3}, 1.565×10^{-3}, 0.535×10^{-3}]

9. Three strain gauges *a*, *b*, and *c* in a rectangular strain rosette are attached at a point on the surface of a body. When the body is strained, the strain gauges register the following direct strains:

$$\epsilon_a = -0.3\times10^{-3}, \epsilon_b = -0.8\times10^{-3}, \epsilon_c = 1.5\times10^{-3}$$

Find the magnitudes and directions of the principal strains at the point concerned.

[1.115×10^{-3}, -2.215×10^{-3}, 28.6°, and 118.6°]

3

Stress Analysis

Learning Objectives

After careful study of this chapter, students should be able to do the following:

LO1: Define stress at a point.

LO2: Describe stresses on an oblique plane.

LO3: Define principal stresses, hydrostatic, and deviatorial stress tensor.

LO4: Calculate shear stresses.

LO5: Construct Mohr's circle.

LO6: Analyze equations of equilibrium.

3.1 STATE OF STRESS AT A POINT [LO1]

When a body is subjected to external forces, its behavior depends on the magnitude and distribution of forces and properties of the body material. Depending on these factors, the body may deform elastically or plastically, or it may fracture. The body may also fail by fatigue when subjected to repetitive loading. Here we are primarily interested in elastic deformation of materials.

In order to establish the concept of stress and stress at a point, let us consider a straight bar of uniform cross-section of area A and subjected to uniaxial force F as shown in Figure 3.1. Stress at a typical section $A-A'$ is normally given as $\sigma = F/A$. This is true only if the force is uniformly distributed over the area A, but this is rarely true. Therefore, definition of stress must be considered by progressively reducing the area until it is small enough such that the force may be considered to be uniformly distributed.

To understand this, consider a body subjected to external forces P_1, P_2, P_3, and P_4 as shown in Figure 3.2. If we now cut the body in two pieces,

Internal forces f_1, f_2, f_3, etc. are developed to keep the pieces in equilibrium. Now consider an infinitesimal element of area ΔA at the cut section and let the resultant force on the element be Δf. Since the element is very small, we may consider that the forces are uniformly distributed and therefore we may define stress σ at the element as

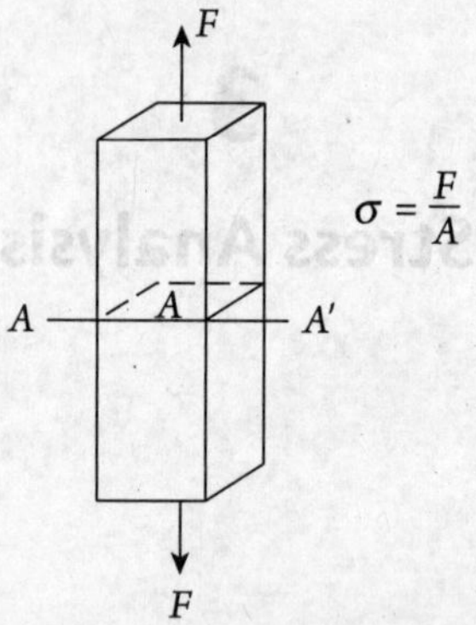

Figure 3.1 Average stress at a section in a body subjected to uniaxial load

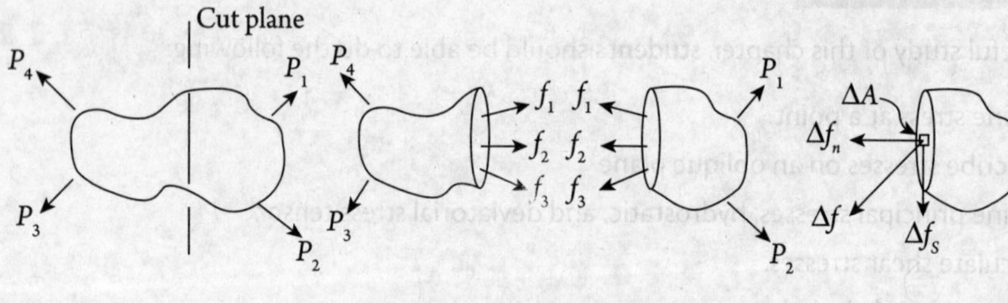

Figure 3.2 Internal forces developed at P_2, the cut plane

$$\sigma = \lim_{\Delta A \to 0} \frac{\Delta f}{\Delta A}. \tag{3.1.1}$$

If we now resolve the force Δf in the perpendicular direction as Δf_n and in the parallel direction as Δf_s, then we may define normal and shear stress as

$$\sigma_n = \lim_{\Delta A \to 0} \frac{\Delta f_n}{\Delta A}$$
$$\tau = \lim_{\Delta A \to 0} \frac{\Delta f_s}{\Delta A}. \tag{3.1.2}$$

Stresses developed within a body may be normal or shear in a three-dimensional (3D) system. There are many different systems of notations for stresses to represent the state of stress at a point. Perhaps the most common is the single suffix for normal stresses and double suffix for shear stress. The state of stress at a point is traditionally shown on the six faces of a cube as represented in Figure 3.3.

The complete stress system at a point may be given by nine stress elements, and the following stress tensor (Cauchy stress tensor) defines this completely.

$$[\sigma] = \begin{bmatrix} \sigma_x & \tau_{xy} & \tau_{xz} \\ \tau_{yx} & \sigma_y & \tau_{yz} \\ \tau_{zx} & \tau_{zy} & \sigma_z \end{bmatrix}. \tag{3.1.3}$$

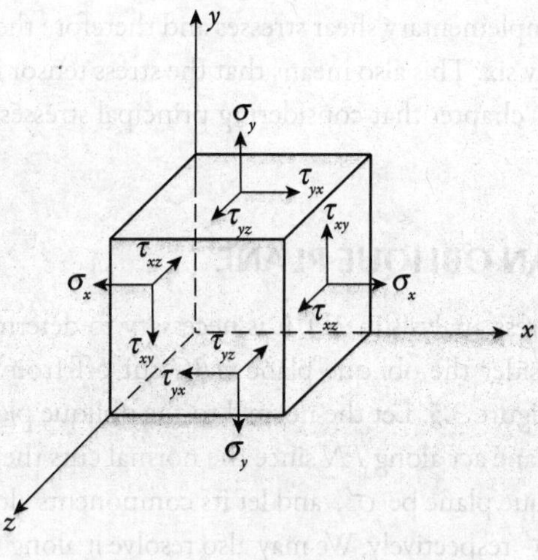

Figure 3.3 State of stress at a point

The complete stress system requires three normal stresses and six shear stresses. However, from equilibrium conditions, the number of shear stresses may be reduced to three. Consider only shear stresses on, say, xy plane as shown in Figure 3.4. Normal stresses σ_x and σ_y on the two vertical and horizontal faces cancel each other, and therefore these stresses are not shown in Figure 3.4. Taking moments of only shear forces acting on the element about A, we have

$$\left[\tau_{xy}\left(\delta y \delta z\right)\right]\delta x = \left[\tau_{yx}\left(\delta x \delta z\right)\right]\delta y.$$

This gives $\tau_{xy} = \tau_{yx}$.

Similarly, considering the equilibrium in yz and xz planes, we get

$\tau_{yz} = \tau_{zy}$ and

$\tau_{zx} = \tau_{xz}$.

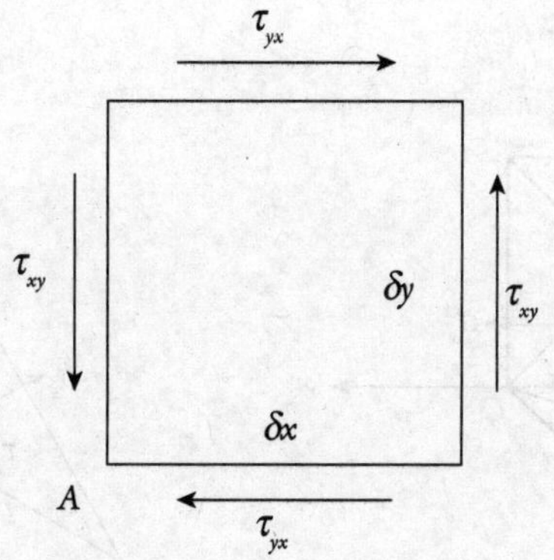

Figure 3.4 Shear stresses in *xy* plane

These are known as complementary shear stresses and therefore the number of stress elements in the tensor reduces to only six. This also means that the stress tensor is symmetric.

We shall see later in this chapter that considering principal stresses, the stress elements would reduce to only three.

3.2 STRESSES ON AN OBLIQUE PLANE [LO2]

To proceed further with stress analysis in 3D, it is necessary to determine the stress components on an oblique plane. Consider the oblique plane *ABC* cut off from the corner of the stressed elemental cube shown in Figure 3.5. Let the normal to the oblique plane be *ON*, and the normal stress σ'_n on the oblique plane act along *PN* since the normal cuts the oblique plane at *P*. Let the resultant stress on the oblique plane be σ', and let its components along the normal and parallel to the surface be σ'_n and σ'_s respectively. We may also resolve it along the three coordinate axes as σ'_x, σ'_y, and σ'_z, and this gives

$$\sigma'^2 = \sigma'^2_n + \sigma'^2_s = \sigma'^2_x + \sigma'^2_y + \sigma'^2_z. \tag{3.2.1}$$

This is shown in Figures 3.5 and 3.6.

Considering the force equilibrium of the tetrahedron shown in Figures 3.5 and 3.6, we may develop the relations between the stress components on the oblique plane along the coordinate axes, σ'_x, σ'_y, and σ'_z and the stresses on the other three faces of the tetrahedron.

For the equilibrium in the *x*-direction,

$$\sigma_x \Delta ABO + \tau_{yx} \Delta AOC + \tau_{xz} \Delta BOC = \sigma'_x \Delta ABC,$$

where ΔABO, ΔAOC, ΔBOC, and ΔABC represent the areas of triangle involved.

Following the definition of direction cosines, *l*, *m*, and *n* in equation (2.3.4), and taking $\tau_{xy} = \tau_{yx}$ we may write:

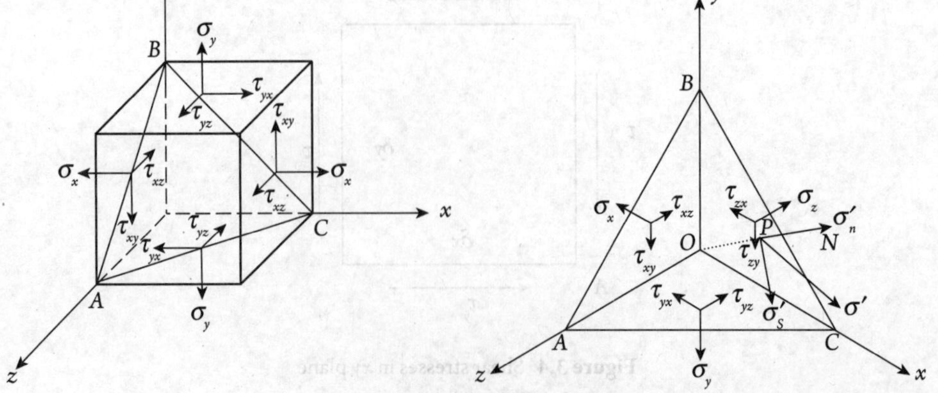

Figure 3.5 Stress on an oblique plane

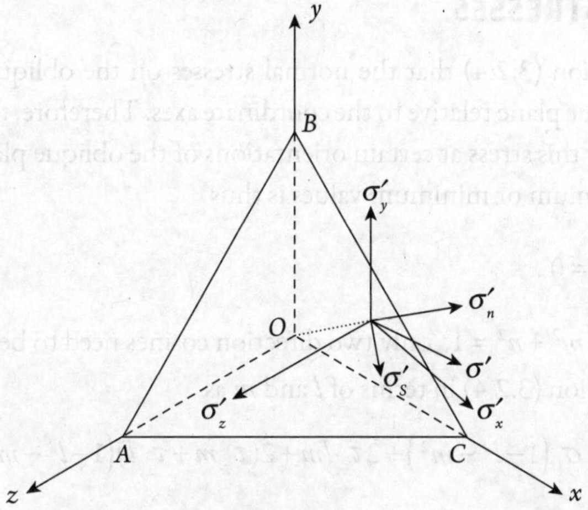

Figure 3.6 Stress components on an oblique plane

$$\sigma'_x = \sigma_x l + \tau_{xy} m + \tau_{xz} n.$$

Considering the equilibrium in the y and z directions, we may write two more equations:

$$\sigma'_y = \tau_{yx} l + \sigma_y m + \tau_{yz} n$$

$$\sigma'_z = \tau_{zx} l + \tau_{zy} m + \sigma_z n.$$

These three equations (Cauchy relations) form the basis of further stress analysis, and we therefore put them under the same bracket as

$$\begin{aligned} \sigma'_x &= \sigma_x l + \tau_{xy} m + \tau_{xz} n \\ \sigma'_y &= \tau_{yx} l + \sigma_y m + \tau_{yz} n \\ \sigma'_z &= \tau_{zx} l + \tau_{zy} m + \sigma_z n. \end{aligned} \tag{3.2.2}$$

Resolving these three stresses along the normal ON, we may write

$$\sigma'_n = \sigma'_x l + \sigma'_y m + \sigma'_z n\,. \tag{3.2.3}$$

On substitution of σ'_x, σ'_y, and σ'_z from equation (3.2.2)

$$\sigma'_n = \sigma_x l^2 + \sigma_y m^2 + \sigma_z n^2 + 2\left(\tau_{xy} lm + \tau_{yz} mn + \tau_{zx} ln\right). \tag{3.2.4}$$

From equation (3.2.1), we may also write

$$\sigma'_s = \sqrt{\sigma'^2_x + \sigma'^2_y + \sigma'^2_z - \sigma'^2_n}\,. \tag{3.2.5}$$

Equation (3.2.4) expresses the normal stress in the oblique plane in terms of the stresses on the other three faces of the tetrahedron. Similarly, equation (3.2.5) may also express the shear stress on the oblique plane in terms of the stresses on the other three faces of the tetrahedron. These two equations are important in further analysis of the state of stress at a point.

3.3 PRINCIPAL STRESSES [LO3]

It can be seen in equation (3.2.4) that the normal stresses on the oblique plane depend on the orientation of the oblique plane relative to the coordinate axes. Therefore, there must be maximum and minimum values of this stress at certain orientations of the oblique plane.

Condition for maximum or minimum values is thus

$$\frac{\partial \sigma'_n}{\partial l} = \frac{\partial \sigma'_n}{\partial m} = \frac{\partial \sigma'_n}{\partial n} = 0 . \tag{3.3.1}$$

However, since $l^2 + m^2 + n^2 = 1$, only two direction cosines need to be considered.

We now write equation (3.2.4) in terms of l and m as

$$\sigma'_n = \sigma_x l^2 + \sigma_y m^2 + \sigma_z \left(1 - l^2 - m^2\right) + 2\tau_{xy} lm + 2(\tau_{yz} m + \tau_{zx} l)\left(1 - l^2 - m^2\right)^{1/2}$$

$$= l^2\left(\sigma_x - \sigma_z\right) + m^2\left(\sigma_y - \sigma_z\right) + \sigma_z + 2\tau_{xy} lm + 2(\tau_{yz} m + \tau_{zx} l)\left(1 - l^2 - m^2\right)^{1/2} .$$

Now differentiating σ'_n w.r.t l, we have

$$\frac{\partial \sigma'_n}{\partial l} = 2l\left(\sigma_x - \sigma_z\right) + 2\tau_{xy} m + 2\tau_{zx}\left(1 - l^2 - m^2\right)^{1/2} + 2(\tau_{yz} m + \tau_{zx} l)\frac{1}{2}\left(1 - l^2 - m^2\right)^{-1/2}(-2l)$$

$$= 2\left(\sigma_x l + \tau_{xy} m + \tau_{zx} n\right) - 2\left(\tau_{yz} m + \tau_{zx} l + \sigma_z n\right)\frac{l}{n} = 0. \tag{3.3.2}$$

From equation (3.2.2), we have:

$$\sigma_x l + \tau_{xy} m + \tau_{zx} n = \sigma'_x$$

$$\tau_{yz} m + \tau_{zx} l + \sigma_z n = \sigma'_z .$$

This gives $\dfrac{\sigma'_x}{l} = \dfrac{\sigma'_z}{n}$.

And on similar analysis, we may show

$$\frac{\sigma'_x}{l} = \frac{\sigma'_y}{m} = \frac{\sigma'_z}{n} = \sigma \text{ (say)}. \tag{3.3.3}$$

Now

$$\sigma'_n = \sigma'_x l + \sigma'_y m + \sigma'_z n.$$

Substituting σ'_x, σ'_y, and σ'_z from equation (3.3.3), we may write

$$\sigma'_n = \sigma\left(l^2 + m^2 + n^2\right) = \sigma.$$

We may also write

$$\sigma'^2 = \sigma'^2_x + \sigma'^2_y + \sigma'^2_z = \sigma^2\left(l^2 + m^2 + n^2\right) = \sigma^2;\ \sigma' = \sigma,$$

we then have $\sigma'_n = \sigma$. (3.3.4)

However, from equation (3.2.1),

$$\sigma'^2 = \sigma_n'^2 + \sigma_s'^2.$$

Therefore, $\sigma_s' = 0$. (3.3.5)

Equation (3.3.1) gives the condition for maximum or minimum normal stress and equation (3.3.5) indicates that at these conditions shear stress is zero.

From the above we may conclude that there exist three orthogonal planes on which the normal stresses are either maximum or minimum, and no shear stress exists on these planes. These planes are known as *principal planes* and normal stresses on these planes are defined as *principal stresses*.

Now, combining equations (3.2.2) and (3.3.3), we may write

$$\begin{aligned} \sigma_x' &= l\sigma = \sigma_x l + \tau_{xy} m + \tau_{xz} n \\ \sigma_y' &= m\sigma = \tau_{yx} l + \sigma_y m + \tau_{yz} n \\ \sigma_z' &= n\sigma = \tau_{zx} l + \tau_{zy} m + \sigma_z l. \end{aligned} \tag{3.3.6}$$

This may be written in the matrix notation as

$$\begin{bmatrix} \sigma_x - \sigma & \tau_{xy} & \tau_{xz} \\ \tau_{yx} & \sigma_y - \sigma & \tau_{yz} \\ \tau_{zx} & \tau_{zy} & \sigma_z - \sigma \end{bmatrix} \begin{bmatrix} l \\ m \\ n \end{bmatrix} = \begin{bmatrix} 0 \\ 0 \\ 0 \end{bmatrix}. \tag{3.3.7}$$

A trivial solution of the above equation is $l = m = n = 0$ but this does not satisfy $l^2 + m^2 + n^2 = 1$. We therefore consider the nontrivial solution, which is

$$\begin{vmatrix} \sigma_x - \sigma & \tau_{xy} & \tau_{xz} \\ \tau_{yx} & \sigma_y - \sigma & \tau_{yz} \\ \tau_{zx} & \tau_{zy} & \sigma_z - \sigma \end{vmatrix} = 0 \tag{3.3.8}$$

and this indicates that σ is the eigenvalue of the stress matrix.

On expansion of equation (3.3.8), we have the following cubic equation

$$\begin{aligned} &\sigma^3 - \left(\sigma_x + \sigma_y + \sigma_z\right)\sigma^2 + \left(\sigma_x\sigma_y + \sigma_y\sigma_z + \sigma_x\sigma_z - \tau_{xy}^{\ 2} - \tau_{yz}^{\ 2} - \tau_{xz}^{\ 2}\right)\sigma \\ &-\left(\sigma_x\sigma_y\sigma_z + 2\tau_{xy}\tau_{yz}\tau_{zx} - \sigma_x\tau_{yz}^{\ 2} - \sigma_y\tau_{xz}^{\ 2} - \sigma_z\tau_{xy}^{\ 2}\right) = 0. \end{aligned} \tag{3.3.9}$$

Three roots of the equation represent the three principal stresses associated with the particular state of stress at a point. Let the roots be σ_1, σ_2, and σ_3, such that $\sigma_1 > \sigma_2 > \sigma_3$. The cubic equation may then be written also as

$$\left(\sigma - \sigma_1\right)\left(\sigma - \sigma_2\right)\left(\sigma - \sigma_3\right) = 0,$$

i.e., $$\sigma^3 - \left(\sigma_1 + \sigma_2 + \sigma_3\right)\sigma + \left(\sigma_1\sigma_2 + \sigma_2\sigma_3 + \sigma_1\sigma_3\right)\sigma - \sigma_1\sigma_2\sigma_3 = 0. \tag{3.3.10}$$

Comparing the coefficients and the constant terms in equations (3.3.9) and (3.3.10), we may also obtain three equations, and by solving them, we find σ_1, σ_2, and σ_3. Although two of the roots of such equations are in general complex, but since here principal stresses are the eigenvalues of a symmetric stress matrix with real coefficients, all three roots are real.

There are many different methods of finding the roots of a cubic equation, such as regula-falsi and Newton–Raphson methods. This will be discussed in more detail while solving problems. Having obtained the eigenvalues, we proceed to find the eigenvectors that represent the directions of the principal stresses. This can be determined by substituting σ_1, σ_2, and σ_3 in the equation (3.3.7). Each substitution would give a set of direction cosines (l_1, m_1, n_1) (say). We may thus finally obtain three sets of direction cosines by substituting σ_1, σ_2, and σ_3 in turn in place of σ in equation (3.3.7): (l_1, m_1, n_1), (l_2, m_2, n_2), (l_3, m_3, n_3). However, since $l^2 + m^2 + n^2 = 1$, some methodology is needed to find the eigenvectors and this was explained in Example 2.1 in Chapter 2. Similar method will be adopted here for stress analysis too. We may also note here that the eigenvectors of a symmetric matrix are mutually orthogonal and thus we may write, for example,

$$l_1 l_2 + m_1 m_2 + n_1 n_2 = 0.$$

This also means that the principal planes are mutually orthogonal.

We may also conclude that there must be an orientation of the stress element where no shear stress acts (Figure 3.7).

This indicates that the state of stress at a point can also be specified by three principal stresses and their directions instead of nine or six stress elements.

It is now convenient to choose the coordinate axes in such a way that they align with the normals to the principal planes. This is shown in Figure 3.8.

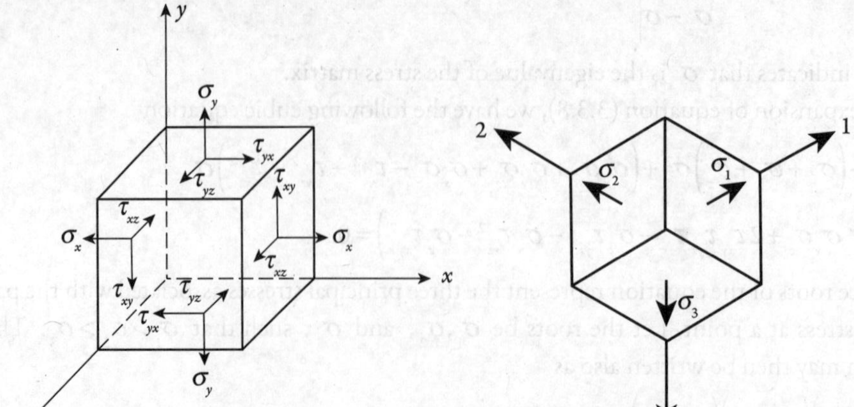

(a) Element with both normal and shear stresses (b) Element with only normal stresses (principal stresses)

Figure 3.7 Two orientations of an element, (a) one showing all the nine stress elements, and (b) the other showing only the principal stresses (no shear stresses)

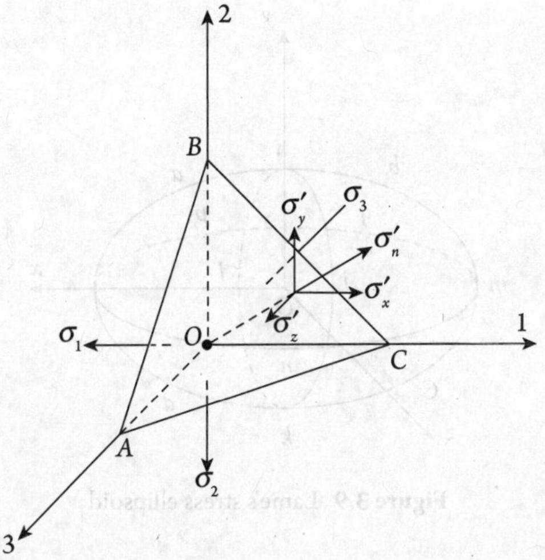

Figure 3.8 Stresses on the faces of a tetrahedron whose axes are aligned with the normal to principal planes

Following equation (3.3.3) and the discussions thereafter, we may say

$$\sigma'_x = l\sigma_1, \sigma'_y = m\sigma_2, \sigma'_z = n\sigma_3, \tag{3.3.11}$$

where l, m, and n are the direction cosines of the normal to oblique plane ABC and the coordinate axes align with the normals to the principal planes.

Lame's Stress Ellipsoid

From the above discussion, we may come up with a graphical representation of state of stress at a point. Since

$$l^2 + m^2 + n^2 = 1$$

from equation (3.3.11), we may write

$$\frac{\sigma'^2_x}{\sigma_1^2} + \frac{\sigma'^2_y}{\sigma_2^2} + \frac{\sigma'^2_z}{\sigma_3^2} = 1. \tag{3.3.12}$$

This is known as *Lame's stress ellipsoid*. This means that the tip of normal stress vector on the inclined plane would trace a 3D ellipsoid as the orientation of the inclined plane is varied, and the semi-axes of the ellipsoid represent the three principal stresses. We may conclude the following from Figure 3.9:

1. Sum of the principal radii = $\sigma_1 + \sigma_2 + \sigma_3 = I_1$, which is the first stress invariant.
2. Sum of the three principal areas, *abcd*, *ijkl*, and *mnpq* is proportional to $\sigma_1\sigma_2 + \sigma_2\sigma_3 + \sigma_3\sigma_1 = I_2$, second stress invariant.
3. Volume of the ellipsoid is proportional to $\sigma_1\sigma_2\sigma_3 = I_3$, the third stress invariant.
4. If $\sigma_1 = \sigma_2 = \sigma_3$, then the ellipsoid turns into a sphere and this represents a hydrostatic state of stress.

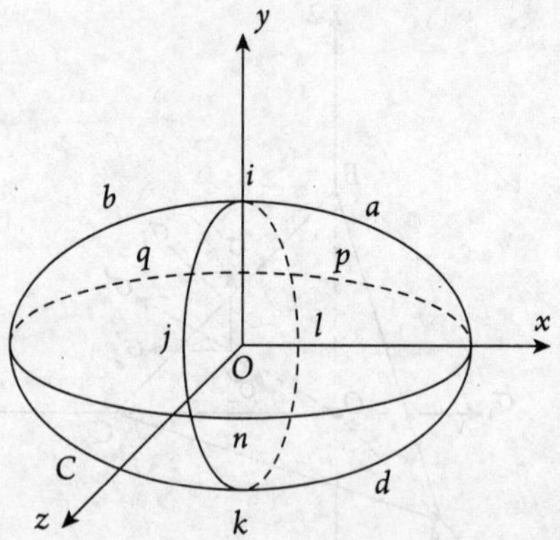

Figure 3.9 Lame's stress ellipsoid

If both σ'_y and σ'_z are zero, $\sigma'_x = \sigma_1$. This represents a simple tension or compression.

Another important issue in defining the state of stress at a point is that it depends only on the physical conditions and not on the coordinate system chosen. Therefore, the state of stress should be defined in terms of invariants with respect to coordinate systems. These were briefly mentioned in Lame's ellipsoid discussion, and here we explain them in detail.

Consider the two cubic equations (3.3.9) and (3.3.10) for evaluating principal stresses. In these equations, no matter what coordinate systems are used the three roots of these two equations must remain the same, and therefore the coefficients must also be the same.

We thus have three *stress invariants* I_1, I_2, and I_3 as follows:

$$I_1 = \sigma_x + \sigma_y + \sigma_z = \sigma_1 + \sigma_2 + \sigma_3$$

$$I_2 = \sigma_x\sigma_y + \sigma_y\sigma_z + \sigma_x\sigma_z - \tau_{xy}^2 - \tau_{yz}^2 - \tau_{zx}^2 = \sigma_1\sigma_2 + \sigma_2\sigma_3 + \sigma_1\sigma_3$$

$$I_3 = \sigma_x\sigma_y\sigma_z + 2\tau_{xy}\tau_{yz}\tau_{xz} - \sigma_x\tau_{yz}^2 - \sigma_y\tau_{xz}^2 - \sigma_z\tau_{xy}^2 = \sigma_1\sigma_2\sigma_3.$$

We may thus write the cubic equation briefly as

$$\sigma^3 - I_1\sigma^2 + I_2\sigma - I_3 = 0. \tag{3.3.13}$$

Some physical phenomena require a description of the state of stress in terms of stress invariants. For example, plastic yielding criterion for metals is often given in terms of von Mises stress, which is given by

$$(\sigma_x - \sigma_y)^2 + (\sigma_y - \sigma_z)^2 + (\sigma_z - \sigma_x)^2 + 6\left(\tau_{xy}^2 + \tau_{yz}^2 + \tau_{zx}^2\right) = (\sigma_1 - \sigma_2)^2 + (\sigma_2 - \sigma_3)^2 + (\sigma_3 - \sigma_1)^2.$$

This may be expressed in terms of I_1 and I_2 as $\left(2I_1^2 - 6I_2\right)$ and this is constant for a given stress tensor.

Hydrostatic and Deviatoric Stress Tensor

The general stress tensor in equation (3.1.3) can be split into two meaningful different tensors. The first one is known as hydrostatic or spherical stress tensor, where all the elements are normal stresses, and they are all equal to a mean stress σ_m defined as $\frac{\sigma_x+\sigma_y+\sigma_z}{3}$. The second one is the remaining part and is known as deviatoric stress tensor. The general stress tensor and its constituents are given as follows:

$$[\sigma]=\begin{bmatrix}\sigma_x & \tau_{xy} & \tau_{xz}\\ \tau_{yx} & \sigma_y & \tau_{yz}\\ \tau_{zx} & \tau_{zy} & \sigma_z\end{bmatrix}=\begin{bmatrix}\sigma_m & 0 & 0\\ 0 & \sigma_m & 0\\ 0 & 0 & \sigma_m\end{bmatrix}+\begin{bmatrix}\sigma_x-\sigma_m & \tau_{xy} & \tau_{xz}\\ \tau_{yx} & \sigma_y-\sigma_m & \tau_{yz}\\ \tau_{zx} & \tau_{zy} & \sigma_z-\sigma_m\end{bmatrix}.$$

Hydrostatic part *Deviatoric part*

The hydrostatic part causes change in volume because it consists of equal mean stresses in all directions whereas the deviatoric part causes distortion and change in shape. This component of the stress tensor is often correlated to yielding.

3.4 SHEAR STRESSES [LO4]

In our earlier discussion, we have seen that three independent shear stresses must be known to define the state of stress at a point. It is also known that three mutually orthogonal coordinate axes exist such that the shear stresses on the planes normal to the axes vanish. The axes are principal axes and the normal stresses on the principal planes are the principal stresses.

Following the analysis of the principal stresses, it is now necessary to find the maximum or minimum shear stresses.

We have from equation (3.2.1)

$$\sigma_s'^2=\sigma_x'^2+\sigma_y'^2+\sigma_z'^2-\sigma_n'^2.$$

Combining this with equations (3.3.11) and (3.2.3), we have

$$\sigma_s'^2=\left(\sigma_1 l\right)^2+\left(\sigma_2 m\right)^2+\left(\sigma_3 n\right)^2-\left(\sigma_1 l^2+\sigma_2 m^2+\sigma_3 n^2\right)^2.$$

This gives

$$\sigma_s'^2=l^2m^2\left(\sigma_1-\sigma_2\right)^2+m^2n^2\left(\sigma_2-\sigma_3\right)^2+n^2l^2\left(\sigma_3-\sigma_1\right)^2. \tag{3.4.1}$$

Again, considering that $l^2+m^2+n^2=1$, equation (3.4.1) needs to be expressed in terms of two direction cosines only, say, for example l and m, and this gives

$$\sigma_s'^2=l^2m^2\left(\sigma_1-\sigma_2\right)^2+m^2(1-l^2-m^2)\left(\sigma_2-\sigma_3\right)^2+l^2(1-l^2-m^2)\left(\sigma_3-\sigma_1\right)^2. \tag{3.4.2}$$

For maximum or minimum values of σ_s', we may write for convenience

$$\frac{\partial\sigma_s'^2}{\partial l}=\frac{\partial\sigma_s'^2}{\partial m}=0.$$

The first condition gives

$$\frac{\partial\sigma_s'^2}{\partial l}=2lm^2\left(\sigma_1-\sigma_2\right)^2-2lm^2\left(\sigma_2-\sigma_3\right)^2+\left(2l-4l^3-2lm^2\right)\left(\sigma_3-\sigma_1\right)^2=0 \tag{a}$$

and the second condition gives

$$\frac{\partial \sigma_s'^2}{\partial m} = 2ml^2\left(\sigma_1 - \sigma_2\right)^2 + \left(\sigma_2 - \sigma_3\right)^2\left(2m - 2ml^2 - 4m^3\right) - 2ml^2\left(\sigma_3 - \sigma_1\right)^2 = 0. \quad \text{(b)}$$

Both the conditions must be satisfied simultaneously.

One solution of equation (a) is $l = 0$ and this gives from (b) $2m - 4m^3 = 0$.

Therefore, solutions are $m = 0;\ m = \pm\frac{1}{\sqrt{2}}$.

We may therefore conclude that

at $l = 0;\ m = 0,\ n = 1$ the condition represents the principal plane.

Also, at $l = 0;\ m = \pm\frac{1}{\sqrt{2}},\ n = \pm\frac{1}{\sqrt{2}}$ the condition represents planes of maximum shear stress.

Similarly, $m = 0$ satisfies equation (b) and this gives from (a)

$$m = 0;\quad l = \pm\frac{1}{\sqrt{2}};\quad n = \pm\frac{1}{\sqrt{2}}$$

$$m = 0;\quad l = 0;\quad n = 1.$$

Here again the first condition represents planes of maximum shear stresses and the second one represents a principal plane.

Since $l = 0$ and $m = 0$ provide solutions as shown above, $n = 0$ must also provide similar solutions.

$$n = 0;\quad l = \pm\frac{1}{\sqrt{2}};\quad m = \pm\frac{1}{\sqrt{2}}$$

$$n = 0;\quad l = 0;\quad m = 1.$$

The same conclusions apply to these conditions too.

In a concise form, the 12 planes of maximum shear stresses are given as follows:

$$\begin{array}{lll} l = 0 & m = \pm\frac{1}{\sqrt{2}} & n = \pm\frac{1}{\sqrt{2}} \\ l = \pm\frac{1}{\sqrt{2}} & m = 0 & n = \pm\frac{1}{\sqrt{2}} \\ l = \pm\frac{1}{\sqrt{2}} & m = \pm\frac{1}{\sqrt{2}} & n = 0. \end{array} \quad (3.4.3)$$

Each set of direction cosines represents four planes, all parallel to one of the principal axes and inclined at 45° with the other two.

Maximum shear stresses and the corresponding normal stresses may be obtained from equations (3.4.1) and (3.2.3) reiterated below.

$$\sigma_s'^2 = l^2m^2\left(\sigma_1 - \sigma_2\right)^2 + m^2n^2\left(\sigma_2 - \sigma_3\right)^2 + n^2l^2\left(\sigma_3 - \sigma_1\right)^2$$

$$\sigma_n' = \sigma_1 l^2 + \sigma_2 m^2 + \sigma_3 n^2. \quad (3.4.4)$$

By substituting the direction cosines from equation (3.4.3), we have the following results:

$$l=0, m=\pm\frac{1}{\sqrt{2}}, n=\pm\frac{1}{\sqrt{2}}: \qquad \sigma'_s=\pm\frac{\sigma_2-\sigma_3}{2}; \qquad \sigma'_n=\frac{\sigma_2+\sigma_3}{2}.$$

$$l=\pm\frac{1}{\sqrt{2}}, m=0, n=\pm\frac{1}{\sqrt{2}}: \qquad \sigma'_s=\pm\frac{\sigma_3-\sigma_1}{2}; \qquad \sigma'_n=\frac{\sigma_3+\sigma_1}{2}. \qquad (3.4.5)$$

$$l=\pm\frac{1}{\sqrt{2}}, m=\pm\frac{1}{\sqrt{2}}, n=0: \qquad \sigma'_s=\pm\frac{\sigma_1-\sigma_2}{2}; \qquad \sigma'_n=\frac{\sigma_1+\sigma_2}{2}.$$

We may recall the two-dimensional (2D) analysis where we get only two maximum shear stresses and a corresponding normal stress. This is simply the third result in equation (3.4.5) and as shown below in the typical form for 2D analysis.

$$\tau_{max}=\pm\frac{\sigma_1-\sigma_2}{2} \text{ and } \sigma_n=\frac{\sigma_1+\sigma_2}{2}.$$

Octahedral Shear Stresses

It is possible to identify eight planes whose normals are equally inclined to the principal planes and they are of practical importance. These planes are known as octahedral planes.

This means $l=m=n=\frac{1}{\sqrt{3}}$ and therefore from equation (3.4.4), we have:

the normal stresses on octahedral planes as

$$\sigma_{n(oct)}=\sigma_1 l^2+\sigma_2 m^2+\sigma_3 n^2=\frac{1}{3}\left(\sigma_1+\sigma_2+\sigma_3\right)=\sigma_{av}. \qquad (3.4.6)$$

Therefore, the normal stresses on the octahedral planes represent the hydrostatic stresses.

Also, from equation (3.4.4), we have the shear stresses on octahedral planes as

$$\sigma_{s(oct)}=\left[l^2m^2\left(\sigma_1-\sigma_2\right)^2+m^2n^2\left(\sigma_2-\sigma_3\right)^2+n^2l^2\left(\sigma_1-\sigma_3\right)^2\right]^{1/2}.$$

With $l=m=n=\frac{1}{\sqrt{3}}$, this gives

$$\sigma_{s(oct)}=\frac{1}{3}\left[\left(\sigma_1-\sigma_2\right)^2+\left(\sigma_2-\sigma_3\right)^2+\left(\sigma_1-\sigma_3\right)^2\right]^{1/2}. \qquad (3.4.7)$$

We may also write shear stress in terms of elemental stresses as follows:

$$\sigma_{s(oct)}=\frac{1}{3}\left[\left(\sigma_x-\sigma_y\right)^2+\left(\sigma_y-\sigma_z\right)^2+\left(\sigma_x-\sigma_z\right)^2+6\left(\tau_{xy}^{\ 2}+\tau_{yz}^{\ 2}+\tau_{zx}^{\ 2}\right)\right]^{1/2}.$$

This shear stress is of practical importance since this gives a criterion for plastic yielding as given in von Mises criterion shown below:

$$\left(\sigma_1-\sigma_2\right)^2+\left(\sigma_2-\sigma_3\right)^2+\left(\sigma_1-\sigma_3\right)^2\geq 2\sigma_Y^{\ 2}.$$

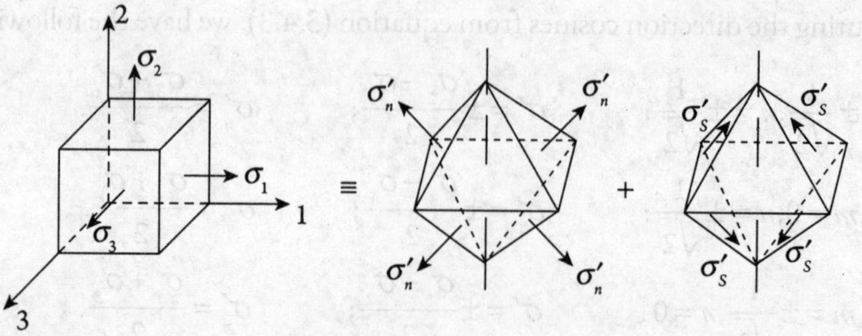

Figure 3.10 Pictorial representation of normal and shear stresses on octahedral planes

Here σ_Y represents yield stress. This means, plastic yielding criterion is simply:

$$\sigma_{s(oct)} \geq \frac{\sqrt{2}}{3}\sigma_Y, \tag{3.4.8}$$

where σ_Y is the tensile yield stress in a standard tensile test.

We may also confirm this by considering a standard tensile yield test where

$\sigma_x = \sigma_Y \qquad \tau_{xy} = \tau_{yz} = \tau_{zx} = 0$

$\sigma_y = 0 \qquad \sigma_z = 0.$

Therefore, the yield criterion is $\sigma_{s(oct)} = \frac{1}{3}\left(2\sigma_Y^2\right)^{1/2} = \frac{\sqrt{2}}{3}\sigma_Y.$

This may be pictorially represented as shown in Figure 3.10.

Octahedral planes are special in a way that the normal stresses on these planes are hydrostatic in nature, whereas the shear stresses represent the deviatoric components used to define the yield criterion.

3.5 GRAPHICAL REPRESENTATION OF STATES OF STRESSES [LO5]

It is useful to represent the state of stress at a point graphically, although it may not be directly useful in solving the elasticity problems.

Combining equations in (3.4.4) and the relation $l^2 + m^2 + n^2 = 1$, we may write by eliminating m and n

$$\left(\sigma'_n - \frac{\sigma_2 + \sigma_3}{2}\right)^2 + \sigma'^2_s = \left(\sigma_1 - \sigma_2\right)\left(\sigma_1 - \sigma_3\right)l^2 + \left(\frac{\sigma_2 - \sigma_3}{2}\right)^2. \tag{3.5.1}$$

This is the equation of a circle with σ'_n and σ'_s as axes, $\frac{\sigma_2 + \sigma_3}{2}$ as the distance of the center from the origin and radius = $\left[\left(\sigma_1 - \sigma_2\right)\left(\sigma_1 - \sigma_3\right)l^2 + \left(\frac{\sigma_2 - \sigma_3}{2}\right)^2\right]^{1/2}$.

If $\sigma_1 \geq \sigma_2 \geq \sigma_3$ and radius is real and its least value at $l = 0$ is

$$\left[(\sigma_1-\sigma_2)(\sigma_1-\sigma_3)\times 0+\left(\frac{\sigma_2-\sigma_3}{2}\right)^2\right]^{1/2}=\frac{1}{2}(\sigma_2-\sigma_3).$$

This gives Mohr's circle (Figure 3.11) for an element in $2-3$ plane with two principal stresses σ_2 and σ_3.

Similarly, we may write by eliminating l and m

$$\left(\sigma_n'-\frac{\sigma_1+\sigma_2}{2}\right)^2+\sigma_s'^2=(\sigma_1-\sigma_3)(\sigma_2-\sigma_3)n^2+\left(\frac{\sigma_1-\sigma_2}{2}\right)^2 \tag{3.5.2}$$

and the least radius at $n = 0$ is $\dfrac{\sigma_1-\sigma_2}{2}$.

This gives Mohr's circle (Figure 3.11) for an element in $1-2$ plane with two principal stresses σ_1 and σ_2.

Another equation may be written by eliminating l and n

$$\left(\sigma_n'-\frac{\sigma_1+\sigma_3}{2}\right)^2+\sigma_s'^2=(\sigma_2-\sigma_1)(\sigma_2-\sigma_3)m^2+\left(\frac{\sigma_1-\sigma_3}{2}\right)^2, \tag{3.5.3}$$

where radius $=\left[(\sigma_2-\sigma_1)(\sigma_2-\sigma_3)m^2+\left(\dfrac{\sigma_1-\sigma_3}{2}\right)^2\right]^{1/2}$.

Since $\sigma_1 > \sigma_2$ the first term in the radius is negative and maximum radius occurs at $m = 0$ and that is $\dfrac{\sigma_1-\sigma_3}{2}$.

Therefore, this gives the outer limit of any combination of direct and shear stresses.

The 3D Mohr's circle is shown in Figure 3.11. The normal and shear stresses acting on any plane fall within the shaded region.

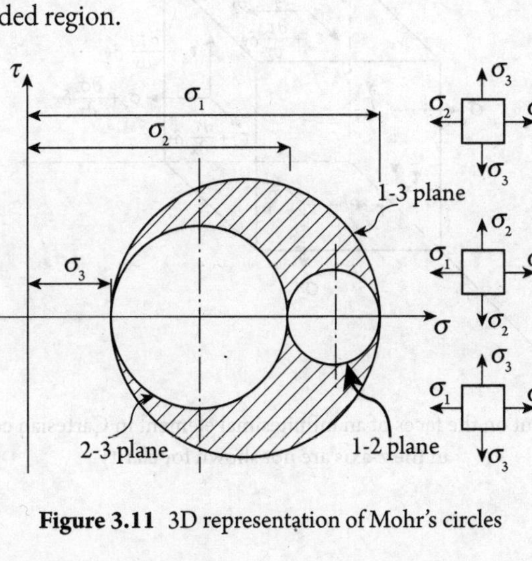

Figure 3.11 3D representation of Mohr's circles

3.6 EQUATIONS OF EQUILIBRIUM [LO6]

So far, we have been discussing the state of stress at a point with invariant stress components over an element. We shall now consider the equilibrium conditions in real situation where stresses vary from point to point. This essentially means that stresses must vary from one end of an infinitesimally small element to its other end. In a Cartesian coordinate system, let us consider the stresses on the faces of a small cubic element as shown in Figure 3.12. In addition to the stresses shown in Figure 3.12, body forces, such as gravitational, electrostatic forces given by X, Y, Z per unit volume in three coordinate axes are also considered.

Net force on the element in the x-direction is given by

$$\left(\frac{\partial \sigma_x}{\partial x}\delta x\right)\delta y \delta z + \left(\frac{\partial \tau_{xy}}{\partial y}\delta y\right)\delta x \delta z + \left(\frac{\partial \tau_{xz}}{\partial z}\delta z\right)\delta x \delta y + X\,\delta x \delta y \delta z\,.$$

For equilibrium, this force must equate to inertial force, i.e., $\rho\,\delta x\,\delta y\,\delta z\,f_x$, where ρ is the density and f_x is the acceleration in the x-direction. We may thus write the equilibrium equation in the x-direction as

$$\left(\frac{\partial \sigma_x}{\partial x}\delta x\right)\delta y \delta z + \left(\frac{\partial \tau_{xy}}{\partial y}\delta y\right)\delta x \delta z + \left(\frac{\partial \tau_{xz}}{\partial z}\delta z\right)\delta x \delta y + X\,\delta x \delta y \delta z = \rho\,\delta x\,\delta y\,\delta z\,f_x.$$

This reduces to $\dfrac{\partial \sigma_x}{\partial x} + \dfrac{\partial \tau_{xy}}{\partial y} + \dfrac{\partial \tau_{xz}}{\partial z} + X = \rho f_x.$

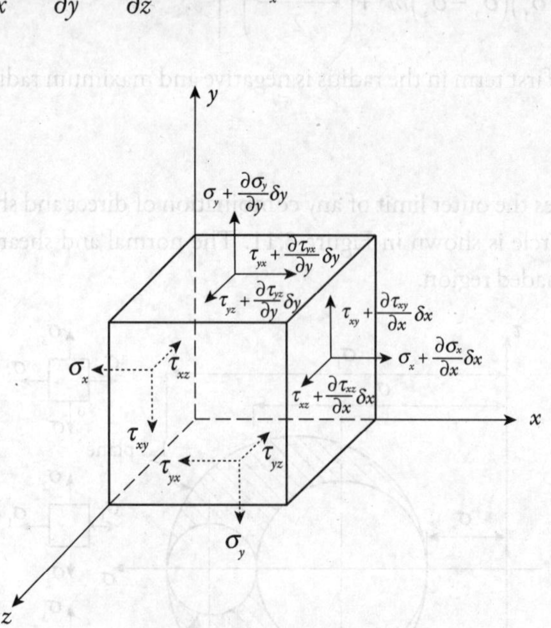

Figure 3.12 Stress increment on the faces of an infinitesimal element in Cartesian coordinate. Stress variations in the z-axis are not shown for clarity

Following this argument, we may write the following three equilibrium equations

$$\begin{aligned}
&\frac{\partial \sigma_x}{\partial x}+\frac{\partial \tau_{xy}}{\partial y}+\frac{\partial \tau_{xz}}{\partial z}+X=\rho f_x \\
&\frac{\partial \tau_{yx}}{\partial x}+\frac{\partial \sigma_y}{\partial y}+\frac{\partial \tau_{yz}}{\partial z}+Y=\rho f_y \\
&\frac{\partial \tau_{zx}}{\partial x}+\frac{\partial \tau_{zy}}{\partial y}+\frac{\partial \sigma_z}{\partial z}+Z=\rho f_z .
\end{aligned} \tag{3.6.1}$$

For static equilibrium, the accelerations in x, y, and z directions are $f_x = f_y = f_z = 0$. Therefore, we have the reduced form of equilibrium equations as

$$\begin{aligned}
&\frac{\partial \sigma_x}{\partial x}+\frac{\partial \tau_{xy}}{\partial y}+\frac{\partial \tau_{xz}}{\partial z}+X=0 \\
&\frac{\partial \tau_{yx}}{\partial x}+\frac{\partial \sigma_y}{\partial y}+\frac{\partial \tau_{yz}}{\partial z}+Y=0 \\
&\frac{\partial \tau_{zx}}{\partial x}+\frac{\partial \tau_{zy}}{\partial y}+\frac{\partial \sigma_z}{\partial z}+Z=0.
\end{aligned} \tag{3.6.2}$$

Example 3.1

The state of stress at a point in a material is given by the following stress tensor

$$[\sigma]=\begin{bmatrix} 100 & -20 & 45 \\ -20 & 50 & 60 \\ 45 & 60 & 40 \end{bmatrix} \text{MPa}.$$

Find the normal and shear stresses on an oblique plane making angles 50°, 80°, and 60° with the coordinate axes x, y, and z respectively.

Solution:

The direction cosines from the given data are

$l = \cos 50° = 0.643 \quad m = \cos 80° = 0.174 \quad n = \cos 60° = 0.5.$

We may now find the components of the resultant stress on the oblique plane along the coordinate axes x, y, and z as

$$\sigma'_x = \sigma_x l + \tau_{xy} m + \tau_{xz} n = (100\times 0.643)+(-20\times 0.174)+(45\times 0.5) = 83.32\,\text{MPa}$$

$$\sigma'_y = \tau_{xy} l + \sigma_y m + \tau_{yz} n = (-20\times 0.643)+(50\times 0.174)+(60\times 0.5) = 25.84\,\text{MPa}$$

$$\sigma'_z = \tau_{zx} l + \tau_{zy} m + \sigma_z n = (45\times 0.643)+(60\times 0.174)+(40\times 0.5) = 59.375\,\text{MPa}$$

σ'_n = Normal stress on oblique plane

$$= \sigma'_x l + \sigma'_y m + \sigma'_z n = (83.32\times 0.643)+(25.84\times 0.174)+(59.375\times 0.5) = 87.76\,\text{MPa}$$

Shear stress on the oblique plane = σ'_s

$$=\left(\sigma_x'^2+\sigma_y'^2+\sigma_z'^2-\sigma_n'^2\right)^{1/2}=\left(83.32^2+25.84^2+59.375^2-87.76^2\right)^{1/2}=58.61\,\text{MPa}.$$

Example 3.2

For the state of stress in Example 3.1, determine the magnitude and direction of principal stresses.

Solution:

We first find the stress invariants as

$$I_1=\sigma_x+\sigma_y+\sigma_z=190\,\text{MPa}$$

$$I_2=\sigma_x\sigma_y+\sigma_y\sigma_z+\sigma_z\sigma_x-\tau_{xy}^2-\tau_{yz}^2-\tau_{zx}^2$$

$$=(100\times50)+(50\times40)+(40\times100)-(-20)^2-(60)^2-(45)^2=4975(\text{MPa})^2$$

$$I_3=\sigma_x\sigma_y\sigma_z+2\tau_{xy}\tau_{yz}\tau_{zx}-\sigma_x\tau_{yz}^2-\sigma_y\tau_{xz}^2-\sigma_z\tau_{xy}^2$$

$$=(100\times50\times40)+2(-20\times60\times45)-100\times60^2-50\times45^2-40\times(-20)^2$$

$$=-385250(\text{MPa})^3.$$

The cubic equation of stress is (from equation 3.3.13)

$$\sigma^3-190\sigma^2+4975\sigma+385250=0$$

Three roots are: $\sigma_1=126.8\,\text{MPa}$, $\sigma_2=95.1\,\text{MPa}$, and $\sigma_3=-32\,\text{MPa}$.

In order to find the direction of the principal stresses, we need to find the eigenvectors of the characteristic equation by substituting the eigenvalues σ_1, σ_2, and σ_3 in return.

First substituting σ_1 for σ in equation (3.3.7), we have

$$(100-126.8)l-20m+45n=0$$
$$-20l+(50-126.8)m+60n=0$$
$$45l+60m+(40-126.8)n=0.$$

In addition to this, we need to satisfy $l^2+m^2+n^2=1$. One simple way of solving this is to assume one of the direction cosines, say, $n=1$ and then solve any two of the equations in the above set and solve for l and m. Consider now the first two equations:

$$-26.8l-20m+45=0$$
$$-20l-76.8m+60=0.$$

This gives $l=1.36$, $m=0.427$, $n=1$.

We now need to normalize this by dividing by $\sqrt{l^2+m^2+n^2}$, which in this case is 1.74, and we have the first set of direction cosine corresponding to $\sigma_1=126.8\,\text{MPa}$ as

$$l_1=0.78,\ m_1=0.245,\ \text{and}\ n_1=0.574.$$

Similarly, for $\sigma_2 = 95.1\,\text{MPa}$

$l_2 = -0.525,\ \ m_2 = 0.755,$ and $n_2 = 0.393.$

And for $\sigma_3 = -32\,\text{MPa}$

$l_3 = -0.337,\ \ m_3 = -0.608,$ and $n_3 = 0.719.$

Example 3.3

For the state of stress in Example 3.1, determine the magnitudes, planes of maximum shear stress, and the corresponding normal stresses.

Solution:

It is known from the discussion in Section 3.4 that in 3D situation there are 12 planes of maximum shear stresses, and the corresponding direction cosines, maximum shear stresses, and normal stresses are shown in equation (3.4.5).

Taking the principal stress values from Example 3.2 as $\sigma_1 = 126.8\,\text{MPa}$, $\sigma_2 = 95.1\,\text{MPa}$, $\sigma_3 = -32\,\text{MPa}$, we have

$$l = 0, m = \pm\frac{1}{\sqrt{2}}, n = \pm\frac{1}{\sqrt{2}}, \tau_{max} = \pm\frac{\sigma_2 - \sigma_3}{2} = \pm 63.55\,\text{MPa} \quad \sigma_n = \frac{\sigma_2 + \sigma_3}{2} = 31.55\,\text{MPa}$$

$$l = \pm\frac{1}{\sqrt{2}}, m = 0, n = \pm\frac{1}{\sqrt{2}}, \tau_{max} = \pm\frac{\sigma_3 - \sigma_1}{2} = \pm 79.4\,\text{MPa} \quad \sigma_n = \frac{\sigma_3 + \sigma_1}{2} = 47.4\,\text{MPa}$$

$$l = \pm\frac{1}{\sqrt{2}}, m = \pm\frac{1}{\sqrt{2}}, n = 0, \tau_{max} = \pm\frac{\sigma_1 - \sigma_2}{2} = \pm 15.85\,\text{MPa} \quad \sigma_n = \frac{\sigma_1 + \sigma_2}{2} = 110.95\,\text{MPa}.$$

Example 3.4

The stress field in a loaded body is as follows:

$$\sigma_x = 10x^2,\ \ \sigma_y = 15y^2 + 5xy,\ \sigma_z = 8\left(x - z^2\right),\ \ \tau_{xy} = -30y + x,\ \ \tau_{yz} = 10yz + 5x,$$

$$\tau_{zx} = 20\left(z^2 + xy\right).$$

Determine the body forces at a point (1, −2, 3) in arbitrary units such that the equation of equilibrium is satisfied.

Solution:

Stress equations of equilibrium are from equation (3.6.2):

$$\frac{\partial \sigma_x}{\partial x} + \frac{\partial \tau_{xy}}{\partial y} + \frac{\partial \tau_{xz}}{\partial z} + X = 0$$

$$\frac{\partial \tau_{yx}}{\partial x} + \frac{\partial \sigma_y}{\partial y} + \frac{\partial \tau_{yz}}{\partial z} + Y = 0$$

$$\frac{\partial \tau_{zx}}{\partial x}+\frac{\partial \tau_{zy}}{\partial y}+\frac{\partial \sigma_z}{\partial z}+Z=0.$$

Substituting the stress field in the above equations, we get $X=-110, Y=74,$ and $Z=58.$

Example 3.5

The state of stress at a point in a loaded machine part is given by the following stress tensor

$$\left[\sigma\right]=\begin{bmatrix}-19 & -4.7 & 6.45\\ -4.7 & 4.6 & 11.8\\ 6.45 & 11.8 & -8.3\end{bmatrix}\text{MPa}.$$

Find the principal stresses and their orientation with respect to the original coordinate system.

Solution:

The stress tensor is given by

$$\left[\sigma\right]=\begin{bmatrix}\sigma_x & \tau_{xy} & \tau_{xz}\\ \tau_{yx} & \sigma_y & \tau_{yz}\\ \tau_{zx} & \tau_{zy} & \sigma_z\end{bmatrix}=\begin{bmatrix}-19 & -4.7 & 6.45\\ -4.7 & 4.6 & 11.8\\ 6.45 & 11.8 & -8.3\end{bmatrix}\text{MPa},$$

where $\sigma_x=-19$ MPa, $\sigma_y=4.6$ MPa, $\sigma_z=-8.3$ MPa,

$\tau_{xy}=-4.7$ MPa, $\tau_{yz}=11.8$ MPa, $\tau_{zx}=6.45$ MPa.

(a) *Computing Principal Stress*

Characteristic equation is

$\sigma^3-I_1\sigma^2+I_2\sigma-I_3=0,$

where the stress invariants are calculated as:

$I_1=\sigma_x+\sigma_y+\sigma_z=-22.7\,\text{MPa}$

$I_2=\sigma_x\sigma_y+\sigma_y\sigma_z+\sigma_z\sigma_x-\left(\tau_{xy}^2+\tau_{yz}^2+\tau_{zx}^2\right)=-170.8125\,(\text{MPa})^2$

$I_3=\det\left|\sigma\right|=2647.52\,(\text{MPa})^3.$

Solving the characteristic equations, we have

$\sigma_1=11.618$ MPa, $\sigma_2=-9$ MPa, $\sigma_3=-25.316$ MPa. ...(Ans.)

(b) *Computing Principal Stress Directions*

For $\sigma_1=11.618$ MPa; following equation (3.3.8), we write

$$\left(\sigma_x-\sigma_1\right)l_1+\tau_{xy}m_1+\tau_{xz}n_1=0$$
$$\tau_{xy}l_1+\left(\sigma_y-\sigma_1\right)m_1+\tau_{yz}n_1=0$$
$$l_1^2+m_1^2+n_1^2=1$$

Solving $l_1=-0.027$, $m_1=0.864$, $n_1=0.503$. ...(Ans.)

For $\sigma_2 = -9$ MPa; following equation (3.3.8), we write

$$(\sigma_x - \sigma_2)l_2 + \tau_{xy} m_2 + \tau_{xz} n_2 = 0$$
$$\tau_{xy} l_2 + (\sigma_y - \sigma_2)m_2 + \tau_{yz} n_2 = 0$$
$$l_2^2 + m_2^2 + n_2^2 = 1$$

Solving, $l_2 = 0.621, \quad m_2 = -0.380, \quad n_2 = 0.685.$ ...(Ans.)

For $\sigma_3 = -25.316$ MPa; following equation (3.3.8), we write

$$(\sigma_x - \sigma_3)l_3 + \tau_{xy} m_3 + \tau_{xz} n_3 = 0$$
$$\tau_{xy} l_3 + (\sigma_y - \sigma_3)m_3 + \tau_{yz} n_3 = 0$$
$$l_3^2 + m_3^2 + n_3^2 = 1$$

Solving, $l_3 = -0.784, \quad m_3 = -0.331, \quad n_3 = 0.526.$ ...(Ans.)

Example 3.6

For the stress state shown in Example 3.5, find the magnitudes and direction cosines of the planes of maximum shear stresses. Find also the corresponding normal stresses.

Solution:

Planes of maximum shear stresses are located at:

(i) $l = 0, \quad m = \pm\frac{1}{\sqrt{2}}, \quad n = \pm\frac{1}{\sqrt{2}}.$

Maximum shear stress and normal stress are given by

$$\tau_{max} = \pm\frac{(\sigma_2 - \sigma_3)}{2} = \pm 8.158\,\text{MPa}$$
$$\sigma_n = \frac{(\sigma_2 + \sigma_3)}{2} = -17.158\,\text{MPa}$$
...(Ans.)

(ii) $l = \pm\frac{1}{\sqrt{2}}, \quad m = 0, \quad n = \pm\frac{1}{\sqrt{2}}.$

$$\tau_{max} = \pm\frac{(\sigma_3 - \sigma_1)}{2} = \pm 18.467\,\text{MPa}$$
$$\sigma_n = \frac{(\sigma_3 + \sigma_1)}{2} = -6.85\,\text{MPa}$$
...(Ans.)

(iii) $l = \pm\frac{1}{\sqrt{2}}, \quad m = \pm\frac{1}{\sqrt{2}}, \quad n = 0.$

$$\tau_{max} = \pm\frac{(\sigma_1 - \sigma_2)}{2} = \pm 10.34\,\text{MPa}$$
$$\sigma_n = \frac{(\sigma_2 + \sigma_3)}{2} = 1.3\,\text{MPa}$$
...(Ans.)

Example 3.7

The state of stress at a point in a material is given as follows:

$$\sigma_x = 109\,\text{MPa},\ \sigma_y = -54\,\text{MPa},\ \sigma_z = 83\,\text{MPa},\ \tau_{xy} = -22\,\text{MPa},\ \tau_{yz} = 63\,\text{MPa},\ \tau_{zx} = 47\,\text{MPa}$$

Find the normal and shear stresses on the plane which is equally inclined to the three coordinate axes. State the significance of these stresses.

Solution:

Direction cosines for equally inclined axes are given as:

$$l^2 + m^2 + n^2 = 1$$

Since $l = m = n$, $l = m = n = \dfrac{1}{\sqrt{3}}$

Normal and shear stresses on an inclined plane defined by the direction cosines (l, m, n) are given by:

Normal stress, $\sigma'_n = \sigma'_x l + \sigma'_y m + \sigma'_z n$

Shear stress, $\sigma'_s = \sqrt{(\sigma'_x)^2 + (\sigma'_y)^2 + (\sigma'_z)^2 - (\sigma'_n)^2}$ (a)

where $\begin{bmatrix} \sigma'_x \\ \sigma'_y \\ \sigma'_z \end{bmatrix} = \begin{bmatrix} \sigma_x & \tau_{xy} & \tau_{xz} \\ \tau_{yx} & \sigma_y & \tau_{yz} \\ \tau_{zx} & \tau_{zy} & \sigma_z \end{bmatrix} \begin{bmatrix} l \\ m \\ n \end{bmatrix}$

$$= \begin{bmatrix} 109 & -22 & 47 \\ -22 & -54 & 63 \\ 47 & 63 & 83 \end{bmatrix} \begin{bmatrix} \frac{1}{\sqrt{3}} \\ \frac{1}{\sqrt{3}} \\ \frac{1}{\sqrt{3}} \end{bmatrix} = \begin{bmatrix} 77.365 \\ -7.505 \\ 111.427 \end{bmatrix} \text{MPa.}$$

Substituting the value of σ'_x, σ'_y, and σ'_z in (a), we have

$\sigma'_n = 104.67\,\text{MPa}$, $\sigma'_s = 86.62\,\text{MPa}$. ...(Ans.)

A plane that is equally inclined to the three axes of reference is called *octahedral plane*; the normal and shear stresses acting on this plane are *octahedral normal* and *octahedral shear stress* respectively. Octahedral normal stress refers to hydrostatic state of stress and octahedral shear stress is useful in estimating the yield of a material using the von Mises yield criteria, widely used in the design of machine components.

Example 3.8

The state of stress at a point in a material is given as follows

$$[\sigma]=\begin{bmatrix}115 & 50 & 15\\ 50 & 75 & -20\\ 15 & -20 & 145\end{bmatrix}\text{MPa.}$$

Determine the mean stress, the deviatoric principal stresses, and the deviatoric stress invariants.

Solution:

$$[\sigma]=\begin{bmatrix}\sigma_x & \tau_{xy} & \tau_{xz}\\ \tau_{yx} & \sigma_y & \tau_{yz}\\ \tau_{zx} & \tau_{zy} & \sigma_z\end{bmatrix}=\begin{bmatrix}115 & 50 & 15\\ 50 & 75 & -20\\ 15 & -20 & 145\end{bmatrix}\text{MPa.}$$

(a) Mean stress: $\sigma_m=\dfrac{(\sigma_x+\sigma_y+\sigma_z)}{3}=111.67\,\text{MPa.}$ ...(Ans.)

Hydrostatic stress tensor:

$$[\sigma]_{hydrostatic}=\begin{bmatrix}\sigma_m & 0 & 0\\ 0 & \sigma_m & 0\\ 0 & 0 & \sigma_m\end{bmatrix}=\begin{bmatrix}111.67 & 0 & 0\\ 0 & 111.67 & 0\\ 0 & 0 & 111.67\end{bmatrix}\text{MPa.}$$

(b) Deviatoric stress tensor:

$$[\sigma_D]=\begin{bmatrix}(\sigma_x-\sigma_m) & \tau_{xy} & \tau_{xz}\\ \tau_{yx} & (\sigma_y-\sigma_m) & \tau_{yz}\\ \tau_{zx} & \tau_{zy} & (\sigma_z-\sigma_m)\end{bmatrix}$$

$$=\begin{bmatrix}3.33 & 50 & 15\\ 50 & -36.67 & -20\\ 15 & -20 & 33.33\end{bmatrix}\text{MPa.}$$ ...(Ans.)

(c) Deviatoric stress invariants:

$$I_1=(\sigma_x+\sigma_y+\sigma_z)\Big|_{[\sigma_D]}=0$$

$$I_2=\left\{\sigma_x\sigma_y+\sigma_y\sigma_z+\sigma_z\sigma_x-\left(\tau_{xy}^2+\tau_{yz}^2+\tau_{zx}^2\right)\right\}\Big|_{[\sigma_D]}=-4358.33(\text{MPa})^2$$

$$I_3=\det|\sigma_D|=-110476.21(\text{MPa})^3.$$

(d) Principal stresses corresponding to the deviatoric stress tensor:

Characteristic equation is $\sigma^3 - I_1\sigma^2 + I_2\sigma - I_3 = 0$

Solving, $\begin{cases} \sigma_1 = 39.64\,\text{MPa} \\ \sigma_2 = 36.57\,\text{MPa} \\ \sigma_3 = -76.21\,\text{MPa} \end{cases}$ ...(Ans.)

Example 3.9

Stress field in a body is as follows:

$$\sigma_x = 80x^3 + y, \qquad \sigma_y = 100\left(x^3 + 10\right), \qquad \sigma_z = 10\left(9y^2 + 100z^3\right),$$

$$\tau_{xy} = 100\left(1 + y^2\right), \quad \tau_{yz} = 0, \quad \tau_{zx} = x\left(z^3 + 100xy\right).$$

Find the body forces in arbitrary units at the point (1,1,5) that satisfies the equations of equilibrium.

Solution:

Using the stress equilibrium equations in the x, y, and z directions

$$\begin{cases} \dfrac{\partial \sigma_x}{\partial x} + \dfrac{\partial \tau_{xy}}{\partial y} + \dfrac{\partial \tau_{xz}}{\partial z} + X = 0 \\ \dfrac{\partial \tau_{yx}}{\partial x} + \dfrac{\partial \sigma_y}{\partial y} + \dfrac{\partial \tau_{yz}}{\partial z} + Y = 0. \\ \dfrac{\partial \tau_{zx}}{\partial x} + \dfrac{\partial \tau_{zy}}{\partial y} + \dfrac{\partial \sigma_z}{\partial z} + Z = 0 \end{cases}$$

Using the given expressions of the stresses in the above equilibrium equations, the body forces are obtained as:

$$X = -\left(240x^2 + 200y + 3xz^2\right)$$

$$Y = 0$$

$$Z = -\left(z^3 + 200xy + 300z^2\right)$$

At $P(1,1,5)$; $X = -515, \quad Y = 0, \quad Z = -7825.$ ...(Ans.)

Exercises

1. In general, there are nine stress elements in a stress tensor. Show how the number of stress elements may be reduced to six and then to only three.

   Write the stress tensors in the following cases:

   (a) In the principal coordinate system

   (b) In the case of simple tension test with a sample subjected to uniaxial tension

   (c) In the case of a thin walled cylinder subjected to torsional loading to pure shear τ on all faces

(d) In the case of hydrostatic compression of a cubical element under uniform p on all faces

$$\left[\begin{bmatrix}\sigma_1 & 0 & 0\\ 0 & \sigma_2 & 0\\ 0 & 0 & \sigma_3\end{bmatrix}, \begin{bmatrix}\sigma & 0 & 0\\ 0 & 0 & 0\\ 0 & 0 & 0\end{bmatrix}, \begin{bmatrix}0 & \tau & 0\\ \tau & 0 & 0\\ 0 & 0 & 0\end{bmatrix}, \begin{bmatrix}-p & 0 & 0\\ 0 & -p & 0\\ 0 & 0 & -p\end{bmatrix}\right]$$

2. Derive expressions for the normal and shear stresses in Cartesian coordinates on an oblique plane defined by the direction cosines (l, m, n). Stress tensors at a point in a body are given by

$$[\sigma] = \begin{bmatrix}40 & 20 & 10\\ 20 & -30 & 35\\ 10 & 35 & 40\end{bmatrix} \text{MPa.}$$

Find the normal and shear stresses on a plane whose normal has the following direction cosines: $l=0.924, m=0.383, n=0$.

[43.9 MPa, 25 MPa]

3. Derive expressions for stress invariants, I_1, I_2, and I_3, and write the characteristic equations in terms of the invariants.

For the following stress tensor at a point in a body, solve the characteristic equations and find the principal stresses

$$[\sigma] = \begin{bmatrix}6 & 4 & 20\\ 4 & 25 & 3\\ 20 & 3 & -3\end{bmatrix} \text{MPa.}$$

[28.7 MPa, 18.28 MPa, −19 MPa]

4. If all the elements of a stress tensor at a point in a material are the same, what is the state of stress at that point.

[Hint: Find the stress invariants, write the characteristic equation and solve it.]

5. Derive the equation of equilibrium in the y direction considering the equilibrium of an infinitesimal element of size $\delta x, \delta y, \delta z$ in Cartesian coordinate system and show that

$$\frac{\partial \tau_{yx}}{\partial x} + \frac{\partial \sigma_y}{\partial y} + \frac{\partial \tau_{yz}}{\partial z} + Y = \rho f_y,$$

where Y is the body force per unit volume, ρ is the density, and f_y is the acceleration in the y direction.

6. For a rectangular beam of cross-sectional depth d and breadth b, elementary strength of material approach yields the following results

$$\sigma_x = \frac{My}{I}, \tau_{xy} = \frac{V(d^2 - 4y^2)}{8I}, \sigma_y = \sigma_z = \tau_{xz} = \tau_{yz} = 0.$$

Do the results satisfy the equations of equilibrium? If not, what could be the possible reasons?

[no, assumptions related to the strength of materials approach]

7. In a 3D configuration, there are 12 planes of maximum shear stresses. From the basic equations (3.4.4) of shear stress on an arbitrary oblique plane, show that there are only two planes of maximum shear stresses in 2D configurations.

[Hint: Consider only σ_1 and σ_2 and $l=0$ in equation (3.4.4).]

4

Constitutive Equations, Boundary Value Problems, and Concepts of Uniqueness and Superposition

Learning Objectives

After careful study of this chapter, students should be able to do the following:

LO1: Describe constitutive equations.

LO2: Relate the elastic constants.

LO3: Recognize boundary value problems.

LO4: Explain St. Venant's principle.

LO5: Describe the principle of superposition.

LO6: Illustrate the uniqueness theorem.

LO7: Develop stress function approach.

4.1 CONSTITUTIVE EQUATIONS [LO1]

So far, we have discussed the strain and stress analysis in detail. In this chapter, we shall link the stress and strain by considering the material properties in order to completely describe the elastic, plastic, elasto-plastic, visco-elastic, or other such deformation characteristics of solids. These are known as constitutive equations, or in simpler terms the stress–strain relations. There are endless varieties of materials and loading conditions, and therefore development of a general form of constitutive equation may be challenging. Here we mainly consider linear elastic solids along with their mechanical properties and deformation behavior.

Fundamental relation between stress and strain was first given by Robert Hooke in 1676 in the most simplified manner as, "Force varies as the stretch". This implies a load–deflection relation that was later interpreted as a stress–strain relation. Following this, we can write $P = k\delta$, where P is the force, δ is the stretch or elongation, and k is the spring constant. This can also be written for linear elastic materials as $\sigma = E\epsilon$, where σ is the stress, ϵ is the strain, and E is the modulus of elasticity. For nonlinear elasticity, we may write in a simplistic manner $\sigma = E\epsilon^n$, where $n \neq 1$.

Hooke's Law based on this fundamental relation is given as the stress–strain relation, and in its most general form, stresses are functions of all the strain components as shown in equation (4.1.1).

$$
\begin{aligned}
\sigma_x &= D_{11}\epsilon_x + D_{12}\epsilon_y + D_{13}\epsilon_z + D_{14}\gamma_{xy} + D_{15}\gamma_{yz} + D_{16}\gamma_{zx}\\
\sigma_y &= D_{21}\epsilon_x + D_{22}\epsilon_y + D_{23}\epsilon_z + D_{24}\gamma_{xy} + D_{25}\gamma_{yz} + D_{26}\gamma_{zx}\\
\sigma_z &= D_{31}\epsilon_x + D_{32}\epsilon_y + D_{33}\epsilon_z + D_{34}\gamma_{xy} + D_{35}\gamma_{yz} + D_{36}\gamma_{zx}\\
\tau_{xy} &= D_{41}\epsilon_x + D_{42}\epsilon_y + D_{43}\epsilon_z + D_{44}\gamma_{xy} + D_{45}\gamma_{yz} + D_{46}\gamma_{zx}\\
\tau_{yz} &= D_{51}\epsilon_x + D_{52}\epsilon_y + D_{53}\epsilon_z + D_{54}\gamma_{xy} + D_{55}\gamma_{yz} + D_{56}\gamma_{zx}\\
\tau_{zx} &= D_{61}\epsilon_x + D_{62}\epsilon_y + D_{63}\epsilon_z + D_{64}\gamma_{xy} + D_{65}\gamma_{yz} + D_{66}\gamma_{zx}
\end{aligned}
\tag{4.1.1}
$$

D_{11} to D_{66} represent material constants relating to stress and strain. This can be expressed in matrix notations as

$$
\begin{bmatrix} \sigma_x \\ \sigma_y \\ \sigma_z \\ \tau_{xy} \\ \tau_{yz} \\ \tau_{zx} \end{bmatrix} =
\begin{bmatrix}
D_{11} & D_{12} & D_{13} & D_{14} & D_{15} & D_{16}\\
D_{21} & D_{22} & D_{23} & D_{24} & D_{25} & D_{26}\\
D_{31} & D_{32} & D_{33} & D_{34} & D_{35} & D_{36}\\
D_{41} & D_{42} & D_{43} & D_{44} & D_{45} & D_{46}\\
D_{51} & D_{52} & D_{53} & D_{54} & D_{55} & D_{56}\\
D_{61} & D_{62} & D_{63} & D_{64} & D_{65} & D_{66}
\end{bmatrix}
\begin{bmatrix} \epsilon_x \\ \epsilon_y \\ \epsilon_z \\ \gamma_{xy} \\ \gamma_{yz} \\ \gamma_{zx} \end{bmatrix}
\tag{4.1.2}
$$

Cauchy formulations for generalized Hooke's Law are given as

$$\sigma_{ij} = \sum_{k=1}^{3}\sum_{l=1}^{3} D_{ijkl}\epsilon_{kl};\ i, j = 1 \text{ to } 3 \tag{4.1.3}$$

In Cauchy general formulation, stress tensor σ_{ij} and strain tensor ϵ_{kl} have 9 components each, and the constitutive tensor D_{ijkl} has 81 elements that represent the independent material constants. The constitutive matrix in equation (4.1.2) is a special case of Cauchy formulation for anisotropic materials, where both stress and strain tensors are symmetrical, i.e., $\sigma_{ij} = \sigma_{ji}$ and $\epsilon_{kl} = \epsilon_{lk}$, and the number of elements in the constitutive matrix reduces to 36. In many analytical approaches to stress–strain analysis, this matrix is taken as stiffness matrix. In simple notation, we may write

$$[\sigma] = [D][\epsilon].$$

The inverse relation is given by

$$[\epsilon] = [c][\sigma],$$

where $[c]$ is the compliance matrix. From this relation, we can also write

$$[c] = [D]^{(-1)}, \tag{4.1.4}$$

which indicates that $[D]$ is the constitutive matrix, as stated above.

There may be other special cases of Cauchy generalized formulation. We may introduce a potential function, namely a strain energy density function, to reduce the number of independent material constants.

Strain energy density function can be defined as a function that relates strain energy density, i.e., strain energy per unit volume to the deformation gradient, and this may be written for an isotropic linear elastic material undergoing small strain as

$$U=\frac{1}{2}\sum_{i=1}^{3}\sum_{j=1}^{3}\sigma_{ij}\in_{ij}=\frac{1}{2}\left(\sigma_x\in_x+\sigma_y\in_y+\sigma_z\in_z+\tau_{xy}\gamma_{xy}+\tau_{yz}\gamma_{yz}+\tau_{zx}\gamma_{zx}\right).$$

We may then write

$$\frac{\partial U}{\partial\in_x}=\sigma_x;\ \frac{\partial U}{\partial\in_y}=\sigma_y$$

and

$$\frac{\partial\sigma_x}{\partial\in_y}=\frac{\partial^2 U}{\partial\in_x\partial\in_y};\ \frac{\partial\sigma_y}{\partial\in_x}=\frac{\partial^2 U}{\partial\in_x\partial\in_y}.$$

Therefore,

$$\frac{\partial\sigma_x}{\partial\in_y}=\frac{\partial\sigma_y}{\partial\in_x}.$$

Combining this with equation (4.1.1), we have

$$D_{12}=D_{21}. \tag{4.1.5}$$

This implies that $D_{ij}=D_{ji}$, where j and i could be 1,2, ..., 6. This means that the number of independent constants in this case (anisotropic materials) is only 21. Constitutive matrix in equation (4.1.2) reduces to

$$\begin{bmatrix}\sigma_x\\ \sigma_y\\ \sigma_z\\ \tau_{xy}\\ \tau_{yz}\\ \tau_{zx}\end{bmatrix}=\begin{bmatrix}D_{11}&D_{12}&D_{13}&D_{14}&D_{15}&D_{16}\\ D_{12}&D_{22}&D_{23}&D_{24}&D_{25}&D_{26}\\ D_{13}&D_{23}&D_{33}&D_{34}&D_{35}&D_{36}\\ D_{14}&D_{24}&D_{34}&D_{44}&D_{45}&D_{46}\\ D_{15}&D_{25}&D_{35}&D_{45}&D_{55}&D_{56}\\ D_{16}&D_{26}&D_{36}&D_{46}&D_{56}&D_{66}\end{bmatrix}\begin{bmatrix}\in_x\\ \in_y\\ \in_z\\ \gamma_{xy}\\ \gamma_{yz}\\ \gamma_{zx}\end{bmatrix}. \tag{4.1.6}$$

The matrix can be split diagonally such that in the bottom portion, elements are repeated, confirming that the independent elements are only 21 in number for a completely anisotropic material. We can also consider orthotropic materials with three planes of elastic symmetry. The number of independent elastic constants in this case can be shown to be only nine and the constitutive matrix can be given as

$$[D]=\begin{bmatrix}D_{11}&D_{12}&D_{13}&0&0&0\\ D_{12}&D_{22}&D_{23}&0&0&0\\ D_{13}&D_{23}&D_{33}&0&0&0\\ 0&0&0&D_{44}&0&0\\ 0&0&0&0&D_{55}&0\\ 0&0&0&0&0&D_{66}\end{bmatrix}. \tag{4.1.7}$$

For homogenous, isotropic, and linearly elastic solids, there are only two independent constants: E, Young's modulus and ν, Poisson's ratio. The constitutive equations then reduce to

$$\begin{aligned}
\epsilon_x &= \frac{1}{E}\left[\sigma_x - \nu\sigma_y - \nu\sigma_z\right] \\
\epsilon_y &= \frac{1}{E}\left[\sigma_y - \nu\sigma_x - \nu\sigma_z\right] \\
\epsilon_z &= \frac{1}{E}\left[\sigma_z - \nu\sigma_x - \nu\sigma_y\right] \\
\gamma_{xy} &= \frac{\tau_{xy}}{G} \\
\gamma_{yz} &= \frac{\tau_{yz}}{G} \\
\gamma_{zx} &= \frac{\tau_{zx}}{G},
\end{aligned} \tag{4.1.8}$$

where G is the modulus of rigidity that is related to E as

$$G = \frac{E}{2(1+\nu)}. \tag{4.1.9}$$

This will be further discussed in the next section. For isotropic materials, the modulus of elasticity is independent of three orthogonal directions, and therefore

$$E = \frac{\sigma_x}{\epsilon_x} = \frac{\sigma_y}{\epsilon_y} = \frac{\sigma_z}{\epsilon_z}, \tag{4.1.10}$$

whereas for orthotropic materials, the modulus of elasticity along the three orthogonal directions is different, and we may write

$$\begin{aligned}
&E_x = \frac{\sigma_x}{\epsilon_x}, \quad E_y = \frac{\sigma_y}{\epsilon_y}, \quad E_z = \frac{\sigma_z}{\epsilon_z} \\
&G_{xy} = \frac{\tau_{xy}}{\gamma_{xy}}, \quad G_{yz} = \frac{\tau_{yz}}{\gamma_{yz}}, \quad G_{xz} = \frac{\tau_{xz}}{\gamma_{xz}}.
\end{aligned} \tag{4.1.11}$$

On similar arguments, we also define Poisson's ratio differently along different orthogonal planes as

$$\begin{aligned}
&\nu_{xy} = -\frac{\epsilon_x}{\epsilon_y}, \quad \nu_{yz} = -\frac{\epsilon_y}{\epsilon_z}, \quad \nu_{zx} = -\frac{\epsilon_z}{\epsilon_x} \\
&\nu_{yx} = -\frac{\epsilon_y}{\epsilon_x}, \quad \nu_{zy} = -\frac{\epsilon_z}{\epsilon_y}, \quad \nu_{xz} = -\frac{\epsilon_x}{\epsilon_z}.
\end{aligned} \tag{4.1.12}$$

4.2 RELATIONS AMONG THE ELASTIC CONSTANTS *E*, *G*, *k*, AND ν [LO2]

In solving elasticity problems, it is important to be able to relate the elastic constants. Firstly, we consider the relation between E and G. Here we recall the elementary strength of the material concept that deformation due to pure shear can be correlated to deformation due to normal stress. To illustrate this, we consider an element subjected to pure shear (Figure 4.1a).

From elementary two-dimensional (2D) analysis, principal stresses are given as

$$\sigma_{1,2} = \frac{\sigma_x + \sigma_y}{2} \pm \sqrt{\left(\frac{\sigma_x - \sigma_y}{2}\right)^2 + \tau_{xy}^{\ 2}}, \tag{4.2.1}$$

where σ_x and σ_y are normal stresses and τ_{xy} is the shear stress on a 2D element in the *xy* plane.

Here the only nonzero shear stress on the element in Figure 4.1(a) is τ_{xy}, and therefore

$$\sigma_1 = +\tau_{xy}; \quad \sigma_2 = -\tau_{xy}$$

and the planes at which they act are

$$\tan 2\theta = \frac{2\tau_{xy}}{(\sigma_x - \sigma_y)}. \tag{4.2.2}$$

Since the only nonzero stress here is τ_{xy}, the two roots of θ are 45° and 135°.

This means that, due to pure shear, normal stress acts along the diagonal planes of the element.

Now we consider the strain transformation for the element in Figure 4.1. From the elementary strength of material, the transformed strain ϵ'_x is given by

$$\epsilon' = \epsilon_x \cos^2\theta + \epsilon_y \sin^2\theta + \gamma_{xy} \sin\theta\cos\theta. \tag{4.2.3}$$

Here again the only nonzero strain is γ_{xy} and $\theta = 45°$

$$\epsilon'_x = \frac{\gamma_{xy}}{2} = \frac{\tau_{xy}}{2G}; \text{ since } \tau_{xy} = G\gamma_{xy}. \tag{4.2.4}$$

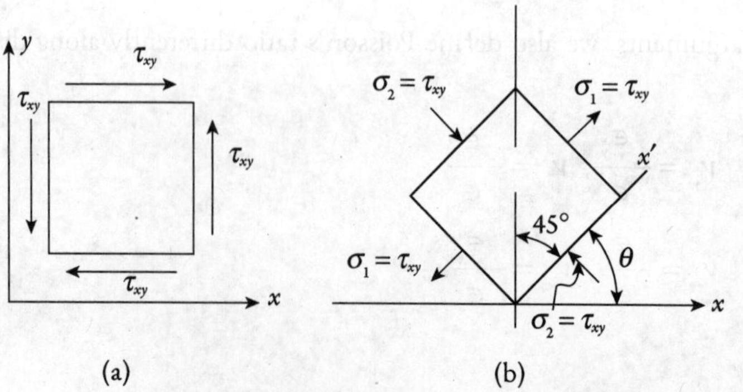

Figure 4.1 Element under pure shear related to normal stress

Now using the generalized Hooke's Law in 2D, we have

$$\epsilon'_x = \epsilon_1 = \frac{\sigma_1}{E} - \frac{\nu\sigma_2}{E} = \frac{\tau_{xy}}{E}(1+\nu). \tag{4.2.5}$$

Since σ_2 is compressive, as seen in Figure 4.1, from equations (4.2.4) and (4.2.5), we have

$$G = \frac{E}{2(1+\nu)}. \tag{4.2.6}$$

This has been stated earlier as equation (4.1.9) without proof.

Now we consider the correlation between bulk modulus k and elastic modulus (Young's modulus) E.

For this purpose, we consider a small element of sides dx, dy, and dz subjected to normal stresses σ_x, σ_y, and σ_z along the three orthogonal coordinate axes that give rise to normal strains ϵ_x, ϵ_y, and ϵ_z. The change in volume is thus

$$(1+\epsilon_x)dx(1+\epsilon_y)dy(1+\epsilon_z)dz - dxdydz = (\epsilon_x + \epsilon_y + \epsilon_z)dxdydz.$$

Here the products of small strains are neglected. The change in volume per unit volume, referred to as dilatation e, is given by

$$e = \epsilon_x + \epsilon_y + \epsilon_z. \tag{4.2.7}$$

It is worth mentioning here that dilatation is invariant and we may write

$$e = \epsilon_x + \epsilon_y + \epsilon_z = \epsilon_1 + \epsilon_2 + \epsilon_3. \tag{4.2.8}$$

Using the generalized Hooke's Law, equation (4.2.8) may be written as

$$e = \epsilon_x + \epsilon_y + \epsilon_z = \frac{(1-2\nu)}{E}(\sigma_x + \sigma_y + \sigma_z). \tag{4.2.9}$$

If we now consider that the elastic body is subjected to a state of hydrostatic stress $\sigma_x = \sigma_y = \sigma_z = -p$, then we have $e = \frac{3(1-2\nu)}{E}(-p)$, which means that $\frac{-p}{e} = \frac{E}{3(1-2\nu)}$ and this is defined as *bulk modulus k*, and we may write

$$k = \frac{E}{3(1-2\nu)}. \tag{4.2.10}$$

4.3 BOUNDARY VALUE PROBLEMS [LO3]

In solving mechanics or elasticity problems, we have 15 basic equations:

1. 6 strain compatibility equations
2. 3 stress equations of equilibrium
3. 6 constitutive equations

These equations must be solved along with the boundary conditions of a problem. The boundary conditions generally consist of data about the displacements and the applied traction on the surface of the body. Combination of these state equations and the boundary conditions forms the boundary value problems, and the solution to these problems constitutes the solution of the state equations satisfying the boundary conditions.

The boundary value problems in elasticity are categorized into three different kinds.

1. *Boundary value problem of the first kind*

   Here displacements at all points on the boundary are specified. This may be given by

   $$u_i = \overline{u}_i \text{ on } S,$$

   where $\overline{u}_i$ is the specified displacement values on the surface, S, of an elastic body.

2. *Boundary value of the second kind*

   Here tractions at all points on the boundary are specified. This may be given by

   $$\sigma_{ij} n_i = \overline{F}_i \text{ on } S,$$

   where $\overline{F}_i$ is the specified surface force value at all points on the surface. $\sigma_{ij} n_i$ refers to the outward normal traction vector.

3. *Boundary value problem of the third kind*

   Here displacements are specified on some parts of the surface, and tractions are specified on the rest of the surface. This may be given by

   $$u_i = \overline{u}_i \text{ on } S_U$$

   $$\sigma_{ij} n_i = \overline{F}_i \text{ on } S_T.$$

1. *Boundary value problem of the first kind*

   Here we express all the stresses and strains in terms of displacements and satisfy the equilibrium equation since the boundary conditions are in terms of displacements (Figure 4.2).

   In order to do this, we start expressing stresses in terms of strains using Hooke's Law and we have from equation (4.2.9)

   $$\epsilon_x + \epsilon_y + \epsilon_z = \frac{1}{E}\left[(1-2\nu)(\sigma_x + \sigma_y + \sigma_z)\right].$$

   We may now express this in terms of first stress and strain invariants I_1 and J_1 as

   $$J_1 = \frac{1-2\nu}{E} I_1,$$

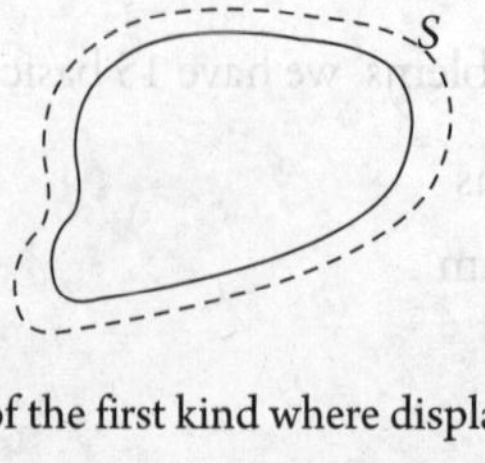

Figure 4.2 Boundary value problems of the first kind where displacements at the boundary are specified

since first strain invariant $J_1 = \epsilon_x + \epsilon_y + \epsilon_z$, and first stress invariant $I_1 = \sigma_x + \sigma_y + \sigma_z$, as discussed in Chapters 2 and 3.

Therefore, $\epsilon_x = \frac{1}{E}\left(\sigma_x - \nu\sigma_y - \nu\sigma_z\right) = \frac{\sigma_x}{E} - \frac{\nu}{E}\left(\sigma_y + \sigma_z\right) = \frac{1+\nu}{E}\sigma_x - \frac{\nu}{E}I_1.$

Combining the above two equations, we have

$$\sigma_x = \epsilon_x \frac{E}{1+\nu} + \frac{\nu E}{(1+\nu)(1-2\nu)}J_1.$$

Defining Lame's constant $\lambda = \frac{\nu E}{(1+\nu)(1-2\nu)}$ and since shear modulus $G = \frac{E}{2(1+\nu)}$, we may write the following three equations:

$$\begin{aligned} \sigma_x &= 2G\epsilon_x + \lambda J_1; \quad \tau_{xy} = G\gamma_{xy} \\ \sigma_y &= 2G\epsilon_y + \lambda J_1; \quad \tau_{yz} = G\gamma_{yz} \\ \sigma_z &= 2G\epsilon_z + \lambda J_1; \quad \tau_{zx} = G\gamma_{zx}. \end{aligned} \tag{4.3.1}$$

We now use these expressions of stresses in terms of strains, in the equation of equilibrium. Starting with the first equation in (3.6.2), we have

$$\frac{\partial}{\partial x}\left(2G\epsilon_x + \lambda J_1\right) + \frac{\partial}{\partial y}\left(G\gamma_{xy}\right) + \frac{\partial}{\partial z}\left(G\gamma_{xz}\right) + X = 0.$$

Now substituting the strains in terms of displacements from equations (2.2.1) and (2.2.2),

$$\frac{\partial}{\partial x}\left(2G\frac{\partial u}{\partial x} + \lambda J_1\right) + \frac{\partial}{\partial y}\left(G\frac{\partial u}{\partial y} + G\frac{\partial v}{\partial x}\right) + \frac{\partial}{\partial z}\left(G\frac{\partial u}{\partial z} + G\frac{\partial w}{\partial x}\right) + X = 0.$$

Rearranging, we have

$$\lambda\frac{\partial J_1}{\partial x} + G\frac{\partial}{\partial x}\left(\frac{\partial u}{\partial x} + \frac{\partial v}{\partial y} + \frac{\partial w}{\partial z}\right) + G\left(\frac{\partial^2 u}{\partial x^2} + \frac{\partial^2 u}{\partial y^2} + \frac{\partial^2 u}{\partial z^2}\right) + X = 0.$$

Since J_1 can be written as $J_1 = \frac{\partial u}{\partial x} + \frac{\partial v}{\partial y} + \frac{\partial w}{\partial z}$, we finally have

$$(\lambda + G)\frac{\partial}{\partial x}\left(\frac{\partial u}{\partial x} + \frac{\partial v}{\partial y} + \frac{\partial w}{\partial z}\right) + G\nabla^2 u + X = 0.$$

We can write two more such equations in y and z coordinate axes and give them in a combined form as

$$\begin{aligned} (\lambda + G)\frac{\partial}{\partial x}\left(\frac{\partial u}{\partial x} + \frac{\partial v}{\partial y} + \frac{\partial w}{\partial z}\right) + G\nabla^2 u + X &= 0 \\ (\lambda + G)\frac{\partial}{\partial y}\left(\frac{\partial u}{\partial x} + \frac{\partial v}{\partial y} + \frac{\partial w}{\partial z}\right) + G\nabla^2 v + Y &= 0 \\ (\lambda + G)\frac{\partial}{\partial z}\left(\frac{\partial u}{\partial x} + \frac{\partial v}{\partial y} + \frac{\partial w}{\partial z}\right) + G\nabla^2 w + Z &= 0. \end{aligned} \tag{4.3.2}$$

here the Laplacian operator is given by $\nabla^2 = \frac{\partial^2}{\partial x^2} + \frac{\partial^2}{\partial y^2} + \frac{\partial^2}{\partial z^2}$.

These equations are known as Navier–Lame equations and are suitable for formulation of displacement boundary value problems. Here three unknown displacements are represented as u, v, and w.

2. *Boundary problems of the second kind*

Here we express all strain terms in stress terms using Hooke's Law as in equation (4.1.8), since the boundary conditions are in terms of stresses (Figure 4.3) and use them in the compatibility equations.

We start with the first compatibility equation in (2.7.2)

$$\frac{\partial^2 \epsilon_x}{\partial y^2} + \frac{\partial^2 \epsilon_y}{\partial x^2} = \frac{\partial^2 \gamma_{xy}}{\partial x \partial y}.$$

From equations (4.1.8) and (4.1.9), we have

$$\epsilon_x = \frac{1}{E}\left[(1+v)\sigma_x - v\left(\sigma_x + \sigma_y + \sigma_z\right)\right]$$
$$\epsilon_y = \frac{1}{E}\left[(1+v)\sigma_y - v\left(\sigma_x + \sigma_y + \sigma_z\right)\right]$$
$$\gamma_{xy} = \frac{2(1+v)}{E}\tau_{xy}.$$

Substituting these relations in the above-mentioned compatibility equation, we have

$$(1+v)\left(\frac{\partial^2 \sigma_x}{\partial y^2} + \frac{\partial^2 \sigma_y}{\partial x^2}\right) - v\left(\frac{\partial^2 I_1}{\partial x^2} + \frac{\partial^2 I_1}{\partial y^2}\right) = 2(1+v)\frac{\partial^2 \tau_{xy}}{\partial x \partial y} \tag{4.3.3}$$

Next, we consider the first two of the equilibrium equations in (3.6.2) without the body forces for simplicity.

$$\frac{\partial \tau_{xy}}{\partial x} = -\frac{\partial \sigma_y}{\partial y} - \frac{\partial \tau_{yz}}{\partial z}$$
$$\frac{\partial \tau_{xy}}{\partial y} = -\frac{\partial \sigma_x}{\partial x} - \frac{\partial \tau_{xz}}{\partial z}.$$

Differentiating the first of these equations with respect to y and the second one with respect to x, and adding them together, we have

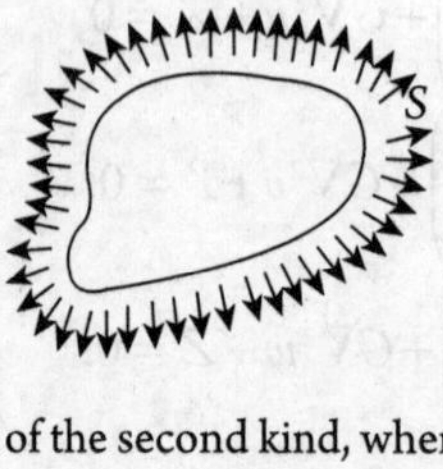

Figure 4.3 Boundary value problems of the second kind, where traction at the boundary is specified

$$2\frac{\partial^2 \tau_{xy}}{\partial x \partial y} = -\frac{\partial^2 \sigma_y}{\partial y^2} - \frac{\partial^2 \sigma_x}{\partial x^2} - \frac{\partial}{\partial z}\left(\frac{\partial \tau_{yz}}{\partial y} + \frac{\partial \tau_{xz}}{\partial x}\right) = -\frac{\partial^2 \sigma_y}{\partial y^2} - \frac{\partial^2 \sigma_x}{\partial x^2} + \frac{\partial^2 \sigma_z}{\partial z^2}. \tag{4.3.4}$$

Substituting equation (4.3.4) in equation (4.3.3), we have

$$(1+v)\left(\nabla^2 I_1 - \nabla^2 \sigma_z - \frac{\partial^2 I_1}{\partial z^2}\right) - v\left(\nabla^2 I_1 - \frac{\partial^2 I_1}{\partial z^2}\right) = 0. \tag{4.3.5}$$

Similar two equations may be obtained by taking the other two compatibility equations. Adding all the three equations of type shown in equation (4.3.5), we get

$$(1-v)\nabla^2 I_1 = 0. \tag{4.3.6}$$

Substituting equation (4.3.6) in equation (4.3.5), we have

$$(1+v)\nabla^2 \sigma_z + \frac{\partial^2}{\partial z^2}(\sigma_x + \sigma_y + \sigma_z) = 0. \tag{4.3.7}$$

We can get two similar equations of type (4.3.7) using the other two compatibility equations in (2.7.3). Similar three equations can be obtained using the last three compatibility equations in (2.7.4), and the first of these equations may be written as

$$(1+v)\nabla^2 \tau_{xy} + \frac{\partial^2}{\partial x \partial y}(\sigma_x + \sigma_y + \sigma_z) = 0. \tag{4.3.8}$$

We can thus concisely write these six equations as

$$\begin{aligned}
&(1+v)\nabla^2 \sigma_x + \frac{\partial^2}{\partial x^2}(\sigma_x + \sigma_y + \sigma_z) = 0 \\
&(1+v)\nabla^2 \sigma_y + \frac{\partial^2}{\partial y^2}(\sigma_x + \sigma_y + \sigma_z) = 0 \\
&(1+v)\nabla^2 \sigma_z + \frac{\partial^2}{\partial z^2}(\sigma_x + \sigma_y + \sigma_z) = 0 \\
&(1+v)\nabla^2 \tau_{xy} + \frac{\partial^2}{\partial x \partial y}(\sigma_x + \sigma_y + \sigma_z) = 0 \\
&(1+v)\nabla^2 \tau_{yz} + \frac{\partial^2}{\partial y \partial z}(\sigma_x + \sigma_y + \sigma_z) = 0 \\
&(1+v)\nabla^2 \tau_{xz} + \frac{\partial^2}{\partial x \partial z}(\sigma_x + \sigma_y + \sigma_z) = 0.
\end{aligned} \tag{4.3.9}$$

These equations are known as Beltrami–Michell compatibility equations in the absence of body forces, and they are suitable for problems where the boundary conditions are in terms of stresses.

The system of equations for stress formulations is still quite complex, and for analytical solutions, the stress function approach is often made.

The boundary value problem of the third kind (Figure 4.4) clearly uses both the above-mentioned displacement and stress formulations for parts of the boundaries.

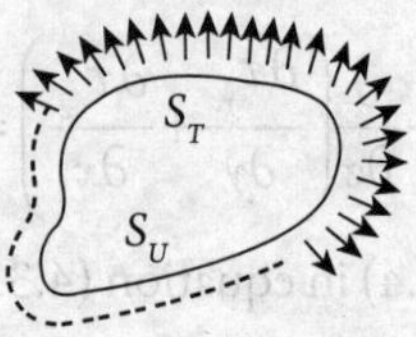

Figure 4.4 Boundary value problems of the third kind, where displacement is specified at some part of the surface and traction at the rest of the surface

4.4 ST. VENANT'S PRINCIPLE [LO4]

Many elasticity problems can be simplified if the system of forces acting on an elastic body can be replaced by a statically equivalent force system with no effect on the solution procedure. St. Venant's principle precisely provides this. The principle states that:

If a system of forces acting on a small portion of the surface of an elastic medium is replaced by another statically equivalent system, then the stress, strain, and displacement field caused by the two systems of forces on parts of the medium far away from the loading point are nearly the same.

The stress, strain, and displacement fields in the vicinity of the loading point would differ, but at remote points, this disturbance would have a negligible effect. For example, consider a distributed and point loading on the surface of an arbitrary body as shown in Figure 4.5.

If the loading in (a) is statically equivalent to the loading in (b), the stress, strain, and displacements at faraway points would not be affected.

A typical example is the loading of the specimen in a simple uniaxial tension test where the specimen is gripped at two ends and is subjected to tensile or compressive force. Gripping produces a distributed load at the, ends but at a remote point, say A, in Figure 4.6, we consider a statically equivalent force P and the stress–strain measurement is based on this force. Experimental results show correct material behavior with this assumption. The local stress distribution near the grip does not affect the stress, strain, or displacements at remote points.

There are many such elasticity problems where the use of St. Venant's principle simplifies the solution procedure. However, there is a point that needs a special mention regarding the

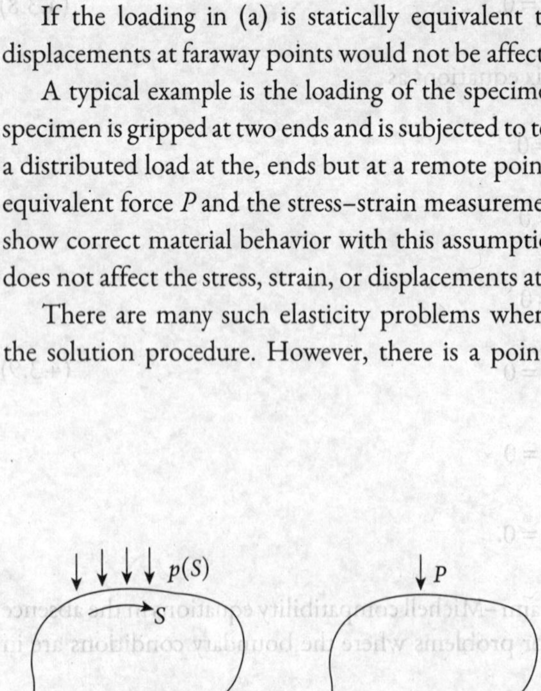

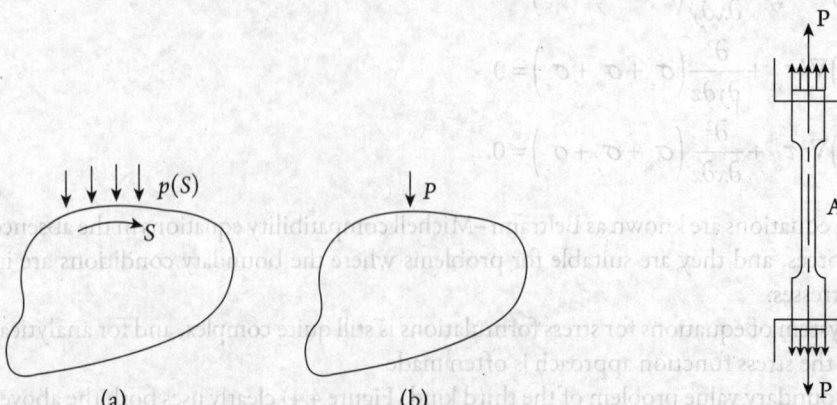

Figure 4.5 Statically equivalent loading on the surface of an arbitrary body

Figure 4.6 A typical loading of a tensile test specimen

statement of the principle. The principle uses terms such as "far away" or "remote" that do not provide any quantitative estimate. Normally, for common usage, the terms are taken as orders greater than the size of the loading area. Von Mises (1945), Sternberg (1954), and Toupin (1965) have discussed quantitative results on this issue.

4.5 PRINCIPLE OF SUPERPOSITION [LO5]

The principle of superposition is a very useful tool in solving many elasticity problems. However, this applies to problems governed by linear equations only. In elasticity formulations, all the 15 major equations and usual boundary conditions are linear, and therefore the principle of superposition holds good for most of the elasticity problems. To illustrate this, let us consider an elastic medium subjected to two different loadings consisting of

(a) Body force $X_i^{(1)}$ and surface force $F_i^{(1)}$.

(b) Body force $X_i^{(2)}$ and surface force $F_i^{(2)}$.

If the stress, strain, and displacement field due to the first loading are $\sigma_{ij}^{(1)}$, $\in_{ij}^{(1)}$, $u_i^{(1)}$ and those due to the second loading are $\sigma_{ij}^{(2)}$, $\in_{ij}^{(2)}$, $u_i^{(2)}$, then the solutions due to the combined effect of the two loadings would be $\sigma_{ij}^{(1)}+\sigma_{ij}^{(2)}$, $\in_{ij}^{(1)}+\in_{ij}^{(2)}$, $u_i^{(1)}+u_i^{(2)}$ for stress, strain, and displacements respectively.

We can also illustrate the principle of superposition in the stress–strain relations using the generalized Hooke's Law. Since in linear elasticity, stress has a linear relationship to all the strain components, we may write as in equation (4.1.1)

$$\sigma_x = D_{11}\in_x + D_{12}\in_y + D_{13}\in_z + D_{14}\gamma_{xy} + D_{15}\gamma_{yz} + D_{16}\gamma_{xz}.$$

This implies that the effects of different strain components can be algebraically added to give the total effect.

The principle of superposition is also valid in summing up the displacement produced due to forces acting at different points in an elastic medium. It is important to recall that Hooke's Law states that, "deflections are proportional to the force which produces them", and thus we can relate the displacement at a point due to forces acting at any point within the elastic body. To illustrate this, consider an arbitrary elastic body acted upon by forces F_1, F_2, and F_3 as shown in Figure 4.7. Consider the displacement at point 2 due to a force at 1. The displacement Δ_2 may be in any direction (different from the direction of F_1) and according to Hooke's Law, this will be proportional to F_1 and we may write $\Delta_2 = kF_1$, where k is an arbitrary constant. Taking the displacement δ_2 in a specific direction (say vertical), we may also write

$$\delta_2 = a_{21}F_1,$$

where a_{21} is known as influence coefficient. Now let us consider the effect of another force F_3. When F_3 acts alone, the displacement at point 2 in the specified direction is $a_{23}F_3$, but when F_3 acts in the presence of F_1 the displacement would be $a'_{23}F_3$, a'_{23} being different from a_{23}. Therefore, the total displacement at point 2, when both F_1 and F_3 act, is

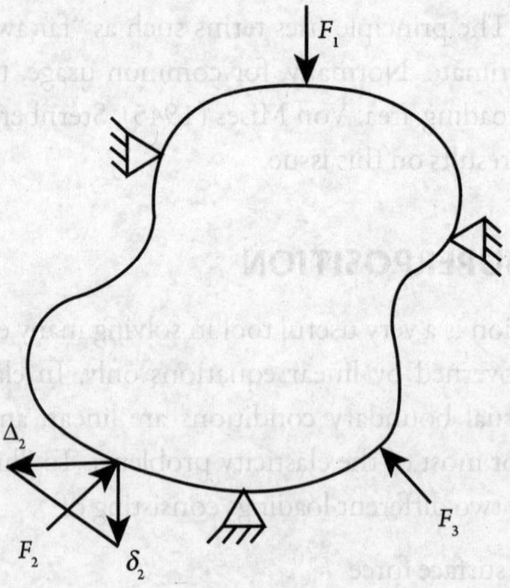

Figure 4.7 An elastic medium acted upon by forces at different locations

$$\delta_2 = a_{21}F_1 + a'_{23}F_3. \tag{4.5.1}$$

Now let us apply $-F_1$ and the displacement at 2 is $-a'_{21}F_1$, a'_{21} being different from a_{21}. Let us also apply $-F_3$ and since now only F_3 is acting, the displacement is $-a_{23}F_3$. Therefore, the total displacement at point 2 in a specified direction is zero as given in equation (4.5.2):

$$\delta_2 = a_{21}F_1 + a'_{23}F_3 - a'_{21}F_1 - a_{23}F_3 = 0. \tag{4.5.2}$$

This gives

$$\frac{a_{21} - a'_{21}}{F_3} = \frac{a_{23} - a'_{23}}{F_1}. \tag{4.5.3}$$

Difference $(a_{21} - a'_{21})$ is due to F_3 and the difference $(a_{23} - a'_{23})$ is due to F_1. Therefore, the LHS of equation (4.5.3) is a function of F_3, and the RHS of the equation is a function of F_1. We may then write

$$\frac{a_{21} - a'_{21}}{F_3} = \frac{a_{23} - a'_{23}}{F_1} = k. \tag{4.5.4}$$

k being a constant.

Combining (4.5.1) and (4.5.4), we have

$$\delta_2 = a_{21}F_1 + a_{23}F_3 - kF_1F_3. \tag{4.5.5}$$

The last term in equation (4.5.5) is nonlinear, and is therefore not compatible with Hooke's Law. We therefore conclude $k = 0$, and this gives, from equation (4.5.4),

$$a_{21} = a'_{21} \text{ and } a_{23} = a'_{23}.$$

Therefore, taking into account the displacement due to F_2 at a point 2 as $a_{22}F_2$, equation (4.5.1) may be written as

$$\delta_2 = a_{21}F_1 + a_{22}F_2 + a_{23}F_3 \ldots \tag{4.5.6}$$

This implies that displacements due to forces at any point on an elastic body may be superposed. Component of the total displacement at a point in the direction of the applied force at that point is called *corresponding displacement*. If we consider the corresponding displacements in an elastic body, we may easily find the total strain energy using the principle of superposition as

$$U = \frac{1}{2}\left(F_1\delta_1 + F_2\delta_2 + F_3\delta_3 \ldots\right). \tag{4.5.7}$$

4.6 UNIQUENESS THEOREM [LO6]

Elasticity problems are governed by linear equations and boundary conditions. The theorem states that for every problem in elasticity there exists one and only one solution. For a simple proof, consider two sets of solutions $\sigma'_x, \sigma'_y, \ldots, \tau'_{zx}$ and $\sigma''_x, \sigma''_y, \ldots, \tau''_{zx}$ for the same body forces X, Y, Z and surface forces F_x, F_y, and F_z. Therefore, we may write

$$\frac{\partial \sigma'_x}{\partial x} + \frac{\partial \tau'_{xy}}{\partial y} + \frac{\partial \tau'_{xz}}{\partial z} + X = 0$$
$$\sigma'_x l + \tau'_{xy} m + \tau'_{xz} n = F_x. \tag{4.6.1}$$

$$\frac{\partial \sigma''_x}{\partial x} + \frac{\partial \tau''_{xy}}{\partial y} + \frac{\partial \tau''_{xz}}{\partial z} + X = 0$$
$$\sigma''_x l + \tau''_{xy} m + \tau''_{xz} n = F_x. \tag{4.6.2}$$

Similarly, we can write the same equations for the difference of the solutions

$$\frac{\partial(\sigma'_x - \sigma''_x)}{\partial x} + \frac{\partial(\tau'_{xy} - \tau''_{xy})}{\partial y} + \frac{\partial(\tau'_{xz} - \tau''_{xz})}{\partial z} = 0$$
$$(\sigma'_x - \sigma''_x)l + (\tau'_{xy} - \tau''_{xy})m + (\tau'_{xz} - \tau''_{xz})n = 0. \tag{4.6.3}$$

Similar arguments hold for the strain components and compatibility equations.

Now equation (4.6.3) shows that for the solutions with stress differences, both body forces and surface forces are zero. We can now consider that the integrated strain energy represents the work done during loading, which must be zero, and we may write

$$\iiint U dx dy dz = 0,$$

where $U = \left(\sigma_x \epsilon_x + \sigma_y \epsilon_y + \sigma_z \epsilon_z + \tau_{xy} \epsilon_{xy} + \tau_{yz} \epsilon_{yz} + \tau_{zx} \epsilon_{zx}\right)$.

For U to be zero, all the stress and strain difference components must vanish at all points, implying

$$\sigma'_x - \sigma''_x = \sigma'_y - \sigma''_y = \sigma'_z - \sigma''_z = \tau'_{xy} - \tau''_{xy} = \tau'_{yz} - \tau''_{yz} = \tau'_{zx} - \tau''_{zx} = 0.$$

This means that both solutions are identical, implying that there is one and only one solution for an elasticity problem.

4.7 STRESS FUNCTION APPROACH [LO7]

In Section 4.3, we discussed boundary value problems with displacement and stress boundary conditions, and it was shown that for problems with stress boundary conditions, Beltrami–Michell equations are most suitable. However, the system of equations is difficult to solve analytically. In such situations, stress functions approach offers a simpler solution methodology. In this section, we shall discuss some elementary stress functions and solution techniques using these functions.

It is possible to identify some functions that automatically satisfy the equations of equilibrium and thus reduce the number of equations to be solved. In a three-dimensional system, stress components are defined in terms of three independent stress functions ϕ_1, ϕ_2, and ϕ_3 as

$$\begin{aligned}
\sigma_x &= \frac{\partial^2 \phi_3}{\partial y^2} + \frac{\partial^2 \phi_2}{\partial z^2}; \quad & \tau_{xy} &= -\frac{\partial^2 \phi_3}{\partial x \partial y} \\
\sigma_y &= \frac{\partial^2 \phi_3}{\partial x^2} + \frac{\partial^2 \phi_1}{\partial z^2}; \quad & \tau_{yz} &= -\frac{\partial^2 \phi_1}{\partial y \partial z} \\
\sigma_z &= \frac{\partial^2 \phi_1}{\partial y^2} + \frac{\partial^2 \phi_2}{\partial x^2}; \quad & \tau_{zx} &= -\frac{\partial^2 \phi_2}{\partial z \partial x}.
\end{aligned} \tag{4.7.1}$$

These are known as Maxwell stress functions.

It is now necessary to find the strain components using the constitutive equations and then satisfy the compatibility equations. Even here, the resulting differential equations may be difficult to solve analytically. Therefore, problems are often reduced to 2D ones wherever possible.

If $\phi_2 = \phi_1 = 0$, the stress functions reduce to Airy stress function ϕ and the two-dimensional stresses are defined as

$$\sigma_x = \frac{\partial^2 \phi}{\partial y^2}; \; \sigma_y = \frac{\partial^2 \phi}{\partial x^2}; \; \tau_{xy} = -\frac{\partial^2 \phi}{\partial x \partial y}. \tag{4.7.2}$$

2D problems are generally considered as either plane stress or plane strain problems. A detailed discussion on these problems will be given in Chapter 5. Here we shall briefly consider the use of stress functions in plane stress conditions.

In both plane stress and plane strain conditions, the three-dimensional elasticity problems are reduced to 2D cases. In plane stress conditions, it is assumed that stresses along one of the coordinate axes are zero. We may then write

$$\sigma_z = \tau_{zy} = \tau_{zx} = 0. \tag{4.7.3}$$

This also means $\gamma_{yz} = \gamma_{zx} = 0$.

This assumption is valid when the dimension of an elastic body in the z-direction is small compared to the dimensions in the other two directions. Now it can be shown that the functions defined in equation (4.7.2) automatically satisfy the equilibrium equations in (3.6.1).

We now turn our attention to the constitutive equations (4.1.8). Combining equations (4.7.2) and (4.1.8), we may write

$$\begin{aligned}
\epsilon_x &= \frac{1}{E}\left[\sigma_x - \nu\sigma_y\right] = \frac{1}{E}\left[\frac{\partial^2\phi}{\partial y^2} - \nu\frac{\partial^2\phi}{\partial x^2}\right]\\
\epsilon_y &= \frac{1}{E}\left[\sigma_y - \nu\sigma_x\right] = \frac{1}{E}\left[\frac{\partial^2\phi}{\partial x^2} - \nu\frac{\partial^2\phi}{\partial y^2}\right]\\
\epsilon_z &= -\frac{\nu}{E}\left[\sigma_x + \sigma_y\right] = -\frac{\nu}{E}\left[\frac{\partial^2\phi}{\partial y^2} + \frac{\partial^2\phi}{\partial x^2}\right]\\
\gamma_{xy} &= \frac{\tau_{xy}}{G} = \frac{2(1+\nu)}{E}\tau_{xy} = -\frac{2(1+\nu)}{E}\frac{\partial^2\phi}{\partial x\partial y}.
\end{aligned} \tag{4.7.4}$$

Substituting equation (4.7.4) in the first compatibility equation as shown below

$$\frac{\partial^2\epsilon_x}{\partial y^2} + \frac{\partial^2\epsilon_y}{\partial x^2} = \frac{\partial^2\gamma_{xy}}{\partial x\partial y}.$$

We have

$$\frac{1}{E}\left[\frac{\partial^4\phi}{\partial y^4} - \nu\frac{\partial^4\phi}{\partial x^2\partial y^2} + \frac{\partial^4\phi}{\partial x^4} - \nu\frac{\partial^4\phi}{\partial x^2\partial y^2}\right] = -\frac{2(1+\nu)}{E}\frac{\partial^4\phi}{\partial x^2\partial y^2}$$

$$\frac{\partial^4\phi}{\partial x^4} + 2\frac{\partial^4\phi}{\partial x^2\partial y^2} + \frac{\partial^4\phi}{\partial y^4} = 0,$$

which gives

$$\nabla^4\phi = 0. \tag{4.7.5}$$

Equation (4.7.5) is known as *biharmonic equation*, and this needs to be satisfied for the solution of elasticity problems with stress boundary conditions. However, since ϕ is a function of x and y, all the compatibility equations must be satisfied. Substituting equations (4.7.4) in the compatibility equations (2.7.2) and (2.7.4), it can be shown that all the other five equations are satisfied.

Biharmonic equation $\nabla^4\phi = 0$ is valid if body forces are absent. In the presence of body forces, the governing equation for plane stress problems is

$$\nabla^4\phi + (1-\nu)\nabla^2\Omega = 0,$$

where Ω is the potential function and the body forces are given as

$$X = -\frac{\partial\Omega}{\partial x},\ Y = -\frac{\partial\Omega}{\partial y},\ \text{and}\ Z = -\frac{\partial\Omega}{\partial z}.$$

Numerical solutions may be needed for elasticity problems using stress functions. However, there exist some simple forms of stress functions that may be used to solve some relatively simple problems analytically. The simplest form of a stress function is a polynomial involving x and y.

Consider the following polynomials:

1. $\phi = x$: This does not represent a stress distribution since stresses are given by second derivatives of ϕ.

2. $\phi = a_2 x^2$: This satisfies $\nabla^4 = 0$ and

$$\sigma_x = 0 \qquad \sigma_y = 2a_2 \qquad \tau_{xy} = 0.$$

This represents a pure tension situation as shown in Figure 4.8.

3. $\phi = Bxy$: Satisfies $\nabla^4 \phi = 0$ and

$$\sigma_x = \sigma_y = 0 \qquad \tau_{xy} = -B.$$

This represents a pure shear situation as shown in Figure 4.9.

4. $\phi = d_3 y^3$: Satisfies $\nabla^4 \phi = 0$ and

$$\sigma_x = 6d_3 y \qquad \sigma_y = \tau_{xy} = 0.$$

This represents a pure bending as shown in Figure 4.10.

Depending on applications, we may choose stress functions and solve the problem. However, there are some general forms of stress function in terms of polynomials of varying degrees.

Let us consider first a second-degree polynomial as shown below:

$$\phi = a_2 x^2 + b_2 xy + c_2 y^2.$$

This satisfies $\nabla^4 \phi = 0$, and it is therefore a valid stress function.

This gives

$$\sigma_x = \frac{\partial^2 \phi}{\partial y^2} = 2c_2$$

$$\sigma_y = \frac{\partial^2 \phi}{\partial x^2} = 2a_2$$

$$\tau_{xy} = -\frac{\partial^2 \phi}{\partial x \partial y} = -b_2.$$

This represents a stress distribution as shown in Figure 4.11.

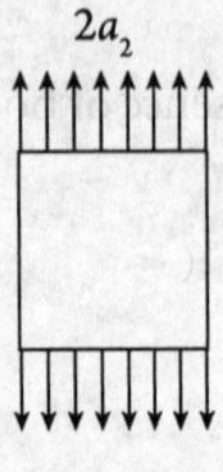

Figure 4.8 A case of pure tension

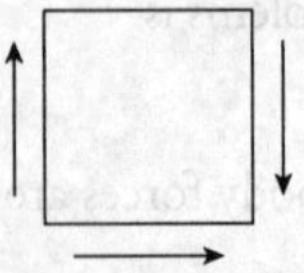

Figure 4.9 A case of pure shear

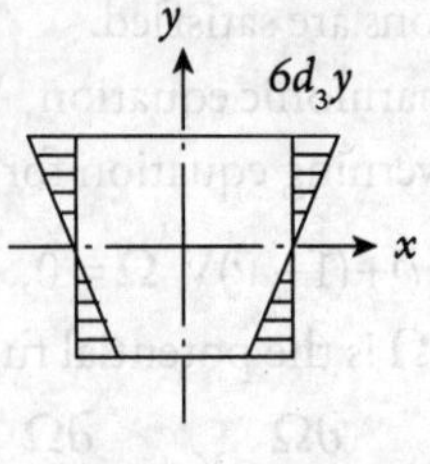

Figure 4.10 A case of pure bending

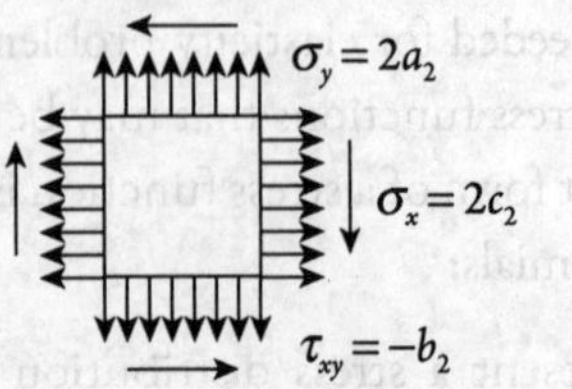

Figure 4.11 Stress distribution as derived from a second-degree polynomial

Now we consider a third-degree polynomial as shown below:

$$\phi = a_3x^3 + b_3x^2y + c_3xy^2 + d_3y^3.$$

This satisfies $\nabla^4\phi = 0$ for any value of a_3, b_3, c_3, and d_3 and the stresses are

$$\sigma_x = \frac{\partial^2\phi}{\partial y^2} = 2c_3x + 6d_3y$$

$$\sigma_y = \frac{\partial^2\phi}{\partial x^2} = 6a_3x + 2b_3y$$

$$\tau_{xy} = -\frac{\partial^2\phi}{\partial x\partial y} = -2b_3x - 2c_3y.$$

Stresses are linear functions of coordinate positions, and they offer several cases:

(a) If only $d_3 \neq 0$, then we have

$$\sigma_x = 6d_3y$$

$$\sigma_y = 0$$

$$\tau_{xy} = 0.$$

This implies pure bending, as discussed before in Figure 4.10, reproduced here in Figure 4.12 for convenience.

(b) If only $a_3 \neq 0$, then we have

$$\sigma_x = 0$$

$$\sigma_y = 6a_3x$$

$$\tau_{xy} = 0.$$

This also implies bending as shown in Figure 4.13.

(c) If only $b_3 \neq 0$, then we have

$$\sigma_x = 0.$$

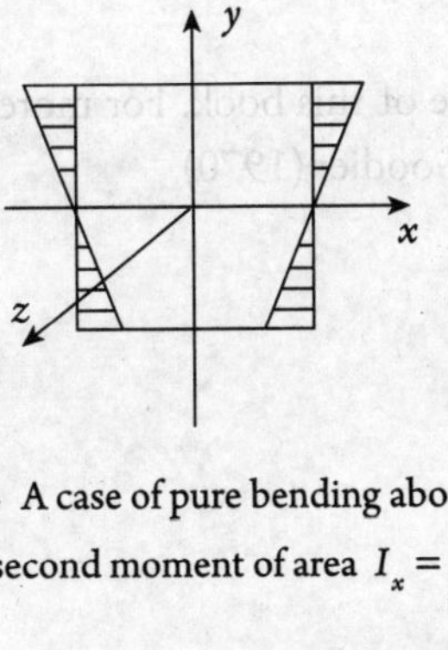

Figure 4.12 A case of pure bending about the z-axis with a second moment of area $I_x = \int_A y^2 dA$

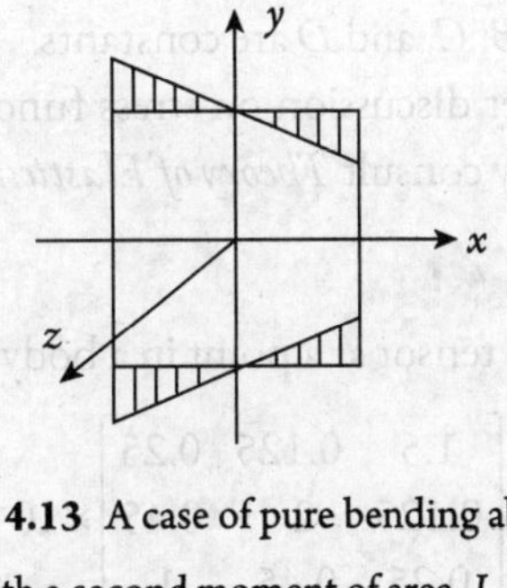

Figure 4.13 A case of pure bending about the z-axis with a second moment of area $I_y = \int_A x^2 dA$

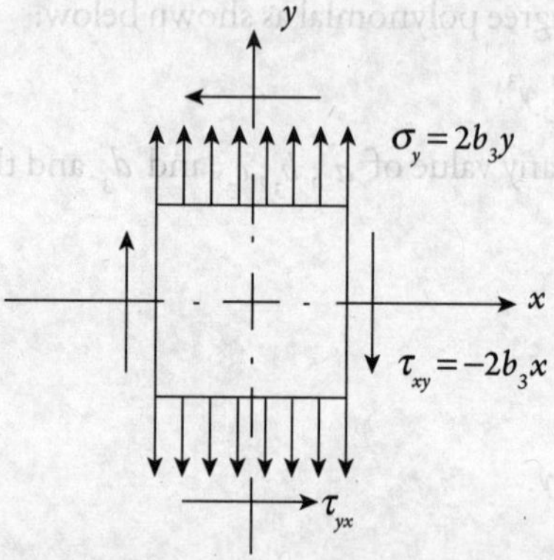

Figure 4.14 Tension in the *y* direction and shear

$$\sigma_y = 2b_3 y$$

$$\tau_{xy} = -2b_3 x$$

This represents a tension in the *y* direction and also shear, as shown in Figure 4.14.

Polynomials of higher degree and combinations of polynomials of different degrees may be used as stress function for different elasticity problems. Some problems require stress functions in polar coordinates. Detailed discussions of governing equations in the cylindrical and polar coordinates will be given in the next chapter, but here we briefly mention a general form of stress function as

$$\phi = f_1(r) f_2(\theta).$$

Certain simplifications may be introduced by assuming axi-symmetric problems with no variation with θ. This gives

$$\phi = f_1(r).$$

Substituting this in the biharmonic equation and on repeated integration, we get the general form of stress function as

$$f_1(r) = A \ln r + Br^2 \ln r + Cr^2 + D,$$

where A, B, C, and D are constants.

Further discussion on stress function is beyond the scope of this book. For more details, the reader may consult *Theory of Elasticity* by Timoshenko and Goodier (1970).

Example 4.1

The strain tensor at a point in a body is given by

$$[\epsilon] = \begin{bmatrix} 1.5 & 0.125 & 0.25 \\ 0.125 & 2 & 0.45 \\ 0.25 & 0.45 & 1 \end{bmatrix} \times 10^{-3}.$$

If Young's modulus of elasticity E and Poisson's ratio ν are 200 GPa and 0.3 respectively, find the stress tensor at the point.

Solution:

Using equation (4.3.1), we have

$$\sigma_x = 2G\in_x + \lambda(\in_x + \in_y + \in_z);\ \tau_{xy} = G\gamma_{xy}$$

$$\sigma_y = 2G\in_y + \lambda(\in_x + \in_y + \in_z);\ \tau_{yz} = G\gamma_{yz}$$

$$\sigma_z = 2G\in_z + \lambda(\in_x + \in_y + \in_z);\ \tau_{zx} = G\gamma_{zx}.$$

Now

$$G = \frac{E}{2(1+\nu)} = \frac{200\times10^9}{2\times1.3} = 77\times10^9\ \text{Pa}$$

$$\lambda = \frac{\nu E}{(1+\nu)(1-2\nu)} = \frac{0.3\times200\times10^9}{1.3\times0.4} = 115.4\times10^9\ \text{Pa}.$$

Substituting

$$\in_x = 0.0015,\ \in_y = 0.002,\ \in_z = 0.001$$

$$\gamma_{xy} = 0.00025,\ \gamma_{yz} = 0.0009,\ \gamma_{zx} = 0.0005.$$

We have, from equation (4.3.1),

$$\sigma_x = 750.3\,\text{MPa},\ \sigma_y = 827.3\,\text{MPa},\ \sigma_z = 673.3\,\text{MPa}$$

$$\tau_{xy} = 19.25\,\text{MPa},\ \tau_{yz} = 69.3\,\text{MPa},\ \tau_{zx} = 38.5\,\text{MPa}.$$

Therefore, the stress tensor at the point is

$$[\sigma] = \begin{bmatrix} 750.3 & 19.25 & 38.5 \\ 19.25 & 827.3 & 69.3 \\ 38.5 & 69.3 & 673.3 \end{bmatrix} \text{MPa}.$$

Example 4.2

Derive the flexure formula for pure bending using the stress function approach.

Solution

As discussed in Section 4.7, we shall first consider a third-degree polynomial for this problem. This is given as $\phi = a_3x^3 + b_3x^2y + c_3xy^2 + d_3y^3$.

Consider a beam of constant cross-section of width b, and depth d, subjected to a constant bending moment M.

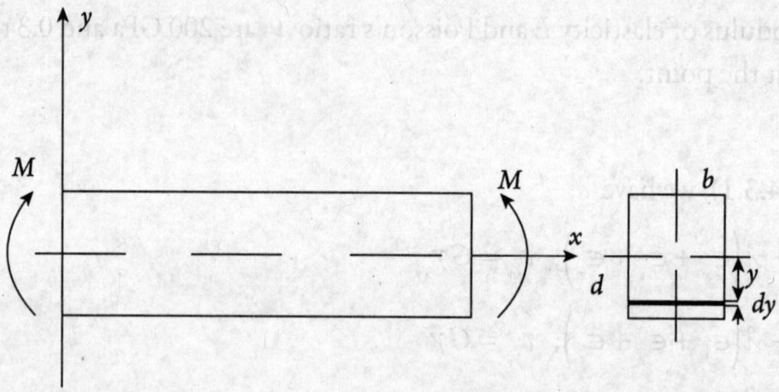

Since stresses are independent of x, $a_3 = b_3 = c_3 = 0$ and the stress function reduces to

$\phi = d_3 y^3$.

Therefore, the stresses are

$$\sigma_x = \frac{\partial^2 \phi}{\partial y^2} = 6 d_3 y$$
$$\sigma_y = \frac{\partial^2 \phi}{\partial x^2} = 0$$
$$\tau_{xy} = \frac{\partial^2 \phi}{\partial x \partial y} = 0.$$

Boundary conditions for this problem are:

(a) $\tau_{xy} = \sigma_y = 0$ at $y = \pm \frac{d}{2}$.

(b) At any cross-section,

$$M = b \int_{-d/2}^{d/2} \sigma_x y dy.$$

This can be seen by considering a small element of width of b and depth dy at a distance of y from the neutral axis. Force acting on the element along the x-direction is $(\sigma_x bdy)$ and the moment arm is y. This gives the moment M as shown above.

(c) Since there is no resultant axial force at any cross-section, we may write

$$b \int_{-d/2}^{d/2} \sigma_x dy = 0.$$

Boundary condition (c) is satisfied, since

$$b \int_{-d/2}^{d/2} \sigma_x dy = \int_{-d/2}^{d/2} 6 d_3 y dy = 0.$$

Now, from boundary condition (b), we have

$$b \int_{-d/2}^{d/2} \sigma_x y dy = b \int_{-d/2}^{d/2} 6 d_3 y^2 dy = d_3 \frac{bd^3}{2} = M.$$

This gives

$$d_3 = \frac{M}{\frac{bd^3}{2}} = \frac{M}{6I},$$

where I = Second moment of area $= \frac{bd^3}{12}$.

Substituting $d_3 = \frac{M}{6I}$ in the expression for $\sigma_x = 6d_3 y$, we have

$$\sigma_x = \frac{My}{I},$$

which is valid for the 2D pure bending case.

Example 4.3

In the most general stress–strain relation, the constitutive matrix has 81 elements. Show that for an anisotropic material, the maximum number of elastic constants is only 21, whereas for an isotropic material, this reduces to only 2.

Solution:

The generalized stress–strain relations are given by

$$\sigma_{ij} = D_{ijkl} \in_{kl},$$

where: $\sigma_{ij} = \left[\sigma_x \sigma_y \sigma_z \tau_{xy} \tau_{yz} \tau_{zx} \tau_{xz} \tau_{zy} \tau_{yx}\right]^T$

$$\in_{kl} = \left[\in_x \in_y \in_z \gamma_{xy} \gamma_{yz} \gamma_{zx} \gamma_{xz} \gamma_{zy} \gamma_{yx}\right]^T$$

Both the stress (σ_{ij}) and strain $(\in_{kl})$ have 9 components each.

Therefore, the stress–strain transformation matrix (constitutive matrix) D_{ijkl} has $(9\times9) = 81$ components.

For anisotropic materials, the stress–strain symmetry exists such that

$$\tau_{xz} = \tau_{zx};\ \tau_{yz} = \tau_{zy};\ \tau_{xy} = \tau_{yx}$$

$$\gamma_{xz} = \gamma_{zx};\ \gamma_{yz} = \gamma_{zy};\ \gamma_{xy} = \gamma_{yx}.$$

Thus, the relation reduces to

$$\begin{bmatrix} \sigma_x \\ \sigma_y \\ \sigma_z \\ \tau_{xy} \\ \tau_{yz} \\ \tau_{zx} \end{bmatrix} = \begin{bmatrix} D_{11} & D_{12} & D_{13} & D_{14} & D_{15} & D_{16} \\ - & D_{22} & D_{23} & D_{24} & D_{25} & D_{26} \\ - & - & D_{33} & D_{34} & D_{35} & D_{36} \\ - & - & - & D_{44} & D_{45} & D_{46} \\ [Sym] & - & - & - & D_{55} & D_{56} \\ - & - & - & - & - & D_{66} \end{bmatrix} \begin{bmatrix} \in_x \\ \in_y \\ \in_z \\ \gamma_{xy} \\ \gamma_{yz} \\ \gamma_{zx} \end{bmatrix} \quad \text{Anisotropic or Hyperelastic}$$

——— 21 components ———

Further, with one plane of symmetry (let the *xz* plane be the symmetric plane)

$$[D]=\begin{bmatrix} \boldsymbol{D}_{11} & \boldsymbol{D}_{12} & \boldsymbol{D}_{13} & 0 & 0 & \boldsymbol{D}_{16} \\ D_{12} & \boldsymbol{D}_{22} & \boldsymbol{D}_{23} & 0 & 0 & \boldsymbol{D}_{26} \\ D_{13} & D_{23} & \boldsymbol{D}_{33} & 0 & 0 & \boldsymbol{D}_{36} \\ 0 & 0 & 0 & \boldsymbol{D}_{44} & \boldsymbol{D}_{45} & 0 \\ 0 & 0 & 0 & D_{45} & \boldsymbol{D}_{55} & 0 \\ D_{16} & D_{26} & D_{36} & 0 & 0 & \boldsymbol{D}_{66} \end{bmatrix} \text{ Monoclinic material}$$

——— 13 independent components ———

Considering the orthogonal plane of symmetry (two orthogonal planes of symmetry), then the third plane also becomes symmetric: Orthotropic material.

Where

$$\left[D\right]=\begin{bmatrix} D_{11} & D_{12} & D_{13} & 0 & 0 & 0 \\ D_{12} & D_{22} & D_{23} & 0 & 0 & 0 \\ D_{13} & D_{23} & D_{33} & 0 & 0 & 0 \\ 0 & 0 & 0 & D_{44} & 0 & 0 \\ 0 & 0 & 0 & 0 & D_{55} & 0 \\ 0 & 0 & 0 & 0 & 0 & D_{66} \end{bmatrix} \text{ Orthotropic material}$$

For isotropic material, the modulus of elasticity is independent of the three orthogonal directions. Therefore,

$$E=\frac{\sigma_x}{\epsilon_x}=\frac{\sigma_y}{\epsilon_y}=\frac{\sigma_z}{\epsilon_z} \quad \text{Isotropic material constants}$$

$$G=\frac{\tau_{xy}}{\gamma_{xy}}=\frac{\tau_{yz}}{\gamma_{yz}}=\frac{\tau_{xz}}{\gamma_{xz}}$$

where $G=\dfrac{E}{2\left(1+\nu\right)}$.

Therefore, two independent constants $\left(E,\nu\right)$. ... (Proved).

Example 4.4

The displacement field in a homogenous, isotropic, and linearly elastic body is as follows:

$$u=5x^2z+100x,\ v=3z^2+6xy,\ w=12z^2+4xyz$$

Determine the stress field and show if the stress field satisfies the Beltrami–Michell compatibility equations.

Solution:

From the given displacement field, we have

$$\epsilon_x=\frac{\partial u}{\partial x}=10xz+100,\quad \epsilon_y=\frac{\partial v}{\partial y}=6x,\quad \epsilon_z=\frac{\partial w}{\partial z}=24z+4xy$$

$$\gamma_{xy} = \frac{\partial u}{\partial y} + \frac{\partial v}{\partial x} = 6y, \quad \gamma_{yz} = \frac{\partial v}{\partial z} + \frac{\partial w}{\partial y} = 6z + 4xz, \quad \gamma_{zx} = \frac{\partial u}{\partial z} + \frac{\partial w}{\partial x} = 5x^2 + 4yz.$$

Using equation (4.3.1), we have

$$\sigma_x = 2G(10xz + 100) + \lambda\left[2x(5z + 2y + 3) + 24z + 100\right]$$

$$\sigma_y = 2G(6x) + \lambda\left[2x(5z + 2y + 3) + 24z + 100\right]$$

$$\sigma_z = 2G(24z + 4xy) + \lambda\left[2x(5z + 2y + 3) + 24z + 100\right]$$

$$\tau_{xy} = G.6y, \; \tau_{yz} = G(6z + 4xz), \; \tau_{xz} = G(5x^2 + 4yz).$$

Substituting these in equation (4.3.9), it can be seen that the stress field satisfies the Beltrami–Michell equations.

Example 4.5

In some cases of 2D problems that deal with long prismatic bodies subjected to uniformly distributed lateral loads, say, in the z direction, deformation of the body at points remote from the ends is assumed to be independent of z. In these cases, only nonzero strain components are: $\in_x$, $\in_y$, and γ_{xy}. Show that under such circumstances, the stress–strain relation may be given as

$$\begin{bmatrix} \sigma_x \\ \sigma_y \\ \tau_{xy} \end{bmatrix} = \lambda \begin{bmatrix} \frac{1-\nu}{\nu} & 1 & 0 \\ 1 & \frac{1-\nu}{\nu} & 0 \\ 0 & 0 & \frac{1-2\nu}{2\nu} \end{bmatrix} \begin{bmatrix} \in_x \\ \in_y \\ \gamma_{xy} \end{bmatrix}.$$

Where Lame's constant λ is defined as $\dfrac{E\nu}{(1+\nu)(1-2\nu)}$.

Solution:

Here $\in_z = \gamma_{xz} = \gamma_{yz} = 0$. Therefore, we may write

$$\in_x = \frac{1}{E}\left[\sigma_x - \nu(\sigma_y + \sigma_z)\right]$$

$$\in_y = \frac{1}{E}\left[\sigma_y - \nu(\sigma_x + \sigma_z)\right]$$

$$0 = \frac{1}{E}\left[\sigma_z - \nu(\sigma_x + \sigma_y)\right].$$

Solving these, we have

$$\sigma_x = \frac{E(1-\nu)}{(1+\nu)(1-2\nu)} \in_x + \frac{E\nu}{(1+\nu)(1-2\nu)} \in_y$$

$$\sigma_y = \frac{E\nu}{(1+\nu)(1-2\nu)}\in_x + \frac{E(1-\nu)}{(1+\nu)(1-2\nu)}\in_y$$

and

$$\tau_{xy} = \frac{E}{2(1+\nu)}\gamma_{xy}.$$

We may write this in matrix form

$$\begin{Bmatrix}\sigma_x\\ \sigma_y\\ \tau_{xy}\end{Bmatrix} = \lambda\begin{bmatrix}\frac{1-\nu}{\nu} & 1 & 0\\ 1 & \frac{1-\nu}{\nu} & 0\\ 0 & 0 & \frac{1-2\nu}{2\nu}\end{bmatrix}\begin{Bmatrix}\in_x\\ \in_y\\ \gamma_{xy}\end{Bmatrix}.$$

Example 4.6

The stress tensor of the torsional problem of a circular cylinder of radius r is given by

$$[\sigma] = \begin{bmatrix}0 & 0 & -G'y\\ 0 & 0 & G'x\\ -G'y & G'x & 0\end{bmatrix},$$

where $G' = CG$, G being the shear modulus of elasticity of the material and C a constant.

Here the longitudinal axis of the cylinder coincides with the z-axis in Cartesian coordinate system.

Find the strain tensor at a point $x = y$ on the lateral surface of the cylinder.

Solution:

Since, here $x^2 + y^2 = r^2$, $x = y = \frac{r}{\sqrt{2}}$.

Now using equation (4.3.1), we may write

first stress invariant $I_1 = (2G + 3\lambda)J_1$, J_1 being the first strain invariant.

Since, here $I_1 = 0$, J_1 is also zero if $(2G + 3\lambda) \neq 0$.

Again, since $\sigma_x = \sigma_y = \sigma_z = 0$, from equation (4.3.1), we have

$$\in_x = -\frac{\lambda}{2G}J_1 = \in_y = \in_z.$$

Therefore, since $J_1 = 0$, $\in_x = \in_y = \in_z = 0$.

Here $\tau_{xy} = 0$, $\tau_{yz} = G'x$, $\tau_{xz} = -G'y$.

Equation (4.3.1) gives

$$\gamma_{xy} = 0, \quad \gamma_{yz} = \frac{G'x}{G} = Cx, \quad \gamma_{xz} = \frac{-G'y}{G} = -Cy.$$

Substituting $x = y = \frac{r}{\sqrt{2}}$, we have $\gamma_{yz} = \frac{Cr}{\sqrt{2}}$, $\gamma_{xz} = -\frac{Cr}{\sqrt{2}}$.

Therefore, strain tensor is

$$[\sigma] = \begin{bmatrix} 0 & 0 & -\frac{Cr}{2\sqrt{2}} \\ 0 & 0 & \frac{Cr}{2\sqrt{2}} \\ -\frac{Cr}{2\sqrt{2}} & \frac{Cr}{2\sqrt{2}} & 0 \end{bmatrix}.$$

Example 4.7

Consider the problem of distributed load at the face of a semi-infinite solid such that

$\sigma_y = \sigma \cos\theta x.$

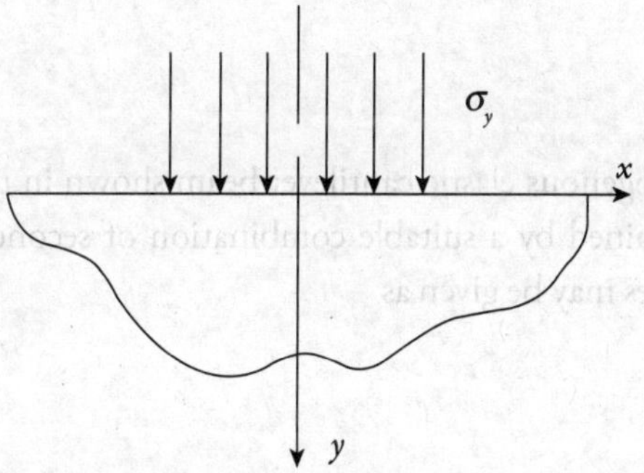

A suitable form of stress function for this problem is

$\phi = f(y)\cos\theta x.$

If the stresses tend to zero remote from the loaded surface, i.e., when $y \to -\infty$, find the function $f(y)$. Also show that along the edge $\sigma_x = -\sigma_y$.

Solution:

For the given stress function to be valid, it must satisfy the biharmonic equation $\nabla^4\phi = 0$.

This gives $f''''(y)\cos\theta x - 2f''(y)\theta^2 \cos\theta x + \theta^4 \cos\theta x \cdot f(y) = 0$

If $\cos\theta x \neq 0$,

$f''''(y) - 2f''(y)\theta^2 + \theta^4 \cdot f(y) = 0.$

To find the roots, we may write

$r^4 - 2r^2\theta^2 + \theta^4 = 0.$

This yields $r = \pm\theta$.

This gives

$f(y) = C_1 e^{\theta y} + C_2 e^{-\theta y}$.

Thus, $\phi = (C_1 e^{\theta y} + C_2 e^{-\theta y})\cos\theta x$

$$\sigma_y = \frac{\partial^2 \phi}{\partial x^2} = -(C_1 e^{\theta y} + C_2 e^{-\theta y})\theta^2 \cos\theta x.$$

Boundary conditions:

At $y = 0$, $\sigma_y = \sigma\cos\theta x$ gives $\sigma = -(C_1 + C_2)\theta^2$.

At $y \to -\infty$, $\sigma_y = 0$ gives $0 = -C_2 e^{\infty}\theta^2 \cos\theta x$.

Since e^{∞} approaches infinity faster than σ_y approaches 0, $C_2 = 0$.

Thus, $C_1 = \dfrac{-\sigma}{\theta^2}$

$$\phi = \frac{-\sigma}{\theta^2}\cos\theta x . e^{\theta y} \text{ and } f(y) = \frac{-\sigma}{\theta^2}e^{\theta y}.$$

$$\text{Now, } \sigma_x = \frac{\partial^2 \phi}{\partial y^2} = -\sigma\cos\theta x \cdot e^{\theta y} \text{ and } \sigma_y = \frac{\partial^2 \phi}{\partial x^2} = \sigma\cos\theta x \cdot e^{\theta y}.$$

This gives $\sigma_x = -\sigma_y$.

Example 4.8

For the isotropic and homogenous elastic cantilever beam shown in the given figure, a workable stress function can be obtained by a suitable combination of second, third, and fourth degree polynomials, and the stresses may be given as

$$\sigma_x = 6d_3 y + 6d_4 xy$$

$$\sigma_y = 0$$

$$\tau_{xy} = -b_2 - 3d_4 y^2,$$

where b_2, d_3, and d_4 are constants.

Write the boundary conditions and find the constants in terms of load W, length l, width b, depth d, and second moment of area I. Finally, write the stresses in terms of these variables.

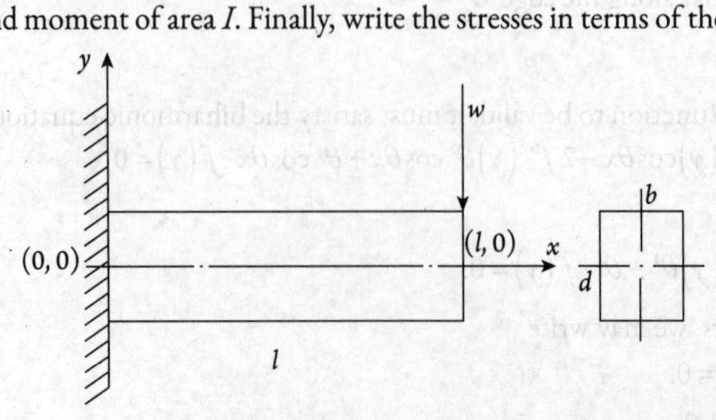

Solution:

Given $\sigma_x = 6d_3 y + 6d_4 xy,\ \sigma_y = 0,\ \tau_{xy} = -b_2 - 3d_4 y^2$

Boundary conditions:

(a) $\tau_{xy} = \sigma_y = 0$ at $y = \pm\dfrac{d}{2}$

Substituting, $-b_2 - 3d_4 (d/2)^2 = 0$. We have $b_2 = -\dfrac{3}{4} d_4 d^2$. ... (i)

(b) In the absence of any external axial force,

$$b \int_{-d/2}^{d/2} \sigma_x dy = 0;\ \text{at } x = l.$$

The condition is invalid at $x = 0$, where a clamping reaction force develops to balance the external load.

$$b \int_{-d/2}^{d/2} \left(6d_3 y + 6d_4 xy\right) dy = 3b\left(d_3 + d_4 l\right)\left[\frac{d^2}{4} - \frac{d^2}{4}\right] = 0.$$

Therefore, $d_3 + d_4 l = 0$. ... (ii)

(c) At $x = 0$, $M = Wl$ (moment)

$$M = b \int_{-d/2}^{d/2} \sigma_x y dy = b \int_{-d/2}^{d/2} \left(6d_3 y + 6d_4 xy\right) y dy$$

$$Wl = 6d_3 b \int_{-d/2}^{d/2} y^2 dy = 2d_3 b\left[\left(\frac{d^3}{4}\right)\right] = 6d_3 (I)$$

$$d_3 = \frac{Wl}{6I}. \quad \text{... (iii)}$$

From (i), (ii), and (iii), we get

$$b_2 = \frac{3}{2}\frac{W}{bd},\ d_3 = \frac{Wl}{6I},\ d_4 = -\frac{W}{6I}. \quad \text{... (Ans.)}$$

Where $I = \dfrac{1}{12} bd^3$: area moment of inertia.

Substituting these relations in the given stress functions,

$$\sigma_x = \frac{Wl}{I} y - \frac{W}{I} xy,\ \sigma_y = 0,\ \tau_{xy} = \frac{W}{2I}\left(y^2 - \left(\frac{d}{2}\right)^2\right). \quad \text{... (Ans.)}$$

Exercises

1. Write the boundary conditions for the following problems:
   (a) A simply supported beam with distributed load as shown below:

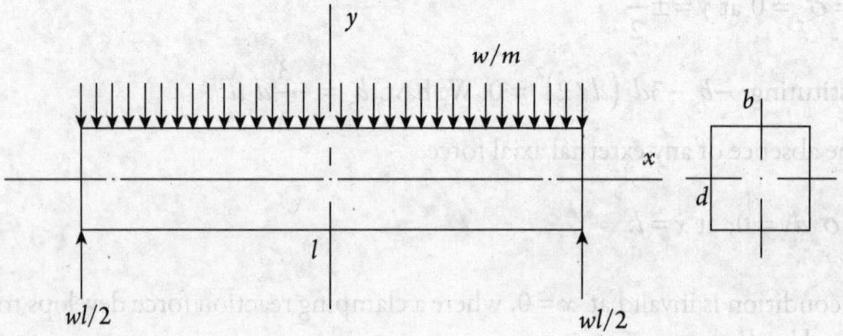

   (b) A cantilever beam with distributed shear stress on the upper face as shown below:

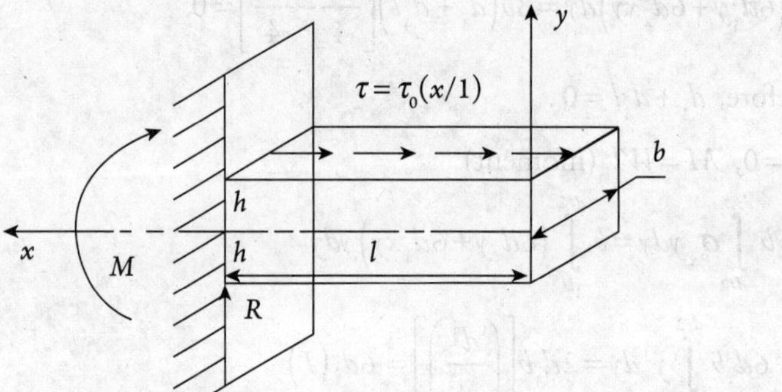

[(a) $\tau_{xy} = 0$ at $y = \pm\dfrac{d}{2}$; $b\displaystyle\int_{-d/2}^{+d/2} \tau_{xy}\,dy = \dfrac{wl}{2}$ at $x = \pm\dfrac{l}{2}$; $\sigma_y = \dfrac{-w}{b}$ at $y = \dfrac{d}{2}$; $\sigma_y = 0$ at $y = -\dfrac{d}{2}$; $b\displaystyle\int \sigma_x dy = 0$ at $x = \pm\dfrac{l}{2}$; $\sigma_x = 0$ at $x = \pm\dfrac{l}{2}$; $b\displaystyle\int \sigma_x y dy = 0$ at $x = \pm\dfrac{l}{2}$

(b) $(\tau_{xy})_{y=h} = \tau_0\left(\dfrac{x}{l}\right)$; $(\tau_{xy})_{y=-h} = 0$; $b\displaystyle\int \tau_{xy} dy = R$ at $x = l$; $\sigma_y = 0$ at $y = \pm h$;

$b\displaystyle\int_{-d/2}^{+d/2} \sigma_x dy = 0$ at $x = 0, l$]

2. From the basic considerations, write the equilibrium equations in terms of displacements in the absence of body forces. [Refer to the boundary value problem of the first kind in Section 4.3.]
3. If an elastic body is acted upon by forces $F_1, F_2, F_3 \ldots F_n$ at points 1, 2, 3, ...*n*, then show that following the principal of superposition displacement at a point, say 2, is the algebraic sum of displacements at point 2 due to all the forces. [Refer to Section 4.5.]

4. For an isotropic linearly elastic material, derive a relation between bulk modulus k and Young's modulus E.
5. Strain tensor at a point in a body is given by

$$[\epsilon]=\begin{bmatrix} -0.25 & 1.5 & -2.5 \\ 1.5 & 0.1 & -4 \\ -2.5 & -4 & 0.5 \end{bmatrix}\times 10^{-3}.$$

For E = 200 GPa and $\nu = 0.3$, find the stress tensor at that point.

$$[\begin{bmatrix} 1.92 & 231 & -385 \\ 231 & 55.8 & -615 \\ -385 & -615 & 117.4 \end{bmatrix}\text{MPa}].$$

5

Two-Dimensional Problems in Elasticity

Learning Objectives

After careful study of this chapter, students should be able to do the following:

LO1: Identify two-dimensional problems in elasticity.

LO2: Illustrate plane stress and plane strain problems.

LO3: Construct governing equations in cylindrical coordinate system.

LO4: Analyze axisymmetric problems.

5.1 INTRODUCTION [LO1]

In any three-dimensional (3D) elasticity problem, there are 15 unknown parameters: 6 stress components, 6 strain components, and 3 displacements. There are 15 related equations: 3 equations of equilibrium, 6 compatibility equations, and 6 constitutive equations. Solutions to a particular elasticity problem require evaluation of these 15 unknown parameters using 15 equations, satisfying all the boundary conditions. As discussed in the earlier chapters, there may be displacement or stress, or mixed boundary conditions. In many cases, solutions to 3D problems are not easy analytically. Even numerical solutions may be difficult.

There are mainly three methods of simplification of solution techniques:

(a) If the boundary conditions are in terms of stresses, stress function approach may be made as discussed in the earlier chapter. This makes the solution simpler.

(b) Assumptions of plane stress and plane strain reduce 3D problems to two-dimensional (2D) ones and this also makes the solution simpler.

(c) Use of St. Venant's principle and superposition principle also makes the solution of elasticity problems simpler.

An introduction to stress function approach has been discussed in Chapter 4. We therefore start our discussion on plane stress and plane strain approaches.

5.2 PLANE STRESS AND PLANE STRAIN PROBLEMS [LO2]

The idealizations of both plane stress and plane strain states are suitable for certain classes of problems that are made to reduce the complexity of solutions. We shall consider the plane stress state first.

5.2.1 Plane Stress

As discussed earlier, here stresses in one of the coordinate directions are assumed to be zero. Let us consider the stresses in the z-direction to be zero:

$$\sigma_z = \tau_{yz} = \tau_{xz} = 0, \text{ also body force } Z = 0. \tag{5.2.1}$$

Z being the body force per unit volume in the z-direction.

This applies to domains bounded by two parallel plates separated by a small distance in comparison to other dimensions; for example, a thin plate as shown in Figure 5.1. This state of stress reduces the stress tensor to a 2×2 matrix as

$$[\sigma] = \begin{bmatrix} \sigma_x & \tau_{xy} \\ \tau_{xy} & \sigma_y \end{bmatrix}. \tag{5.2.2}$$

Principal stresses reduce to simple 2D forms as

$$\sigma_{1,2} = \frac{\sigma_x + \sigma_y}{2} \pm \sqrt{\left(\frac{\sigma_x - \sigma_y}{2}\right)^2 + \tau_{xy}^{\,2}} \tag{5.2.3}$$

and stress invariants are: $I_1 = \sigma_x + \sigma_y = \sigma_1 + \sigma_2$ and $I_2 = \sigma_x \sigma_y - \tau_{xy}^{\,2} = \sigma_1 \sigma_2$.

Using Hooke's Law, the strain field corresponding to the state of plane stress is given by

$$\begin{aligned} \in_x &= \frac{1}{E}\left(\sigma_x - \nu\sigma_y\right) \\ \in_y &= \frac{1}{E}\left(\sigma_y - \nu\sigma_x\right) \\ \in_z &= -\frac{\upsilon}{E}\left(\sigma_x + \sigma_y\right) = -\frac{\nu}{1-\nu}\left(\in_x + \in_y\right) \\ \gamma_{xy} &= \frac{2(1+\nu)}{E}\tau_{xy} \\ \gamma_{yz} &= \gamma_{xz} = 0. \end{aligned} \tag{5.2.4}$$

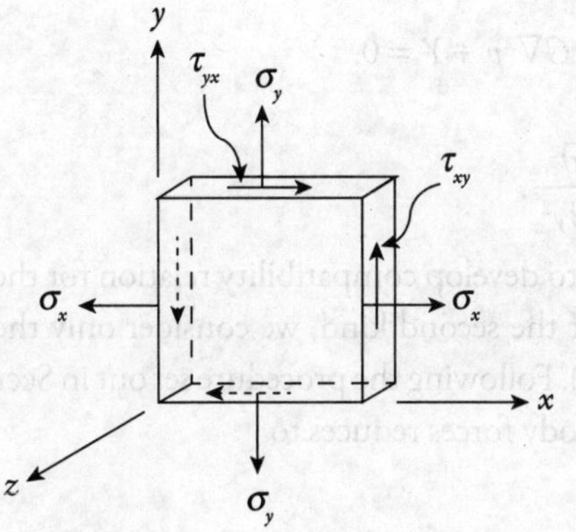

Figure 5.1 State of plane stress

It is important to note that out-of-plane strain $\in_z$ has been written in terms of in-plane components $\in_x$ and $\in_y$. Although in-plane stress state $\sigma_z = 0$, but $\in_z \neq 0$. Following equation (5.2.4), we may write the strain–displacement equations for the plane stress state as

$$\begin{aligned}
&\in_x = \frac{\partial u}{\partial x}, \quad \in_y = \frac{\partial v}{\partial y}, \quad \in_z = \frac{\partial w}{\partial z} \\
&\gamma_{xy} = \left(\frac{\partial u}{\partial y} + \frac{\partial v}{\partial x}\right) \\
&\gamma_{yz} = \left(\frac{\partial v}{\partial z} + \frac{\partial w}{\partial y}\right) = 0 \\
&\gamma_{xz} = \left(\frac{\partial u}{\partial z} + \frac{\partial w}{\partial x}\right) = 0.
\end{aligned} \tag{5.2.5}$$

It is interesting to note that the last two relations in equation (5.2.5) show that the in-plane displacements *u* and *v* are functions of *z* also, and thus the plane stress state is not strictly 2D. It also shows that *w* is a linear function of *z*, and this again implies that the theory is 3D.

In-plane stress state equilibrium equations reduce to a simple form as

$$\begin{aligned}
&\frac{\partial \sigma_x}{\partial x} + \frac{\partial \tau_{xy}}{\partial y} + X = 0 \\
&\frac{\partial \tau_{xy}}{\partial x} + \frac{\partial \sigma_y}{\partial y} + Y = 0.
\end{aligned} \tag{5.2.6}$$

X and *Y* are the body forces per unit volume and functions of *x* and *y* only. As seen in Section 4.3, for boundary value problems of the first kind, we need to express stresses and strains in terms of displacements. This was done by first expressing stresses in terms of strains, as in equation (4.3.1), and then by substituting these in equilibrium equations, we get Navier–Lame equations, as in equation (4.3.2). For the plane stress state, Navier's equations reduce to

$$\begin{aligned}
&\frac{E}{2(1-v)}\frac{\partial}{\partial x}\left(\frac{\partial u}{\partial x} + \frac{\partial v}{\partial y}\right) + G\nabla^2 u + X = 0 \\
&\frac{E}{2(1-v)}\frac{\partial}{\partial y}\left(\frac{\partial u}{\partial x} + \frac{\partial v}{\partial y}\right) + G\nabla^2 v + Y = 0.
\end{aligned} \tag{5.2.7}$$

Note here, $\nabla^2 = \dfrac{\partial^2}{\partial x^2} + \dfrac{\partial^2}{\partial y^2}$.

Again, as in Section 4.3, to develop compatibility relation for the plane stress state, suitable for boundary value problems of the second kind, we consider only the first of the six compatibility equations in equation (2.7.3). Following the procedure set out in Section 4.3, the Beltrami–Michell equation in the absence of body forces reduces to

$$\nabla^2\left(\sigma_x + \sigma_y\right) = 0. \tag{5.2.8}$$

It is interesting to note that the equation of equilibrium in (5.2.6), in the absence of body forces, and Beltrami–Michell equations (5.2.8), also in the absence of body forces, are independent of material properties. This allows us to develop experimental techniques, such as photoelasticity with transparent birefringent materials and the results are applicable to metallic machine parts and other structures.

5.2.2 Plane Strain

In the plane strain state, strains along one of the coordinate axes are considered to be zero; say, strains along the *z* direction, and we may write

$$\epsilon_z = \gamma_{yz} = \gamma_{xz} = 0. \tag{5.2.9}$$

This applies to a long prismatic body as shown in Figure 5.2.

In such cases, the displacements in all cross-sections in the *xy* plane are identical, and therefore this reduces a 3D problem to a 2D one. We may also write

$$\frac{\partial}{\partial z} = 0.$$

In such conditions, the strain tensor reduces to

$$[\epsilon] = \begin{bmatrix} \epsilon_x & \frac{1}{2}\gamma_{xy} \\ \frac{1}{2}\gamma_{xy} & \epsilon_y \end{bmatrix} \tag{5.2.10}$$

and the principal strains are given as in equation (2.8.2)

$$\epsilon_{1,2} = \frac{1}{2}\left(\epsilon_x + \epsilon_y\right) \pm \frac{1}{2}\sqrt{\left(\epsilon_x - \epsilon_y\right)^2 + \gamma_{xy}^{\ 2}}$$

and the principal strain directions are given by

$$\tan 2\theta = \frac{\gamma_{xy}}{\epsilon_x - \epsilon_y}.$$

In the plane strain state, the strain–displacement relations can be given as

$$\epsilon_x = \frac{\partial u}{\partial x},\ \epsilon_y = \frac{\partial v}{\partial y},\ \gamma_{xy} = \left(\frac{\partial u}{\partial y} + \frac{\partial v}{\partial x}\right)$$
$$\epsilon_z = \gamma_{xz} = \gamma_{yz} = 0. \tag{5.2.11}$$

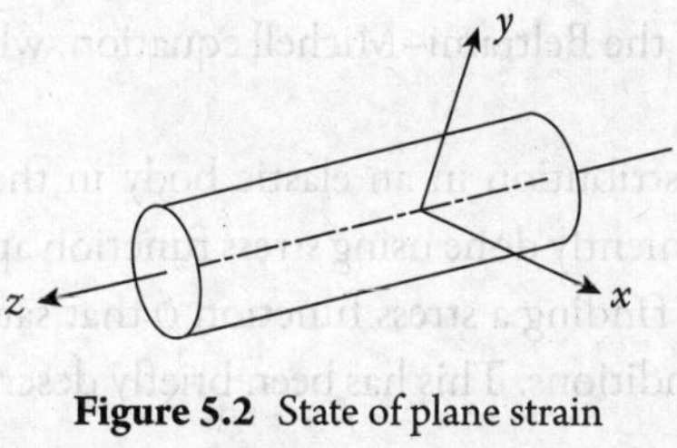

Figure 5.2 State of plane strain

It should be noted that unlike in plane stress, here $\in_z = 0$, but $\sigma_z \neq 0$.

Using Hooke's Law, stresses may be expressed in terms of strains, and from equation (4.3.1), we may write stresses in the plane strain state as

$$\begin{aligned}
&\sigma_x = 2G\in_x + \lambda\left(\in_x + \in_y\right)\\
&\sigma_y = 2G\in_y + \lambda\left(\in_x + \in_y\right)\\
&\sigma_z = \lambda\left(\in_x + \in_y\right)\\
&\tau_{xy} = G\gamma_{xy},\quad \tau_{yz} = \tau_{zx} = 0.
\end{aligned} \tag{5.2.12}$$

For plane strain conditions, equilibrium equations reduce to only two equations as in plane stress state shown in equation (5.2.6). These equations are reproduced here for reference purpose:

$$\frac{\partial \sigma_x}{\partial x} + \frac{\partial \tau_{xy}}{\partial y} + X = 0$$

$$\frac{\partial \tau_{xy}}{\partial x} + \frac{\partial \sigma_y}{\partial y} + Y = 0.$$

Using equations (5.2.11) and (5.2.12), equilibrium equations in terms of displacements are as shown in equation (5.2.13).

$$\begin{aligned}
&(\lambda + G)\frac{\partial}{\partial x}\left(\frac{\partial u}{\partial x} + \frac{\partial v}{\partial y}\right) + G\nabla^2 u + X = 0\\
&(\lambda + G)\frac{\partial}{\partial y}\left(\frac{\partial u}{\partial x} + \frac{\partial v}{\partial y}\right) + G\nabla^2 v + Y = 0.
\end{aligned} \tag{5.2.13}$$

Here $\nabla^2 = \frac{\partial^2}{\partial x^2} + \frac{\partial^2}{\partial y^2}$, 2D Laplacian operator, and equation (5.2.13) is the reduced form of 3D Navier–Lame equations shown in (4.3.2). For the plane strain state, the only compatibility equation that is not automatically satisfied is

$$\frac{\partial^2 \in_x}{\partial y^2} + \frac{\partial^2 \in_y}{\partial x^2} = \frac{\partial^2 \gamma_{xy}}{\partial x \partial y}.$$

Expressing this relation in terms of stresses in the absence of body forces, we have

$$\nabla^2\left(\sigma_x + \sigma_y\right) = 0. \tag{5.2.14}$$

This is the reduced form of the Beltrami–Michell equation, which is the same as in the case of plane stress, equations (5.2.8).

Determination of stress distribution in an elastic body in the case of both plane stress and plane strain states can be conveniently done using stress function approach. In the absence of body forces, the problem reduces to finding a stress function ϕ that satisfies the biharmonic equation $\nabla^4\phi = 0$ and the boundary conditions. This has been briefly described in Section 4.7.

5.3 GOVERNING EQUATIONS IN CYLINDRICAL COORDINATE SYSTEM [LO3]

Solutions to many elasticity problems make use of polar coordinates. The governing equations used in such problems are the reduced form of 3D governing equations in the cylindrical coordinate system.

In cylindrical coordinates, a point in space is given by $(r,\ \theta,\ z)$ coordinates, where r denotes radial distance from the origin, θ denotes azimuthal coordinate, and z denotes the height as shown in Figure 5.3. Relation between the Cartesian and cylindrical coordinate systems is given by

$$x = r\cos\theta,\ \ y = r\sin\theta,\ \ z = z. \tag{5.3.1}$$

Polar coordinate system is a 2D curvilinear system, where a point is given by $(r,\ \theta)$ coordinates as shown in Figure 5.4.

Relation between the Cartesian and polar coordinate systems is simply

$$x = r\cos\theta,\ \ y = r\sin\theta. \tag{5.3.2}$$

In order to develop the governing equations in cylindrical coordinates, we consider a cylindrical element as shown in Figure 5.5. Variations of normal stresses, σ_r, σ_θ, and σ_z, and shear stresses, $\tau_{r\theta}$, τ_{rz}, and $\tau_{z\theta}$, on the six faces of the element are shown in the figure. First, we consider equilibrium in the r-direction:

Net force in the r-direction due to variation of σ_r as shown in Figure 5.5 (a) is

$$\left(\sigma_r + \frac{\partial \sigma_r}{\partial r}\delta r\right)\delta\theta(r+\delta r)\delta z - \sigma_r\delta\theta r\delta z = r\delta r\delta\theta\delta z\left(\frac{\partial \sigma_r}{\partial r} + \frac{\sigma_r}{r}\right).$$

Net force in the r-direction due to variation in $\tau_{r\theta}$ as shown in Figure 5.5 (b) is

$$\left(\tau_{r\theta} + \frac{\partial \tau_{r\theta}}{\partial \theta}\delta\theta - \tau_{r\theta}\right)\delta r\delta z\cos\left(\frac{\delta\theta}{2}\right) = \frac{\partial \tau_{r\theta}}{\partial \theta}\delta\theta\delta r\delta z\cos\left(\frac{\delta\theta}{2}\right) = r\delta\theta\delta r\delta z\left(\frac{1}{r}\frac{\partial \tau_{r\theta}}{\partial \theta}\right).$$

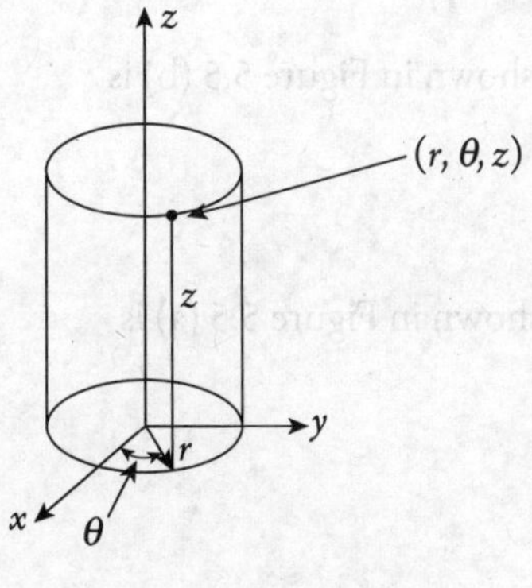

Figure 5.3 Cylindrical coordinate system

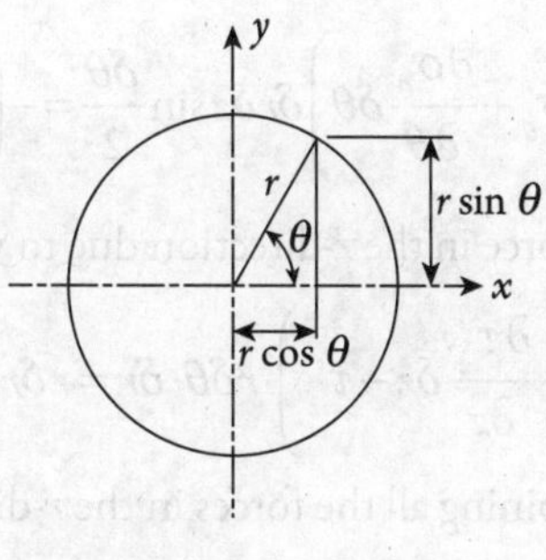

Figure 5.4 Polar coordinate system

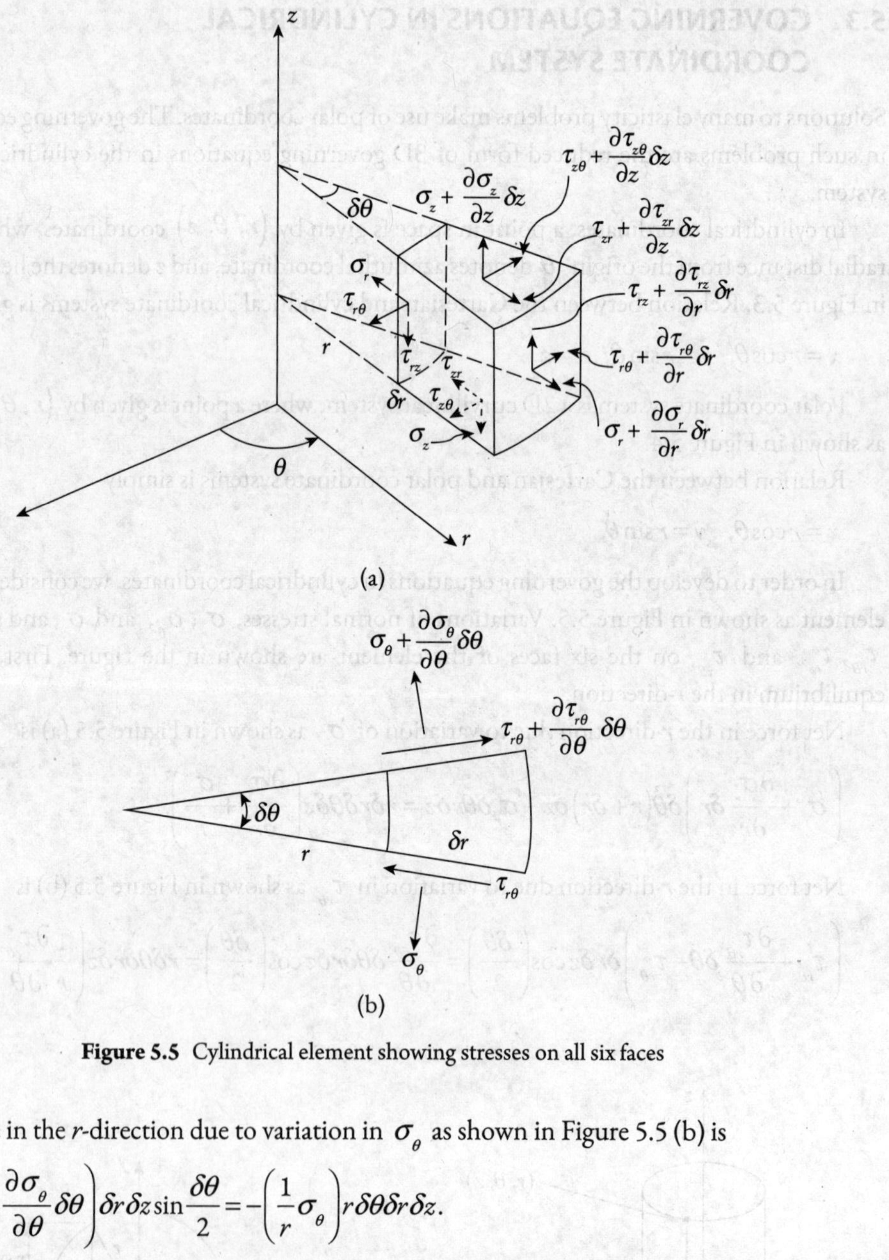

Figure 5.5 Cylindrical element showing stresses on all six faces

Net force in the r-direction due to variation in σ_θ as shown in Figure 5.5 (b) is

$$-\left(2\sigma_\theta+\frac{\partial\sigma_\theta}{\partial\theta}\delta\theta\right)\delta r\,\delta z\sin\frac{\delta\theta}{2}=-\left(\frac{1}{r}\sigma_\theta\right)r\delta\theta\delta r\delta z.$$

Net force in the r-direction due to variation in τ_{rz} as shown in Figure 5.5 (a) is

$$\left(\tau_{rz}+\frac{\partial\tau_{rz}}{\partial z}\delta z-\tau_{rz}\right)\cdot r\delta\theta\cdot\delta r=r\delta r\cdot\delta\theta\cdot\delta z\left(\frac{\partial\tau_{rz}}{\partial z}\right).$$

Combining all the forces in the r-direction, we have

$$r\delta r\delta\theta\delta z\left(\frac{\partial\sigma_r}{\partial r}+\frac{1}{r}\frac{\partial\tau_{r\theta}}{\partial\theta}+\frac{\partial\tau_{rz}}{\partial z}+\frac{\sigma_r-\sigma_\theta}{r}\right).$$

Therefore, for static equilibrium with no body forces, we have

$$\frac{\partial \sigma_r}{\partial r}+\frac{1}{r}\frac{\partial \tau_{r\theta}}{\partial \theta}+\frac{\partial \tau_{rz}}{\partial z}+\frac{\sigma_r-\sigma_\theta}{r}=0.$$

Similarly, two more equations may be written considering equilibrium in the θ and z-directions. Equations of equilibrium in the presence of body forces in cylindrical coordinates system may then be concisely given as follows:

$$\begin{aligned}&\frac{\partial \sigma_r}{\partial r}+\frac{1}{r}\frac{\partial \tau_{r\theta}}{\partial \theta}+\frac{\partial \tau_{rz}}{\partial z}+\frac{\sigma_r-\sigma_\theta}{r}+F_r=0\\&\frac{\partial \tau_{\theta r}}{\partial r}+\frac{1}{r}\frac{\partial \sigma_\theta}{\partial \theta}+\frac{\partial \tau_{\theta z}}{\partial z}+2\frac{\tau_{r\theta}}{r}+F_\theta=0\\&\frac{\partial \tau_{zr}}{\partial r}+\frac{1}{r}\frac{\partial \tau_{z\theta}}{\partial \theta}+\frac{\partial \sigma_z}{\partial z}+\frac{\tau_{rz}}{r}+F_z=0,\end{aligned} \tag{5.3.3}$$

where F_r, F_θ, and F_z are body forces per unit volume in the r, θ, and z-directions respectively.

Solutions to plane stress and plane strain problems in 2D polar coordinates involve determination of displacements in radial and circumferential directions u_r, u_θ, normal strains in radial and circumferential directions $\in_r$, $\in_\theta$ and shear strain in $r\theta$ plane, normal stresses in radial and circumferential stresses σ_r and σ_θ and shear stress $\tau_{r\theta}$ in $r\theta$ plane.

Before proceeding further, we need to establish the basic displacement, stress, and strain relations in 2D polar coordinates. We first consider the strain–displacement relation, similar to equations (2.2.1) and (2.2.2) in Cartesian coordinates. Following Timoshenko and Goodier, we consider an element *abcd* in the polar coordinate system as in Figure 5.6.

Now, let the radial displacement of *ad* be u_r. Therefore, the radial displacement of *bc* is $u_r+\frac{\partial u_r}{\partial r}dr$. The strain of the element *abcd* in the radial direction is

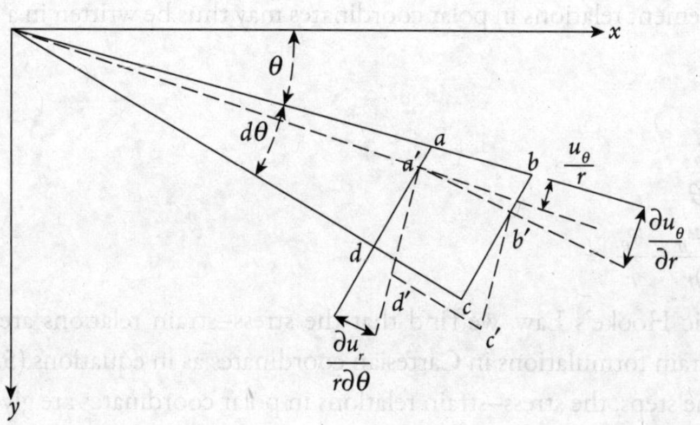

Figure 5.6 An element in the polar coordinate system where the displacements in r and θ directions are indicated

$$\epsilon_r = \frac{u_r + \frac{\partial u_r}{\partial r}dr - u_r}{dr} = \frac{\partial u_r}{\partial r}.$$

Now, the circumferential strain here depends both on the radial and circumferential displacements u and u_θ. If we first consider that ad has only radial displacement, then circumferential strain due to radial displacement alone is

$$\epsilon_{\theta_1} = \frac{(r+u_r)d\theta - rd\theta}{rd\theta} = \frac{u_r}{r}.$$

Consider now the difference in the circumferential displacement of ab and cd of the element $abcd$ in Figure 5.6. This is given by $\frac{\partial u_\theta}{\partial \theta}d\theta$, and therefore the strain due to the circumferential displacement is

$$\epsilon_{\theta_2} = \frac{\frac{\partial u_\theta}{\partial \theta}d\theta}{rd\theta} = \frac{\partial u_\theta}{r\partial \theta}.$$

Total circumferential strain is given by

$$\epsilon_\theta = \epsilon_{\theta_1} + \epsilon_{\theta_2} = \frac{u_r}{r} + \frac{\partial u_\theta}{r\partial \theta}.$$

For shear strain, we consider that the element $abcd$ is deformed to $a'b'c'd'$ as shown in Figure 5.6. The shear strain here is the sum of the angles between ad and $a'd'$ and ab and $a'b'$.

The angle between ad and $a'd'$ is $\frac{\partial u_r}{r\partial \theta}$.

The effective angle between ab and $a'b'$ is $\frac{\partial u_\theta}{\partial r} - \frac{u_\theta}{r}$.

Therefore, the shear strain is given by $\gamma_{r\theta} = \frac{\partial u_r}{r\partial \theta} + \frac{\partial u_\theta}{\partial r} - \frac{u_\theta}{r}$.

Strain–displacement relations in polar coordinates may thus be written in a concise form as

$$\begin{aligned} \epsilon_r &= \frac{\partial u_r}{\partial r} \\ \epsilon_\theta &= \frac{u_r}{r} + \frac{\partial u_\theta}{r\partial \theta} \\ \gamma_{r\theta} &= \frac{\partial u_r}{r\partial \theta} + \frac{\partial u_\theta}{\partial r} - \frac{u_\theta}{r}. \end{aligned} \tag{5.3.4}$$

Using the basic Hooke's Law, we find that the stress–strain relations are similar for plane stress and plane strain formulations in Cartesian coordinates as in equations (5.2.4) and (5.2.12). Following the same steps, the stress–strain relations in polar coordinates are given below for both the plane stress and plane strain states.

For the plane stress state

$$\begin{aligned}
\epsilon_r &= \frac{1}{E}(\sigma_r - \nu\sigma_\theta) \\
\epsilon_\theta &= \frac{1}{E}(\sigma_\theta - \nu\sigma_r) \\
\epsilon_z &= -\frac{\nu}{E}(\sigma_r + \sigma_\theta) = -\frac{\nu}{1-\nu}(\epsilon_r + \epsilon_\theta) \\
\gamma_{r\theta} &= \frac{2(1+\nu)}{E}\tau_{r\theta},\ \ \gamma_{\theta z} = \gamma_{rz} = 0
\end{aligned} \tag{5.3.5}$$

and for the plane strain state

$$\begin{aligned}
\sigma_r &= 2G\epsilon_r + \lambda(\epsilon_r + \epsilon_\theta) \\
\sigma_\theta &= 2G\epsilon_\theta + \lambda(\epsilon_r + \epsilon_\theta) \\
\sigma_z &= \lambda(\epsilon_r + \epsilon_\theta) = \nu(\sigma_r + \sigma_\theta) \\
\tau_{r\theta} &= G\gamma_{r\theta},\ \ \tau_{\theta z} = \tau_{rz} = 0.
\end{aligned} \tag{5.3.6}$$

In 2D polar coordinates, the equilibrium equations reduce to the following equations:

$$\begin{aligned}
&\frac{\partial\sigma_r}{\partial r} + \frac{1}{r}\frac{\partial\tau_{r\theta}}{\partial\theta} + \frac{\sigma_r - \sigma_\theta}{r} + F_r = 0 \\
&\frac{\partial\tau_{\theta r}}{\partial r} + \frac{1}{r}\frac{\partial\sigma_\theta}{\partial\theta} + 2\frac{\tau_{\theta r}}{r} + F_\theta = 0.
\end{aligned} \tag{5.3.7}$$

Equation (5.3.7) applies to both the plane stress and plane strain states.

Following the procedure set out in Section 4.3, equilibrium equations can be written in terms of displacements, which are known as Navier–Lame equations.

For the plane stress state with $\sigma_z = \tau_{rz} = \tau_{\theta z} = 0$, we have

$$\begin{aligned}
&\frac{E}{2(1-\nu)}\frac{\partial}{\partial r}\left(\frac{\partial u_r}{\partial r} + \frac{u_r}{r} + \frac{1}{r}\frac{\partial u_\theta}{\partial\theta}\right) + G\nabla^2 u_r + F_r = 0 \\
&\frac{E}{2(1-\nu)}\frac{1}{r}\frac{\partial}{\partial\theta}\left(\frac{\partial u_r}{\partial r} + \frac{u_r}{r} + \frac{1}{r}\frac{\partial u_\theta}{\partial\theta}\right) + G\nabla^2 u_\theta + F_\theta = 0
\end{aligned} \tag{5.3.8}$$

and for the plane strain state with $\epsilon_z = \gamma_{rz} = \gamma_{\theta z} = 0$, we have

$$\begin{aligned}
&(\lambda+G)\frac{\partial}{\partial r}\left(\frac{\partial u_r}{\partial r} + \frac{u_r}{r} + \frac{1}{r}\frac{\partial u_\theta}{\partial\theta}\right) + G\nabla^2 u_r + F_r = 0 \\
&(\lambda+G)\frac{1}{r}\frac{\partial}{\partial\theta}\left(\frac{\partial u_r}{\partial r} + \frac{u_r}{r} + \frac{1}{r}\frac{\partial u_\theta}{\partial\theta}\right) + G\nabla^2 u_\theta + F_\theta = 0.
\end{aligned} \tag{5.3.9}$$

Here, $\nabla^2 = \frac{\partial^2}{\partial r^2} + \frac{1}{r}\frac{\partial}{\partial r} + \frac{1}{r^2}\frac{\partial^2}{\partial\theta^2}$, the Laplacian operator in the polar coordinate system and body forces in the r and θ directions are F_r and F_θ respectively.

Expressing the compatibility equations in terms of stresses, we get the Beltrami–Michell equations suitable for problems with stress boundary conditions. Since $\sigma_x + \sigma_y = \sigma_r + \sigma_\theta$, the Beltrami–Michell equations in polar coordinates in the absence of body forces reduce to

$$\nabla^2\left(\sigma_r + \sigma_\theta\right) = 0 \tag{5.3.10}$$

for both the plane stress and plane strain conditions.

In Section 4.6, the stress function approach was briefly discussed in Cartesian coordinates for problems with stress boundary conditions. Here, we briefly mention some salient points of stress function approach in polar coordinates.

Stress functions are defined as functions that satisfy the equilibrium equations automatically. In the absence of body forces, the equilibrium equations in polar coordinates, given in equations (5.3.7), are satisfied by the following definition of stress functions:

$$\begin{aligned} \sigma_r &= \frac{1}{r}\frac{\partial \phi}{\partial r} + \frac{1}{r^2}\frac{\partial^2 \phi}{\partial \theta^2} \\ \sigma_\theta &= \frac{\partial^2 \phi}{\partial r^2} \\ \tau_{r\theta} &= -\frac{\partial}{\partial r}\left(\frac{1}{r}\frac{\partial \phi}{\partial \theta}\right), \end{aligned} \tag{5.3.11}$$

where ϕ is the stress function, which is a function of r and θ. It can be seen easily by direct substitution that equations (5.3.11) satisfy the equilibrium equations in (5.3.7). Further details of equations (5.3.11) may be found in *Theory of Elasticity* by Timoshenko and Goodier (1970).

Another important formulation in solving elasticity problems using the stress function is the biharmonic equation. In polar coordinates, this may be given as follows:

$$\nabla^4 \phi = \left(\frac{\partial^2}{\partial r^2} + \frac{1}{r}\frac{\partial}{\partial r} + \frac{1}{r^2}\frac{\partial^2}{\partial \theta^2}\right)\left(\frac{\partial^2}{\partial r^2} + \frac{1}{r}\frac{\partial}{\partial r} + \frac{1}{r^2}\frac{\partial^2}{\partial \theta^2}\right)\phi = 0. \tag{5.3.12}$$

With a defined stress function of the form $\phi(r, \theta)$, equations (5.3.11) and (5.3.12) are sufficient to proceed further with elasticity problems in polar coordinates.

5.4 AXISYMMETRIC PROBLEMS [LO4]

There is another class of problems of some engineering importance, where stress distribution is independent of θ coordinate. These problems are known as axisymmetric problems. If solids of revolutions are deformed symmetrically about the axis of revolution, then they become independent of angular coordinate θ.

Typical examples are thick cylinders, rotating discs, etc. In many such cases, the problem solution with axisymmetric assumption becomes relatively easy. For instance, in thick cylinders, we may consider a plane stress state and we have $\sigma_z = 0$, $\tau_{rz} = \tau_{\theta z} = 0$ also due to symmetry $\tau_{r\theta} = 0$.

This reduces the equilibrium equations in (5.3.3) for axisymmetric cases in a 2D polar coordinate system in the absence of body forces to

$$\frac{d\sigma_r}{dr} + \frac{\sigma_r - \sigma_\theta}{r} = 0. \tag{5.4.1}$$

This will be discussed in detail later in the chapter on thick cylinders.

Another important point needs to be discussed here. The use of stress functions in polar coordinates has been discussed briefly in Section 4.6, but stress functions in axisymmetric problems are only r-dependent. Therefore, the biharmonic equation is reduced to the following, from equation (5.3.12)

$$\nabla^4\phi = \frac{d^4\phi}{dr^4} + \frac{2}{r}\frac{d^3\phi}{dr^3} - \frac{1}{r^2}\frac{d^2\phi}{dr^2} + \frac{1}{r^3}\frac{d\phi}{dr} = 0. \tag{5.4.2}$$

This is an ordinary fourth-order differential equation with variable coefficients. This can be converted into an equation with a constant coefficient by substitution such as $r = e^t$. General solution of equation (5.4.2) may then be obtained as

$$\phi = A\log r + Br^2\log r + Cr^2 + D, \tag{5.4.3}$$

where A, B, C, and D are constants. Many elasticity problems with symmetrical stress distribution, such as thick cylinders, rotating disks, and pure bending of curved beams, can be solved using equations (5.4.1), (5.4.2), and (5.4.3) with suitable boundary conditions. Stress components from equation (5.4.3) can be given as follows:

$$\begin{aligned}
\sigma_r &= \frac{1}{r}\frac{\partial\phi}{\partial r} = \frac{A}{r^2} + 2B\log r + (B + 2C) \\
\sigma_\theta &= \frac{\partial^2\phi}{\partial r^2} = -\frac{A}{r^2} + 2B\log r + (3B + 2C) \\
\tau_{r\theta} &= -\frac{\partial}{\partial r}\left(\frac{1}{r}\frac{\partial\phi}{\partial\theta}\right) = 0.
\end{aligned} \tag{5.4.4}$$

These equations provide one more step forward for solving axisymmetric problems in polar coordinates.

Example 5.1

Show that the displacement components in plane stress problems in Cartesian coordinates satisfy the following equations:

$$\nabla^2 u + \frac{1+\nu}{1-\nu}\frac{\partial}{\partial x}\left(\frac{\partial u}{\partial x} + \frac{\partial v}{\partial y}\right) = 0$$

$$\nabla^2 v + \frac{1+\nu}{1-\nu}\frac{\partial}{\partial y}\left(\frac{\partial u}{\partial x} + \frac{\partial v}{\partial y}\right) = 0,$$

where $\nabla^2 = \dfrac{\partial^2}{\partial x^2} + \dfrac{\partial^2}{\partial y^2}$.

Solution:

From Hooke's Law, we have

$$\epsilon_x = \frac{\partial u}{\partial x} = \frac{1}{E}\left(\sigma_x - \nu\sigma_y\right)$$

$$\epsilon_y = \frac{\partial v}{\partial y} = \frac{1}{E}\left(\sigma_y - \nu\sigma_x\right)$$

$$\gamma_{xy} = \frac{\partial u}{\partial y} + \frac{\partial v}{\partial x} = \frac{2(1+\nu)}{E}\tau_{xy}$$

Manipulating these equations, we have

$$\sigma_x = \frac{E}{1-\nu^2}\left(\frac{\partial u}{\partial x} + \nu\frac{\partial v}{\partial y}\right)$$

$$\sigma_y = \frac{E}{1-\nu^2}\left(\frac{\partial v}{\partial y} + \nu\frac{\partial u}{\partial x}\right)$$

$$\tau_{xy} = \frac{E}{2(1+\nu)}\left(\frac{\partial u}{\partial y} + \frac{\partial v}{\partial x}\right).$$

Equilibrium equations are

$$\frac{\partial \sigma_x}{\partial x} + \frac{\partial \tau_{xy}}{\partial y} = 0$$

$$\frac{\partial \tau_{xy}}{\partial x} + \frac{\partial \sigma_y}{\partial y} = 0.$$

Substituting expressions for σ_x, σ_y, and τ_{xy}, we have

$$\nabla^2 u + \frac{1+\nu}{1-\nu}\frac{\partial}{\partial x}\left(\frac{\partial u}{\partial x} + \frac{\partial v}{\partial y}\right) = 0$$

and $\nabla^2 v + \dfrac{1+\nu}{1-\nu}\dfrac{\partial}{\partial y}\left(\dfrac{\partial u}{\partial x} + \dfrac{\partial v}{\partial y}\right) = 0.$

Example 5.2

The plane stress distribution in a flat plate of unit thickness is given by

$$\sigma_x = x^3 y - 2y^3 x,\ \sigma_y = y^3 x - 2pxy + qx,\ \tau_{xy} = \frac{y^4}{2} - \frac{3}{2}x^2y^2 + px^2 + s.$$

If the body forces are neglected, show that the equilibrium exists. The dimension of the plate is as shown in the figure. The following boundary conditions apply:

At $y = \pm\dfrac{b}{2}$, $\tau_{xy} = 0$

$y = \dfrac{b}{2}$, $\sigma_y = 0.$

Find the constants p, q, and s.

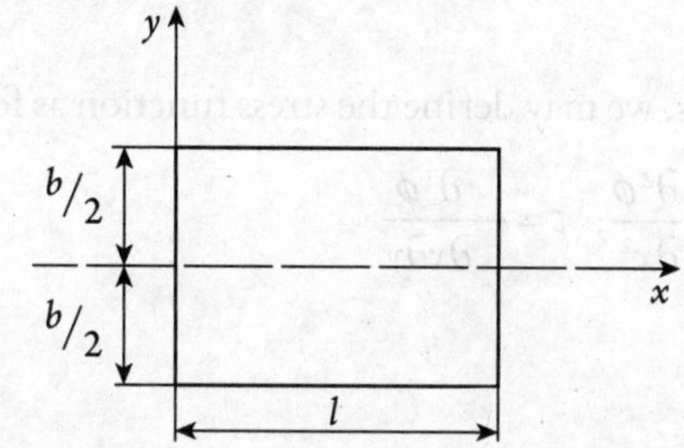

Solution:

First, let us consider the 2D equilibrium equations without the body forces and substitute the given stresses

$$\frac{\partial \sigma_x}{\partial x}+\frac{\partial \tau_{xy}}{\partial y}=0;\ 3x^2y-2y^3+2y^3-3x^2y=0$$

$$\frac{\partial \tau_{xy}}{\partial x}+\frac{\partial \sigma_y}{\partial y}=0;\ -3xy^2+2px+3y^2x-2px=0.$$

This proves that the equilibrium exists.

Now we consider the boundary conditions:

At $y=\pm\frac{b}{2},\ \tau_{xy}=0$

$\frac{b^4}{32}-\frac{3}{8}x^2b^2+px^2+s=0$, the coefficient of x^2 must be zero

$$-\frac{3}{8}b^2+p=0;\ \ p=\frac{3}{8}b^2$$

and $s=-\frac{b^4}{32}$

At $y=\frac{b}{2},\ \sigma_y=0,$

$\frac{b^3x}{8}-2px\frac{b}{2}+qx=0$; the coefficients of x must be zero, thus $-2p\frac{b}{2}+q+\frac{b^3}{8}=0$

$$-2\frac{b}{2}\cdot\frac{3}{8}b^2+q+\frac{b^3}{8}=0;\ \ q=\frac{b^3}{4}.$$

Example 5.3

Show that plane stress elasticity problems in the presence of body forces reduce to solutions of the differential equation

$$\nabla^4\phi+(1-\nu)\nabla^2\Omega=0,$$

having regard to the boundary condition. Here ϕ and Ω are the stress and potential function respectively.

Solution:

In the presence of body forces, we may define the stress function as follows:

$$\sigma_x = \Omega + \frac{\partial^2\phi}{\partial y^2};\ \sigma_y = \Omega + \frac{\partial^2\phi}{\partial x^2};\ \tau = -\frac{\partial^2\phi}{\partial x\partial y}$$

and body forces are given by

$$X = -\frac{\partial\Omega}{\partial x};\ Y = -\frac{\partial\Omega}{\partial y}.$$

2D equilibrium equations reduce to

$$\frac{\partial\sigma_x}{\partial x} + \frac{\partial\tau_{xy}}{\partial y} + X = 0$$

$$\frac{\partial\tau_{xy}}{\partial x} + \frac{\partial\sigma_y}{\partial y} + Y = 0$$

and using Hooke's Law, we can write

$$\epsilon_x = \frac{1}{E}\left(\sigma_x - \nu\sigma_y\right)$$

$$\epsilon_y = \frac{1}{E}\left(\sigma_y - \nu\sigma_x\right).$$

Substituting expressions for stresses as stated above, we have

$$\epsilon_x = \frac{1}{E}\left[\Omega + \frac{\partial^2\phi}{\partial y^2} - \nu\Omega - \nu\frac{\partial^2\phi}{\partial x^2}\right] = \frac{1}{E}\left[(1-\nu)\Omega + \frac{\partial^2\phi}{\partial y^2} - \nu\frac{\partial^2\phi}{\partial x^2}\right]$$

$$\epsilon_y = \frac{1}{E}\left[\Omega + \frac{\partial^2\phi}{\partial x^2} - \nu\Omega - \nu\frac{\partial^2\phi}{\partial y^2}\right] = \frac{1}{E}\left[(1-\nu)\Omega + \frac{\partial^2\phi}{\partial x^2} - \nu\frac{\partial^2\phi}{\partial y^2}\right].$$

γ_{xy} is not affected by the presence of body forces.

Substituting the expression for ϵ_x, ϵ_y, and γ_{xy} in the first compatibility equation, we have

$$\frac{\partial^2\epsilon_x}{\partial y^2} + \frac{\partial^2\epsilon_y}{\partial x^2} = \frac{\partial^2\gamma_{xy}}{\partial x\partial y}$$

$$\frac{1}{E}\left[\frac{\partial^2\Omega}{\partial y^2} + \frac{\partial^4\phi}{\partial y^4} - \nu\frac{\partial^2\Omega}{\partial y^2} - \nu\frac{\partial^4\phi}{\partial x^2\partial y^2}\right] + \frac{1}{E}\left[\frac{\partial^2\Omega}{\partial x^2} + \frac{\partial^4\phi}{\partial x^4} - \nu\frac{\partial^2\Omega}{\partial x^2} - \nu\frac{\partial^4\phi}{\partial x^2\partial y^2}\right]$$

$$= -\frac{\partial^4\phi}{\partial x^2\partial y^2}\frac{2(1+\nu)}{E}.$$

This gives

$\nabla^4\phi + (1-\nu)\nabla^2\Omega = 0\cdot$ ... (Proved).

Example 5.4

Show by the stress function approach that the stress field for the cantilever beam shown in the figure is

$$\sigma_x = \frac{Pxy}{I},\ \sigma_y = 0,\ \tau_{xy} = \frac{P}{2I}\left(\frac{d^2}{4} - y^2\right).$$

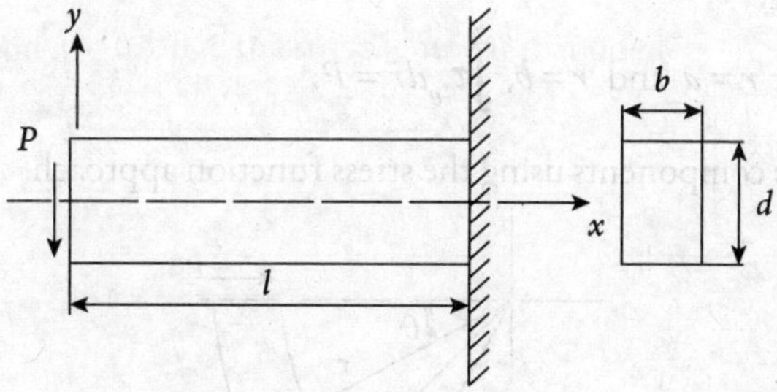

Solution:

By inspecting first-, second-, and third-degree polynomials, we choose the following stress function to be suitable for the present problem:

$$\phi = Axy + Bxy^3.$$

This gives

$$\sigma_x = \frac{\partial^2 \phi}{\partial y^2} = 6Bxy$$

$$\sigma_y = \frac{\partial^2 \phi}{\partial x^2} = 0$$

$$\tau_{xy} = -\frac{\partial^2 \phi}{\partial x \partial y} = -A - 3By^2.$$

Constants can now be determined using the boundary conditions.

1. There are no shear stresses on the top and bottom surfaces of the beam. This means

   At $y = \pm\frac{d}{2}$, $\tau_{xy} = 0$, i.e., $A = -\frac{3}{4}Bd^2$. ...(Ans.)

2. Shear force at the fixed end is P. This means

   at $x = l$

   $$b\int_{-d/2}^{d/2} \tau_{xy}\, dy = P.$$

   This gives $B = \frac{2P}{bd^3}$ and $A = -\frac{3}{2}\frac{P}{bd}$.

   Substituting the constants in the expression for σ_x, σ_y, and τ_{xy}, we get

   $$\sigma_x = \frac{Pxy}{I},\ \sigma_y = 0,\ \tau_{xy} = \frac{P}{2I}\left(\frac{d^2}{4} - y^2\right),$$

where I is the second moment of area = $\frac{bd^3}{12}$ for a rectangular cross-section of width b and depth d.

Example 5.5

A curved cantilever beam is subjected to an end shear force P as shown in the figure. The boundary conditions are as follows:

$$\sigma_r = \sigma_\theta = \tau_{r\theta} = 0 \text{ at } r = a \text{ and } r = b, \int_a^b \tau_{r\theta} dr = P.$$

Determine the stress components using the stress function approach.

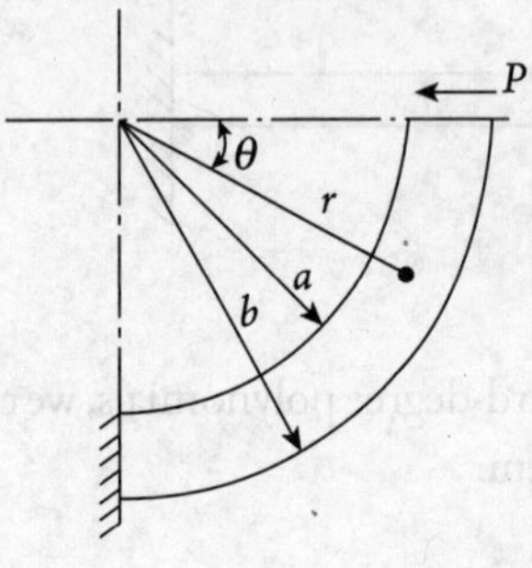

Solution:

Following the procedure similar to the one used in establishing equation (5.4.3), it was shown in *Theory of Elasticity* by Timoshenko and Goodier that, by taking into account the θ dependence, the suitable stress function for this problem is

$$\phi = \left(Ar^3 + \frac{B}{r} + Cr + Dr \ln r \right) \sin\theta.$$

This gives, from (5.3.11),

$$\sigma_r = \frac{1}{r}\frac{\partial \phi}{\partial r} + \frac{1}{r^2}\frac{\partial^2 \phi}{\partial \theta^2} = \left(2Ar - \frac{2B}{r^3} + \frac{D}{r} \right) \sin\theta$$

$$\sigma_\theta = \frac{\partial^2 \phi}{\partial r^2} = \left(6Ar + \frac{2B}{r^3} + \frac{D}{r} \right) \sin\theta$$

$$\tau_{r\theta} = -\frac{\partial}{\partial r}\left(\frac{1}{r}\frac{\partial \phi}{\partial \theta} \right) = -\left(2Ar - \frac{2B}{r^3} + \frac{D}{r} \right) \cos\theta.$$

Boundary conditions for the inner and outer boundaries are

$\sigma_r = 0$ at $r = a, b$ and $\tau_{r\theta} = 0$ at $r = a, b$.

This gives

$$2Aa - \frac{2B}{a^3} + \frac{D}{a} = 0 \qquad \text{... (a)}$$

$$2Ab - \frac{2B}{b^3} + \frac{D}{b} = 0. \qquad \text{... (b)}$$

We also have

$$\int_a^b \tau_{r\theta} dr = -\int_a^b \left(2Ar - \frac{2B}{r^3} + \frac{D}{r}\right)\cos\theta dr = P. \qquad \text{... (c)}$$

Solving equations (a), (b), and (c), we have

$$A = \frac{P'}{2};\ B = -\frac{P'a^2b^2}{2};\ D = -P'\left(a^2 + b^2\right),$$

where

$$P' = \frac{P}{\left(a^2 - b^2\right) + \left(a^2 + b^2\right)\log\frac{b}{a}}.$$

Substituting expressions for A, B, and D, we can write the stress field as

$$\sigma_r = P'\left[r + \frac{a^2b^2}{r^3} - \left(\frac{a^2 + b^2}{r}\right)\right]\sin\theta$$

$$\sigma_\theta = P'\left[3r - \frac{a^2b^2}{r^3} - \left(\frac{a^2 + b^2}{r}\right)\right]\sin\theta$$

$$\tau_{r\theta} = -P'\left[r + \frac{a^2b^2}{r^3} - \left(\frac{a^2 + b^2}{r}\right)\right]\cos\theta.$$

For the upper end of the bar, $\theta = 0$, we have

$$\sigma_\theta = 0,\ \tau_{r\theta} = -P'\left[r + \frac{a^2b^2}{r^3} - \left(\frac{a^2 + b^2}{r}\right)\right].$$

For the lower end, $\theta = \frac{\pi}{2}$, we have

$$\tau_{r\theta} = 0,\ \sigma_\theta = P'\left[3r - \frac{a^2b^2}{r^3} - \left(\frac{a^2 + b^2}{r}\right)\right].$$

These results vary from the strength of materials approach, which gives some overestimation of the stress field.

Example 5.6

An elastic body with the plane stress field has the following displacement field

$$u(x, y) = 30x^2 - 10x^3y + 20y^3 \text{ mm}$$

$$v(x, y) = 10x^3 + 20xy^3 + 5y^2 \text{ mm}.$$

Find the stress components at a point $(x = 0.05\,\text{m},\ y = 0.02\,\text{m})$.

Solution:

The strain field may be determined from the displacement field as follows:

$$\in_x = \frac{\partial u}{\partial x} = 60x - 30x^2 y; \ \in_y = \frac{\partial v}{\partial y} = 60xy^2 + 10y;$$

$$\gamma_{xy} = \frac{\partial u}{\partial y} + \frac{\partial v}{\partial x} = -10x^3 + 60y^2 + 30x^2 + 20y^3.$$

For $E = 200\,\text{GPa}$, $\nu = 0.3$, the stress field at a point $(x = 0.05\,\text{m},\ y = 0.02\,\text{m})$ may be obtained from the stress–strain relationship as follows:

$$\sigma_x = \frac{E}{1-\nu^2}\left(\frac{\partial u}{\partial x} + \nu\frac{\partial v}{\partial y}\right) = 672.28\,\text{MPa}$$

$$\sigma_y = \frac{E}{1-\nu^2}\left(\frac{\partial v}{\partial y} + \nu\frac{\partial u}{\partial x}\right) = 241.92\,\text{MPa}$$

$$\tau_{xy} = \frac{E}{2(1+\nu)}\left(\frac{\partial u}{\partial y} + \frac{\partial v}{\partial x}\right) = 7.53\,\text{MPa}.$$ ... (Ans.)

Example 5.7

Show that the solutions to plane strain problems in the presence of body forces reduce to solving the differential equation

$$\nabla^4\phi + \frac{1-2\nu}{1-\nu}\nabla^2\Omega = 0,$$

where ϕ and Ω are the stress and potential function respectively.

Solution:

For plane strain conditions:

$$\sigma_x = 2G\in_x + \lambda(\in_x + \in_y); \ \sigma_y = 2G\in_y + \lambda(\in_x + \in_y); \ \sigma_z = +\lambda(\in_x + \in_y);$$

$$\tau_{xy} = G\gamma_{xy}; \ \tau_{yz} = \tau_{zx} = 0.$$

Equilibrium equations in the presence of body force:

$$\frac{\partial \sigma_x}{\partial x} + \frac{\partial \tau_{xy}}{\partial y} + X = 0$$

$$\frac{\partial \tau_{xy}}{\partial x} + \frac{\partial \sigma_y}{\partial y} + Y = 0.$$

Combining the above equations, we get

$$\left.\begin{aligned}(\lambda+G)\frac{\partial}{\partial x}\left(\frac{\partial u}{\partial x}+\frac{\partial v}{\partial y}\right)+G\nabla^2 u+X=0\\(\lambda+G)\frac{\partial}{\partial y}\left(\frac{\partial u}{\partial x}+\frac{\partial v}{\partial y}\right)+G\nabla^2 v+Y=0.\end{aligned}\right\} \quad \text{... (i)}$$

The stresses in terms of stress function are given as

$$\sigma_x=\frac{\partial^2\phi}{\partial y^2}+\Omega;\ \sigma_y=\frac{\partial^2\phi}{\partial x^2}+\Omega;\ \tau_{xy}=-\frac{\partial^2\phi}{\partial x\partial y}.$$

where $\phi(x, y)$ – airy stress function; Ω – potential function

Body forces: $X=-\frac{\partial\Omega}{\partial x};\ Y=-\frac{\partial\Omega}{\partial y}.$

The strain compatibility relation for plane strain:

$$\frac{\partial^2 \in_x}{\partial y^2}+\frac{\partial^2 \in_y}{\partial x^2}=\frac{\partial^2\gamma_{xy}}{\partial x\partial y}. \quad \text{... (ii)}$$

Combining the equations (i) and (ii)

$$\nabla^2\left(\sigma_x+\sigma_y\right)=-\frac{1}{1-\nu}\left(\frac{\partial X}{\partial x}+\frac{\partial Y}{\partial y}\right), \quad \text{... (iii)}$$

where the Laplacian operator is defined as: $\nabla^2=\frac{\partial^2}{\partial x^2}+\frac{\partial^2}{\partial y^2}.$

Substituting σ_x, σ_y, X, Y in equation (iii)

$$\left(\frac{\partial^2}{\partial x^2}+\frac{\partial^2}{\partial y^2}\right)\left(\frac{\partial^2\phi}{\partial y^2}+\frac{\partial^2\phi}{\partial x^2}+2\Omega\right)=-\frac{1}{1-\nu}\left(\frac{\partial}{\partial x}\left(\frac{-\partial\Omega}{\partial x}\right)+\frac{\partial}{\partial y}\left(\frac{-\partial\Omega}{\partial y}\right)\right).$$

Simplifying this, we get

$$\nabla^4\phi+\left(\frac{1-2\nu}{1-\nu}\right)\nabla^2\Omega=0. \quad \text{... (Proved).}$$

Example 5.8

Show that the strain field given by

$$\in_x=k\left(x^2+y^2\right),\ \in_y=k\left(y^2+z^2\right),\ \gamma_{xy}=k_1\,xyz,\ \in_z=\gamma_{zx}=\gamma_{zy}=0$$

is not permissible if k and k_1 are constants.

Solution:

Substituting the strain field in the strain compatibility relations, we have

$$\frac{\partial^2 \in_x}{\partial y^2}+\frac{\partial^2 \in_y}{\partial x^2}=\frac{\partial^2 \gamma_{xy}}{\partial x \partial y}; \qquad 2k=k_1 z$$

$$\frac{\partial^2 \in_y}{\partial z^2}+\frac{\partial^2 \in_z}{\partial y^2}=\frac{\partial^2 \gamma_{yz}}{\partial y \partial z}; \qquad k=0$$

$$\frac{\partial}{\partial x}\left(\frac{\partial \gamma_{xy}}{\partial z}-\frac{\partial \gamma_{yz}}{\partial x}+\frac{\partial \gamma_{xz}}{\partial y}\right)=2\left(\frac{\partial^2 \in_x}{\partial y \partial z}\right) \qquad k_1 y=0$$

$$\frac{\partial}{\partial y}\left(\frac{\partial \gamma_{yz}}{\partial x}-\frac{\partial \gamma_{zx}}{\partial y}+\frac{\partial \gamma_{xy}}{\partial z}\right)=2\left(\frac{\partial^2 \in_y}{\partial x \partial z}\right); \qquad k_1 x=0.$$

Hence, for the strain field to be permissible, k and k_1 cannot be constants.

Example 5.9

For thick cylinder analysis, we may assume the plane stress state and axisymmetric situation. Show that equilibrium equations in cylindrical polar coordinates may be reduced to

$$\frac{\partial \sigma_r}{\partial r}+\frac{\sigma_r-\sigma_\theta}{r}=0.$$

Solution:

For the plane stress–strain relations in polar coordinates, we may write

$$\in_r=\frac{1}{E}\left(\sigma_r-\nu\sigma_\theta\right)$$

$$\in_\theta=\frac{1}{E}\left(\sigma_\theta-\nu\sigma_r\right)$$

$$\in_z=-\frac{\nu}{E}\left(\sigma_r+\sigma_\theta\right)=-\frac{\nu}{1-\nu}\left(\in_r+\in_\theta\right)$$

$$\in_{r\theta}=\frac{1+\nu}{E}\gamma_{r\theta}.$$

The 2D polar coordinates equilibrium equation for the plane stress condition is given by

$$\frac{\partial \sigma_r}{\partial r}+\frac{1}{r}\frac{\partial \tau_{r\theta}}{\partial \theta}+\frac{\sigma_r-\sigma_\theta}{r}+F_r=0 \qquad \text{... (i)}$$

$$\frac{\partial \tau_{\theta r}}{\partial r}+\frac{1}{r}\frac{\partial \sigma_\theta}{\partial \theta}+2\frac{\tau_{\theta r}}{r}+F_\theta=0 \qquad \text{... (ii)}$$

In the absence of external body force,

$$F_r=F_\theta=0.$$

Considering axisymmetry, equation (ii) is no longer applicable and $\tau_{\theta r}$ becomes 0. Hence, equation (i) reduces to

$$\frac{\partial \sigma_r}{\partial r}+\frac{\sigma_r-\sigma_\theta}{r}=0. \qquad \text{... (Proved).}$$

Example 5.10

Verify that the stress field obtained for the cantilever beam in Example 5.4 satisfies the equilibrium and compatibility conditions for the plane stress.

Solution:

The stress field obtained in Example 5.4 is given by

$$\sigma_x=\frac{Pxy}{I};\ \sigma_y=0;\ \tau_{xy}=\frac{P}{2I}\left(\frac{d^2}{4}-y^2\right),$$

where P is the applied load, $I=\frac{1}{12}bd^3$; (b, d) are the lateral dimensions.

For the plane stress, the equilibrium equations are given as:

$$\frac{\partial \sigma_x}{\partial x}+\frac{\partial \tau_{xy}}{\partial y}=0$$

$$\Rightarrow \quad \frac{\partial}{\partial x}\left\{\frac{Pxy}{I}\right\}+\frac{\partial}{\partial y}\left\{\frac{P}{2I}\left(\frac{d^2}{4}-y^2\right)\right\}=0$$

$$\Rightarrow \quad \frac{Py}{I}-\frac{Py}{I}=0.$$

Hence, the equation is satisfied.
Similarly,

$$\frac{\partial \sigma_y}{\partial y}+\frac{\partial \tau_{xy}}{\partial x}=0$$

is also satisfied; hence, both equilibrium conditions are satisfied.

Compatibility equation, $\nabla^2\left(\sigma_x+\sigma_y\right)=0,$

where $\nabla^2=\frac{\partial^2}{\partial x^2}+\frac{\partial^2}{\partial x^2}$ is Laplacian operator.

We get $\frac{\partial^2 \sigma_x}{\partial x^2}+\frac{\partial^2 \sigma_y}{\partial x^2}+\frac{\partial^2 \sigma_x}{\partial y^2}+\frac{\partial^2 \sigma_y}{\partial y^2}=\frac{\partial^2}{\partial x^2}\left(\frac{Pxy}{I}\right)+\frac{\partial^2}{\partial x^2}(0)+\frac{\partial^2}{\partial y^2}\left(\frac{Pxy}{I}\right)+\frac{\partial^2}{\partial y^2}(0)=0.$

Hence, the stress function satisfies both the equilibrium and compatibility equations.

Exercises

1. Considering that in the plane strain state $\epsilon_z = 0$ but $\sigma_z \neq 0$, show that

$$\epsilon_x = \frac{1}{E}\left[\left(1-\nu^2\right)\sigma_x - \left(1+\nu\right)\nu\sigma_y\right]$$

$$\epsilon_y = \frac{1}{E}\left[\left(1-\nu^2\right)\sigma_y - \left(1+\nu\right)\nu\sigma_x\right]$$

$$\gamma_{xy} = \frac{2(1+\nu)}{E}\tau_{xy}.$$

2. In the plane stress state, Hooke's Law may be given as

$$\sigma_x = \frac{E}{1-\nu^2}\left(\epsilon_x + \nu\epsilon_y\right)$$

$$\sigma_y = \frac{E}{1-\nu^2}\left(\epsilon_y + \nu\epsilon_x\right)$$

$$\tau_{xy} = \frac{E}{2(1+\nu)}\gamma_{xy}.$$

Eliminating the stresses and using the plane stress equilibrium equations, develop 2D Navier's equations.

3. Find the 2D stress field due to the bending of a simply supported beam by uniform transverse load, as shown in the figure below

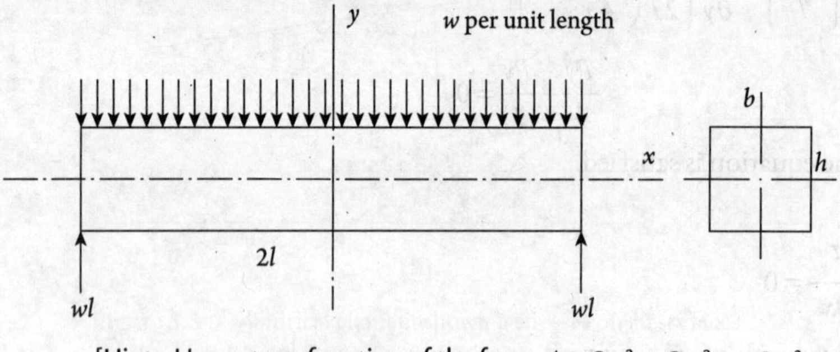

[Hints: Use a stress function of the form $\phi = C_1x^2 + C_2x^2y + C_3y^3 + C_4x^2y^3 - C_5y^5$

$$\sigma_x = \frac{w}{2I}\left(l^2 - x^2\right)y + \frac{w}{I}\left(\frac{y^3}{3} - \frac{h^2y}{20}\right).$$

$$\sigma_y = -\frac{w}{2I}\left(\frac{y^3}{3} - \frac{h^2y}{4} + \frac{h^3}{12}\right)$$

$$\tau_{xy} = -\frac{w}{2I}\left(\frac{h^2}{4} - y^2\right)x].$$

4. Using the stress–strain relations in the plane strain and plane stress states in polar coordinates as given in equations (5.3.5) and (5.3.6), and the equilibrium equations in polar coordinates as given in (5.3.7), develop 2D Navier's equations for both stress states in polar coordinates.

5. Find the 2D stress field in polar coordinates due to a line load w per unit length applied at the right angle to the surface of a semi-infinite body (2D) as shown in the figure below:

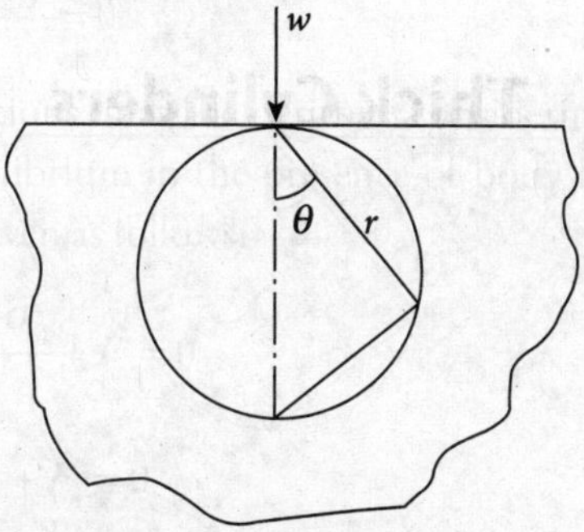

r and θ in the figure represent the polar coordinates.

[Hints: Use a stress function of the form: $\phi = cr\theta \sin\theta$; $\sigma_r = \dfrac{2w}{\pi}\dfrac{\cos\theta}{r}$; $\sigma_\theta = \tau_{r\theta} = 0$.]

6

Thick Cylinders

Learning Objectives

After careful study of this chapter, students should be able to do the following:

LO1: Describe stress equations in thick cylinders.

LO2: Explain stress distribution in pressurized cylinders.

LO3: Analyze compound cylinders.

LO4: Analyze autofrettage.

LO5: Analyze failure theories for thick cylinders.

6.1 INTRODUCTION [LO1]

In earlier chapters, we have discussed axisymmetric problems in two-dimensional (2D) polar coordinate systems. Thick cylinders fall into this class of problems. Cylindrical pressure vessels, hydraulic cylinders, gun-barrels, and pipes carrying fluids at high pressure develop radial and tangential stresses (circumferential). Longitudinal stresses can also be developed if the ends are closed. Therefore, ideally, this is a triaxial stress system as shown in Figure 6.1.

(a) Circumferential or hoop stress $\left(\sigma_\theta\right)$

(b) Longitudinal stress $\left(\sigma_z\right)$

(c) Radial stress $\left(\sigma_r\right)$

If the wall thickness of a hollow cylinder is less than about 10% of its radius, it may be treated as a thin cylinder. Cylinders with higher wall thickness are considered to be thick cylinders. Before analyzing the stress in a thick cylinder, we should briefly consider the stress state in thin cylinders, where radial stress is small compared to the other stresses, and this can be neglected. Stress variation across the thin wall is also negligible. Analysis of thin-walled pressure vessels may therefore be carried out on the basis of biaxial stress system. Since the presence of shear stress at the cut section would lead to incompatible distortion, the longitudinal and circumferential stresses in this case are

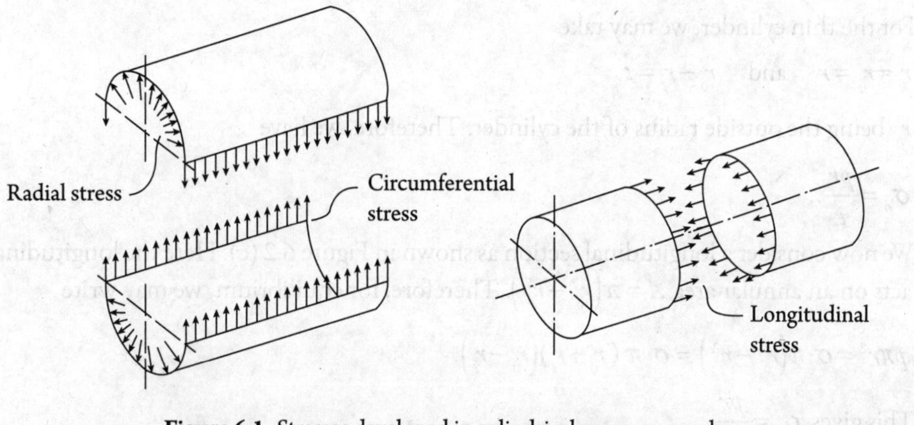

Figure 6.1 Stresses developed in cylindrical pressure vessels

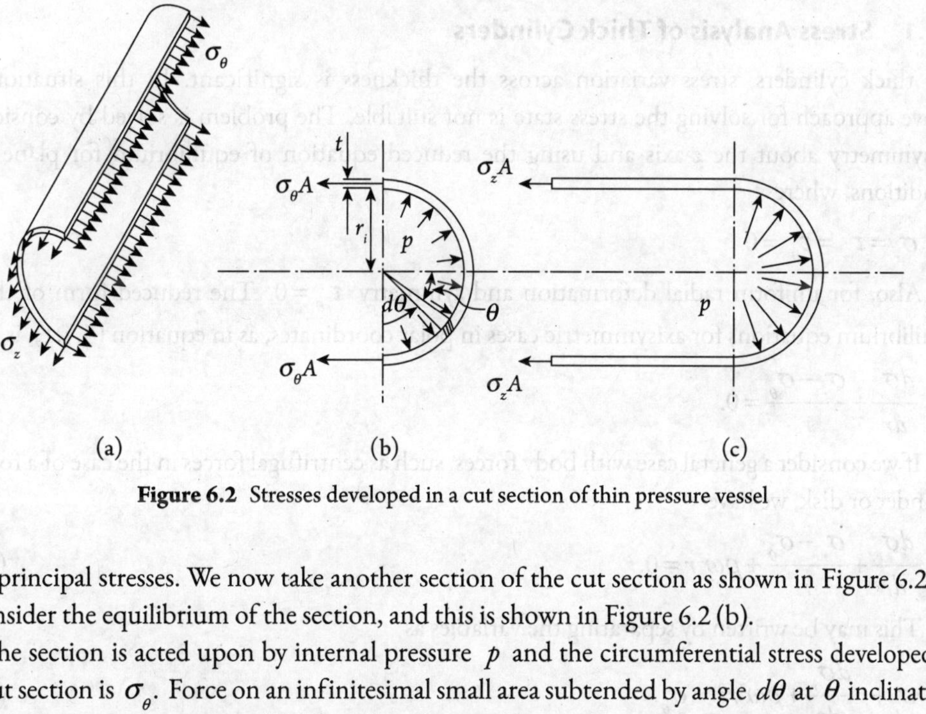

Figure 6.2 Stresses developed in a cut section of thin pressure vessel

both principal stresses. We now take another section of the cut section as shown in Figure 6.2 (a) to consider the equilibrium of the section, and this is shown in Figure 6.2 (b).

The section is acted upon by internal pressure p and the circumferential stress developed at the cut section is σ_θ. Force on an infinitesimal small area subtended by angle $d\theta$ at θ inclination from the horizontal axis is $pr_i d\theta$. Therefore, for equilibrium, we may write

$$2\sigma_\theta A = 2\int_0^{\pi/2} pr_i l\cos\theta d\theta,$$

where r_i is the internal radius of the cylinder and A is the cut section area. Here $A = lt$, l being the cut section length and t being the wall thickness. This gives

$$\sigma_\theta = \frac{pr_i}{t}.$$

For the thin cylinder, we may take

$$r_i \approx r_0 = r \quad \text{and} \quad r_0 - r_i = t.$$

r_0 being the outside radius of the cylinder. Therefore, we have

$$\sigma_\theta = \frac{pr}{t}. \tag{6.1.1}$$

We now consider a longitudinal section as shown in Figure 6.2 (c). Here the longitudinal stress σ_z acts on an annular area $A = \pi\left(r_0^2 - r_i^2\right)$. Therefore, for equilibrium, we may write

$$p\pi r_i^2 = \sigma_z \pi\left(r_0^2 - r_i^2\right) = \sigma_z \pi\,(r_0 + r_i)\left(r_0 - r_i\right).$$

This gives $\sigma_z = \dfrac{pr}{2t}$. (6.1.2)

6.1.1 Stress Analysis of Thick Cylinders

For thick cylinders, stress variation across the thickness is significant. In this situation, the above approach for solving the stress state is not suitable. The problem is solved by considering axisymmetry about the *z*-axis and using the reduced equation of equilibrium for plane stress conditions, where

$$\sigma_z = \tau_{rz} = \tau_{\theta z} = 0.$$

Also, for uniform radial deformation and symmetry $\tau_{r\theta} = 0$. The reduced form of the 2D equilibrium equations for axisymmetric cases in polar coordinates, as in equation (5.4.1), is

$$\frac{d\sigma_r}{dr} + \frac{\sigma_r - \sigma_\theta}{r} = 0.$$

If we consider a general case with body forces, such as centrifugal forces in the case of a rotating cylinder or disk, we have

$$\frac{d\sigma_r}{dr} + \frac{\sigma_r - \sigma_\theta}{r} + \rho\omega^2 r = 0, \tag{6.1.3}$$

This may be written by separating the variables as

$$\sigma_r + r\frac{d\sigma_r}{dr} + \rho\omega^2 r^2 = \sigma_\theta. \tag{6.1.4}$$

Here ρ is the density and ω is the angular velocity.

It is convenient to solve the general equation so that a variety of problems may be solved. It is now necessary to develop another equation relating σ_θ and σ_r and this may be obtained using strain–displacement relation. Now, considering the linear strain of an element of length δr, as shown in Figure 6.3(a), we have

$$\epsilon_r = \frac{du_r}{dr} = \frac{1}{E}\left(\sigma_r - \nu\sigma_\theta\right). \tag{a}$$

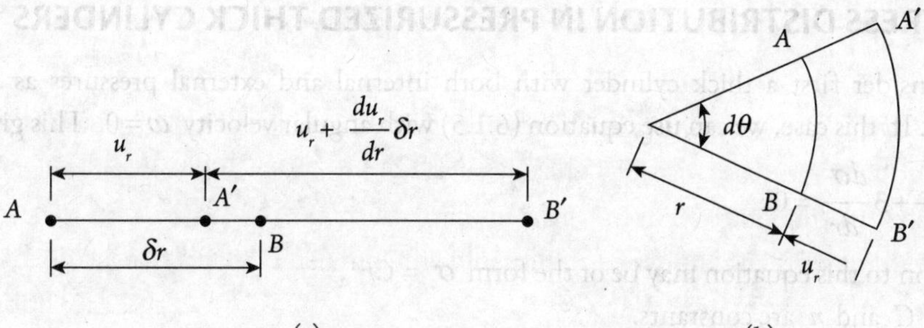

(a) (b)

Figure 6.3 Radial and angular strain

Similarly, considering the angular strain, as shown in Figure 6.3 (b), we have

$$\epsilon_\theta = \frac{(r+u_r)d\theta - rd\theta}{rd\theta} = \frac{u_r}{r} = \frac{1}{E}\left(\sigma_\theta - \nu\sigma_r\right). \tag{b}$$

From equation (b), we have $u_r = \frac{1}{E}\left(r\sigma_\theta - \nu r\sigma_r\right)$.

Differentiating u_r in the above equation with respect to r and combining with equation (a)

$$\frac{du_r}{dr} = \frac{1}{E}\left(\sigma_\theta + r\frac{d\sigma_\theta}{dr} - \nu\sigma_r - \nu r\frac{d\sigma_r}{dr}\right) = \frac{1}{E}\left(\sigma_r - \nu\sigma_\theta\right).$$

This gives

$$r\frac{d\sigma_\theta}{dr} - \nu r\frac{d\sigma_r}{dr} + (1+\nu)\left(\sigma_\theta - \sigma_r\right) = 0. \tag{c}$$

Now differentiating σ_θ with respect to r in equation (6.1.4), we have

$$\frac{d\sigma_\theta}{dr} = r\frac{d^2\sigma_r}{dr^2} + 2\frac{d\sigma_r}{dr} + 2\rho\omega^2 r. \tag{d}$$

Combining equations (c) and (d)

$$r^2\frac{d^2\sigma_r}{dr^2} + 2r\frac{d\sigma_r}{dr} + 2\rho\omega^2 r^2 - \nu r\frac{d\sigma_r}{dr} + (1+\nu)\left(\sigma_\theta - \sigma_r\right) = 0.$$

Substituting σ_θ from equation (6.1.4) and simplifying, we have

$$r\frac{d^2\sigma_r}{dr^2} + 3\frac{d\sigma_r}{dr} + (3+\nu)\rho\omega^2 r^2 = 0. \tag{6.1.5}$$

This is a general equation that can be used to determine the stress distribution in a variety of axisymmetric problems, such as thick cylinders, rotating cylinders, and disks.

6.2 STRESS DISTRIBUTION IN PRESSURIZED THICK CYLINDERS [LO2]

Let us consider first a thick cylinder with both internal and external pressures as shown in Figure 6.4. In this case, we can use equation (6.1.5) with angular velocity $\omega = 0$. This gives

$$r\frac{d^2\sigma_r}{dr^2}+3\frac{d\sigma_r}{dr}=0. \tag{6.2.1}$$

Solution to this equation may be of the form $\sigma_r = Cr^n$, (6.2.2)

where C and n are constants.

Substituting equation (6.2.2) in equation (6.2.1), we have

$$Cr^{n-1}\left(n^2+2n\right)=0.$$

This gives $n = 0$ or -2, and the solution for σ_r is

$$\sigma_r = C_1 + \frac{C_2}{r^2}.$$

From equation (6.1.4) with $\omega = 0$, we may find σ_θ as

$$\sigma_\theta = C_1 - \frac{C_2}{r^2}.$$

We may therefore write in the combined form, the two equations, known as *Lame's equation* for thick cylinders as

$$\left.\begin{aligned}\sigma_r &= C_1 + \frac{C_2}{r^2}\\ \sigma_\theta &= C_1 - \frac{C_2}{r^2}\end{aligned}\right\}. \tag{6.2.3}$$

This forms the basis of any further analysis of stress distribution in thick cylinders. Boundary conditions for a cylinder with both internal and external pressures are shown in Figure 6.4.

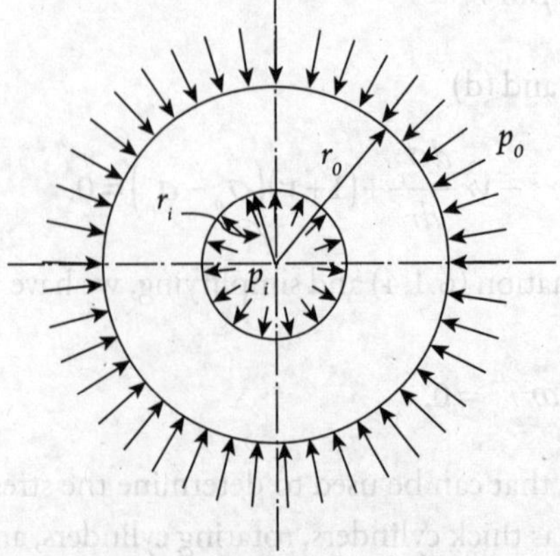

Figure 6.4 A thick cylinder with internal and external radii r_i and r_0 respectively, is subjected to both internal and external pressures p_i and p_0 respectively

At $r = r_i$ $\quad \sigma_r = -p_i$

$r = r_0$ $\quad \sigma_r = -p_0$.

It may be noted that radial stresses in both cases are compressive. This gives

$$C_1 = \frac{p_i r_i^2 - p_0 r_0^2}{\left(r_0^2 - r_i^2\right)}; \qquad C_2 = \frac{r_i^2 r_0^2 \left(p_0 - p_i\right)}{\left(r_0^2 - r_i^2\right)}; \tag{6.2.4}$$

$$\sigma_r = \frac{p_i r_i^2 - p_0 r_0^2}{\left(r_0^2 - r_i^2\right)} + \frac{r_i^2 r_0^2 (p_0 - p_i)}{\left(r_0^2 - r_i^2\right)} \frac{1}{r^2}$$

and

$$\sigma_\theta = \frac{p_i r_i^2 - p_0 r_0^2}{\left(r_0^2 - r_i^2\right)} - \frac{r_i^2 r_0^2 (p_0 - p_i)}{\left(r_0^2 - r_i^2\right)} \frac{1}{r^2}. \tag{6.2.5}$$

σ_θ is tensile if it works out to be *+ve* and is compressive if it is *−ve* but σ_r is always compressive.

Substituting different conditions in equation (6.2.5), for example, $p_0 = 0$, i.e., no external pressure is acting on the cylinder, we have the stress distribution across the wall thickness of a cylinder with internal pressure alone as

$$\sigma_r = \frac{p_i r_i^2}{\left(r_0^2 - r_i^2\right)} \left(1 - \frac{r_0^2}{r^2}\right)$$

$$\sigma_\theta = \frac{p_i r_i^2}{\left(r_0^2 - r_i^2\right)} \left(1 + \frac{r_0^2}{r^2}\right). \tag{6.2.6}$$

The stress distributions are given in Figure 6.5.

It may be noted in equation (6.2.5)

$\sigma_r + \sigma_\theta = constant.$

This indicates that the deformation in the *z*-direction is uniform, and cross-sections perpendicular to the cylinder axis remain plane. Hence, the deformation in an element cutout by two

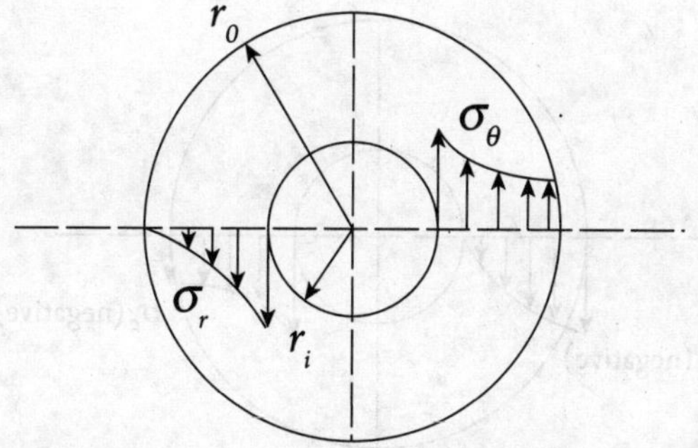

Figure 6.5 Stress distribution in a thick cylinder when there is no external pressure

adjacent cross-sections does not interfere with the adjacent element. Therefore, the assumption of plane stress is justified.

If now we take $p_i = 0$, i.e., no internal pressure, then the stress distribution for an externally pressurized cylinder is given as

$$\sigma_r = \frac{p_o r_o^2}{\left(r_0^2 - r_i^2\right)}\left(\frac{r_i^2}{r^2} - 1\right)$$
$$\sigma_\theta = -\frac{p_0 r_0^2}{\left(r_0^2 - r_i^2\right)}\left(\frac{r_i^2}{r^2} + 1\right). \tag{6.2.7}$$

Stress distributions are shown in Figure 6.6. Here both radial and circumferential stresses are compressive, and the circumferential stress is maximum at the inner radius.

Plane strain conditions:

If the cylinder is fairly long, plane strain conditions may be used. Here we assume

$\in_z = \gamma_{rz} = \gamma_{\theta z} = 0,\quad \tau_{rz} = \tau_{\theta z} = 0$, and also $\frac{\partial}{\partial \theta} = 0$ due to axisymmetry.

Analysis from the basic equilibrium equations would yield the same expressions for σ_θ and σ_r as in the case of plane stress. The analysis would, in addition, give

$$\sigma_z = 2\nu \frac{p_i r_i^2 - p_0 r_0^2}{\left(r_0^2 - r_i^2\right)}. \tag{6.2.8}$$

Although in this case there are three principal stresses σ_r, σ_θ, and σ_z, the σ_z is small in most cases compared to σ_r and σ_θ.

There is yet another method of obtaining σ_z if we consider that the longitudinal stresses are uniformly distributed across the transverse section. From equilibrium consideration (Figure 6.7),

$$\sigma_z \pi\left(r_0^2 - r_i^2\right) = \pi p_i r_i^2 - \pi p_0 r_0^2$$

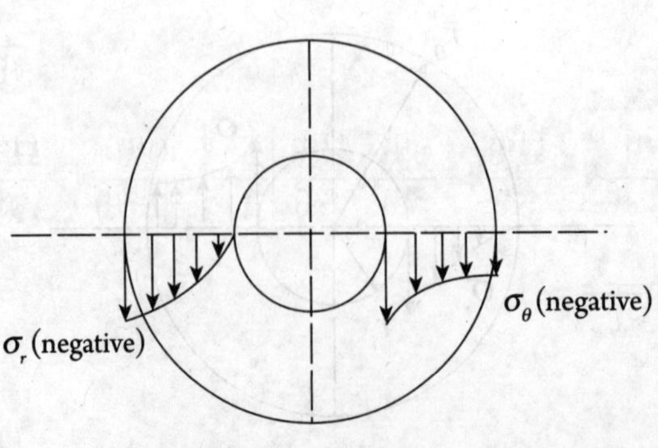

Figure 6.6 Stress distribution in a thick cylinder when there is no internal pressure

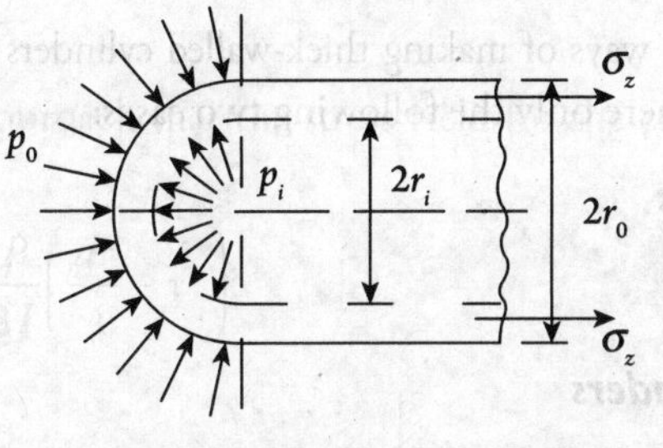

Figure 6.7 Longitudinal section of a pressure vessel subjected to both internal and external pressure p_i and p_0

$$\sigma_z = \frac{p_i r_i^2 - p_0 r_0^2}{\left(r_0^2 - r_i^2\right)}.$$

The difference in σ_z value is due to different assumptions; however, as stated above, σ_z is small in most cases.

6.2.1 Prestressing of Thick Cylinders

To understand the need for prestressing thick cylinders, let us consider the stress distribution for $p_0 = 0$, i.e., for the case where only internal pressure exists. This is given in equation (6.2.6) and Figure 6.5. It can be seen there that maximum values of both σ_r and σ_θ occur at the inner surface of the cylinder, and this is given as

$$\sigma_r(max)\Big|_{r=r_i} = -p_i$$
$$\sigma_\theta(max)\Big|_{r=r_i} = p_i\left(\frac{r_0^2 + r_i^2}{r_0^2 - r_i^2}\right) \tag{6.2.9}$$

This means that as the internal pressure p_i increases, σ_θ may increase more than the yield stress σ_y even when $p_i < \sigma_y$.

Furthermore, it will be shown later that for the large internal pressure in thick cylinders, the wall thickness is required to be very large. Since the stresses σ_r, σ_θ are the maximum near the inner wall and decrease near the outer wall, the material near the outer edge is ineffectively used. A typical pressure versus wall thickness variation is shown in Figure 6.8.

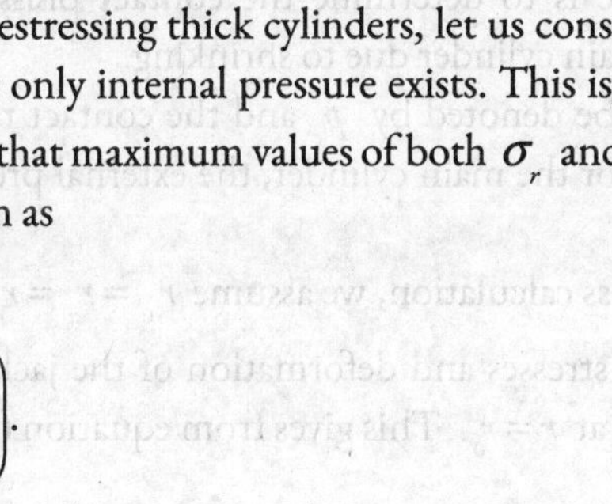

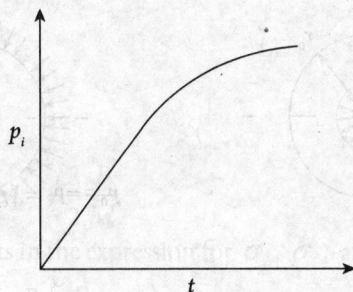

Figure 6.8 A typical variation of wall thickness t with the internal pressure in thick-walled cylinders

There are many different ways of making thick-walled cylinders to resist elastically large internal pressure. We shall consider here only the following two cases:

(a) Compound cylinders
(b) Autofrettage

6.2.1.1 *Compound cylinders* *[LO3]*

These are obtained by shrink fitting an outer cylinder (jacket) with an internal diameter slightly smaller than the outer diameter of the main cylinder. The jacket is heated and fitted onto the main cylinder. When the assembly cools down to room temperature, a composite cylinder is obtained. In this process, the main cylinder is subjected to an external pressure leading to a compressive radial stress at the interface, and the outer cylinder or the jacket is subjected to an internal pressure leading to tensile circumferential stress. Under this condition, as the internal pressure in the main cylinder increases, the compression in the main cylinder is first released and then only the cylinder begins to act in tension. Gun barrels are normally prestressed by hooping, since very large internal pressures are generated here.

Here the important issue is to determine the contact pressure developed at the interface between the jacket and the main cylinder due to shrinking.

Let the contact pressure be denoted by p_s and the contact radius by r_s. Therefore, for the jacket pressure $p_i = p_s$ and for the main cylinder, the external pressure $p_0 = p_s$. This is shown in Figure 6.9.

Here $r_{s0} < r_{si}$, but for stress calculation, we assume $r_{s0} = r_{si} = r_s$ (say).

Now, let us consider the stresses and deformation of the jacket or the outer cylinder. Here, $p_i = p_s$ at $r_i = r_s$ and $p_0 = 0$ at $r = r_0$. This gives from equation (6.2.6).

$$\sigma_r = \frac{p_s r_s^2}{\left(r_0^2 - r_s^2\right)}\left(1 - \frac{r_0^2}{r_s^2}\right) = -p_s$$

$$\sigma_\theta = \frac{p_s r_s^2}{\left(r_0^2 - r_s^2\right)}\left(1 + \frac{r_0^2}{r_s^2}\right) = p_s \frac{r_0^2 + r_s^2}{r_0^2 - r_s^2} \tag{6.2.10}$$

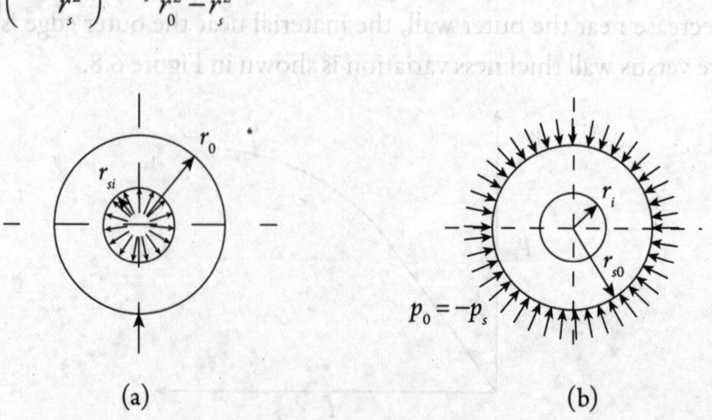

Figure 6.9 (a) Contact pressure developed at the internal radius of the jacket. Contact radius = r_{si} (b) Contact pressure developed at the external radius of the main cylinder. Contact radius = r_{s0}

To find the deformation, we consider the following strain displacement relation:

$$\epsilon_\theta = \frac{u_r}{r} = \frac{1}{E}\left(\sigma_\theta - \nu\sigma_r\right). \tag{6.2.11}$$

Substituting σ_r and σ_θ from equation (6.2.10) to equation (6.2.11), we have radial displacement (say u_{r1}) at the inside radius of the jacket.

$$u_{r1} = \frac{p_s r_s}{E}\left(\frac{r_0^2 + r_s^2}{r_0^2 - r_s^2} + \nu\right). \tag{6.2.12}$$

Now for the inner cylinder, $p_i = 0$ at $r = r_i$ and $p_0 = -p_s$ at $r_0 = r_s$. This gives, from equation (6.2.7),

$$\sigma_r\Big|_{r=r_s} = -p_s$$

$$\sigma_\theta\Big|_{r=r_s} = -p_s\left(\frac{r_s^2 + r_i^2}{r_s^2 - r_i^2}\right). \tag{6.2.13}$$

Again substituting σ_r and σ_θ from equation (6.2.13) in equation (6.2.11), we have the radial displacement, say u_{r2}, at the outside radius of the main cylinder, which is

$$u_{r2} = -\frac{p_s r_s}{E}\left(\frac{r_s^2 + r_i^2}{r_s^2 - r_i^2} - \nu\right). \tag{6.2.14}$$

Therefore, the total interference at the contact is

$$\delta = u_{r1} - u_{r2} = \frac{p_s r_s}{E}\left(\frac{r_0^2 + r_s^2}{r_0^2 - r_s^2} + \frac{r_s^2 + r_i^2}{r_s^2 - r_i^2}\right).$$

The contact pressure is therefore

$$p_s = \frac{E\delta}{r_s\left[\frac{r_0^2 + r_s^2}{r_0^2 - r_s^2} + \frac{r_s^2 + r_i^2}{r_s^2 - r_i^2}\right]}. \tag{6.2.15}$$

The combined stress distribution in the composite cylinder is shown in Figure 6.10.

The residual circumferential stress is maximum at $r = r_i$ for the main cylinder, and this is given by

$$\sigma_\theta\Big|_{r=r_i} = -\frac{2p_s r_s^2}{r_s^2 - r_i^2}. \tag{6.2.16}$$

And the residual circumferential stress is maximum at $r = r_s$ for the outer cylinder and this is given by

$$\sigma_\theta\Big|_{r=r_s} = p_s\left(\frac{r_0^2 + r_s^2}{r_0^2 - r_s^2}\right). \tag{6.2.17}$$

Stresses due to fluid pressure in the main cylinder must be superimposed on these stresses to find the complete stress distribution. This is illustrated in Example 6.1.

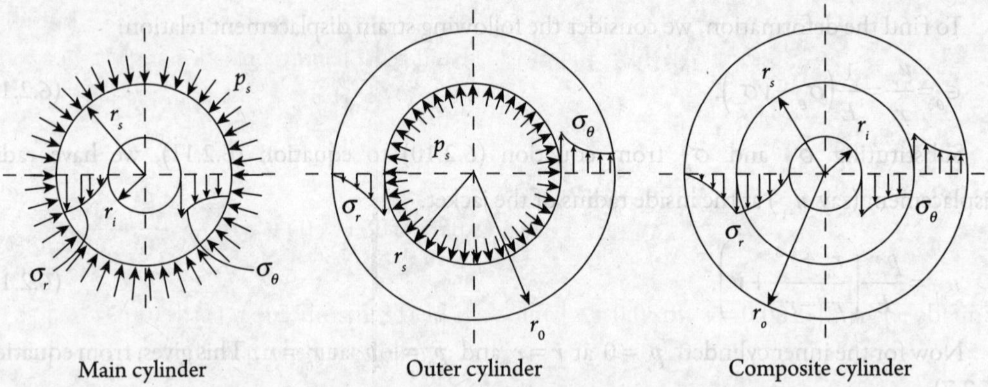

Figure 6.10 Combined stress distribution of the composite cylinder

6.2.1.2 *Autofrettage* **[LO4]**

In most applications of thick cylinders or other components, inelastic deformation is unacceptable in design. However, in some pressure vessel designs, inelastic deformation that starts at the inside radius may be acceptable until plasticity spreads completely over the wall thickness. Here the important step to be taken is to arrest the spread of plasticity beyond a tolerable limit as the fluid pressure inside the thick cylinder rises.

This is uniquely done in practice. As the fluid pressure rises inside the cylinder, inelastic yielding starts spreading from the inside radius, and as the pressure reaches a specified value, it is released. The outer elastic zone would then regain its size and a constant pressure is developed at the yield front. This would exert compression on the inner inelastic layer. This also means that there would be tension on the outer elastic region. This gives the same effect as shrinking a jacket on the inner cylinder. The only difference is that the inner layer remains inelastic allowing the fluid pressure to be higher than even the yield pressure of the cylinder material. This is known as *autofrettage* or *self-hooping*. This is pictorially demonstrated in Figure 6.11.

As the fluid pressure increases, the yield front proceeds through the cylinder wall to the outer radius. When the shear stress at every point in the cylinder wall reaches its yield value, the pressure is described as fully plastic pressure p_f, and the fluid pressure cannot increase any further. This pressure may be determined using the reduced form of the equilibrium equation in polar coordinates used earlier in Section 6.1.1.

$$\frac{d\sigma_r}{dr}+\frac{\sigma_r-\sigma_\theta}{r}=0. \tag{6.2.18}$$

We may also write a yield criterion based on the maximum shear stress at any point within the cylinder wall.

$$\frac{\sigma_\theta-\sigma_r}{2}=\tau_y, \tag{6.2.19}$$

where τ_y is the shear yield stress of the cylinder material.

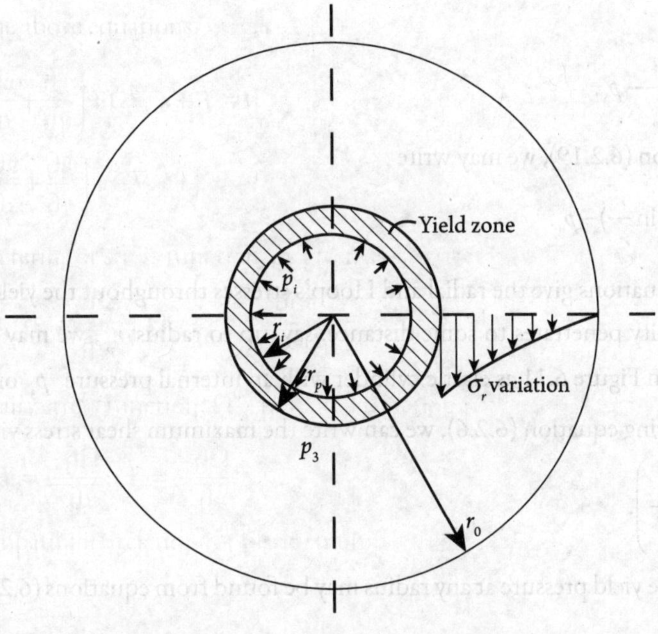

Figure 6.11 Progress of yield front in the cylinder wall as internal pressure rises

Combining equations (6.2.18) and (6.2.19), we have

$$\frac{d\sigma_r}{dr} = \frac{2\tau_y}{r}. \tag{6.2.20}$$

Integrating, we have $\sigma_r = 2\tau_y \ln r + C$.

Since, at $r = r_0,\ \sigma_r = 0 \qquad C = -2\tau_y \ln r_0$ and $\sigma_r = 2\tau_y \ln \frac{r}{r_0}$.

Since for most ductile materials, the tensile yield stress σ_y is twice the shear yield stress, and at $r = r_i$, $\sigma_r = p_f$, we can write

$$\frac{p_f}{\sigma_y} = \ln \frac{r_i}{r_0}. \tag{6.2.21}$$

The equation basically represents the ratio of the fluid pressure necessary to spread inelastic strain throughout the wall to the tensile yield stress in terms of the cylinder geometry. It will be seen later that this is substantially higher than the ratio $\frac{p_i}{\sigma_y}$ according to maximum shear stress and maximum principal stress yielding criteria.

If we now consider the yield zone to extend only up to radius r_p with radial pressure p_3 (as shown in Figure 6.11), then we may write the constant of integration in equation (6.2.20) as

$$C = -2\tau_y \ln r_p - p_3.$$

We thus have

$$\sigma_r = 2\tau_y \ln\frac{r}{r_p} - p_3. \tag{6.2.22}$$

From equation (6.2.19), we may write

$$\sigma_\theta = 2\tau_y (1+\ln\frac{r}{r_p}) - p_3. \tag{6.2.23}$$

These two equations give the radial and Hoop's stresses throughout the yield zone.

When plasticity penetrates to some distance, say, up to radius r_p, we may consider the zone with $r_p < r < r_0$ in Figure 6.11 as elastic cylinder with an internal pressure p_3 only.

Therefore, using equation (6.2.6), we can write the maximum shear stress yield criterion as

$$\sigma_y = \left(\frac{2p_3 r_0^2}{r_0^2 - r_p^2}\right). \tag{6.2.24}$$

Therefore, the yield pressure at any radius may be found from equations (6.2.22), (6.2.23), and (6.2.24).

The initial yield pressure at the inside radius r_i is thus given by

$$p_{yi} = \frac{\sigma_y}{2r_0^2}\left(r_0^2 - r_i^2\right). \tag{6.2.25}$$

And the internal pressure required to cause yielding up to radius r_p is given by

$$p_{yp} = \sigma_y \left[\ln\frac{r_i}{r_p} - \frac{1}{2r_0^2}\left(r_0^2 - r_p^2\right)\right]. \tag{6.2.26}$$

6.3 ANALYSIS OF FAILURE THEORIES FOR THICK CYLINDER [LO5]

Pressure vessels and thick cylinders are designed using certain industrial codes and standards, which are developed based on experimental results and empirical equations. However, we can calculate the wall thickness for a given internal pressure and inside radius based on theoretical failure theories as a guide. Among the failure theories used in the machine design, there are three useful theories: one is used for brittle materials and the other two are for ductile materials. We will now discuss these three theories in some detail for thick cylinders with no external pressure, i.e., $p_0 = 0$.

6.3.1 Maximum Principal Stress Theory

These theories are essentially yield criteria, and according to this theory, if any one of the principal stresses σ_1, σ_2, or σ_3 exceeds the yield stress, then yielding would occur. In the 2D loading, we may develop a yield surface in the $\sigma_1 - \sigma_2$ plane shown in Figure 6.12, and write

$$\sigma_1 = \pm\sigma_y \qquad \sigma_2 = \pm\sigma_y.$$

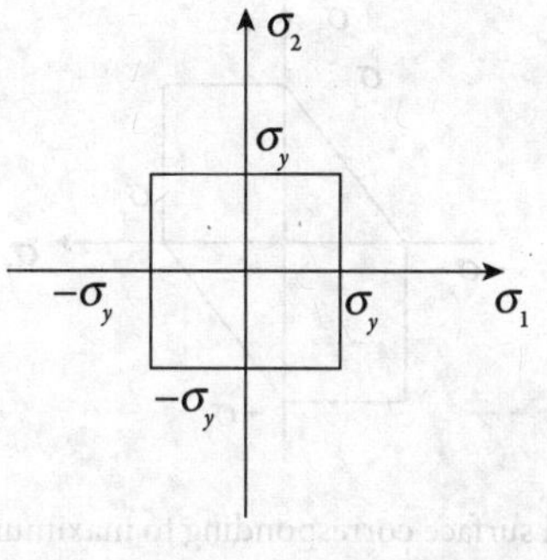

Figure 6.12 Yield surface corresponding to maximum principal stress

This is true considering that yield stress is nearly the same in tension and compression for ductile materials. Now for $p_0 = 0$, σ_r and σ_θ are given by equation (6.2.6). Here both stresses are maximum at the inside radius and σ_θ is always greater than σ_r. We therefore consider the maximum value of σ_θ for further analysis.

$$\sigma_\theta = p_i\left(\frac{r_0^2 + r_i^2}{r_0^2 - r_i^2}\right).$$

This may be rearranged to

$$\frac{t}{r_i} = \sqrt{\frac{1+ p_i/\sigma_y}{1- p_i/\sigma_y}} - 1. \quad (6.3.1)$$

Here $t = r_0 - r_i$, and σ_θ reaches yield stress σ_y.

Although based on yielding, this theory is widely used for brittle materials.

6.3.2 Maximum Shear Stress Theory

According to this theory, yielding would occur when maximum shear stress exceeds shear yield stress $\tau_y \approx \sigma_y/2$ for ductile materials. We may therefore write six yield conditions in three-dimensional (3D) situation.

$$\sigma_1 - \sigma_2 = \pm\sigma_y$$
$$\sigma_2 - \sigma_3 = \pm\sigma_y$$
$$\sigma_1 - \sigma_3 = \pm\sigma_y.$$

The corresponding yield surface in the 2D $\sigma_1 - \sigma_2$ plane is shown in Figure 6.13.

Now considering a 2D case and replacing σ_1 by σ_θ and σ_2 by σ_r from equation (6.2.6) at $r = r_i$, the yield criterion turns out to be

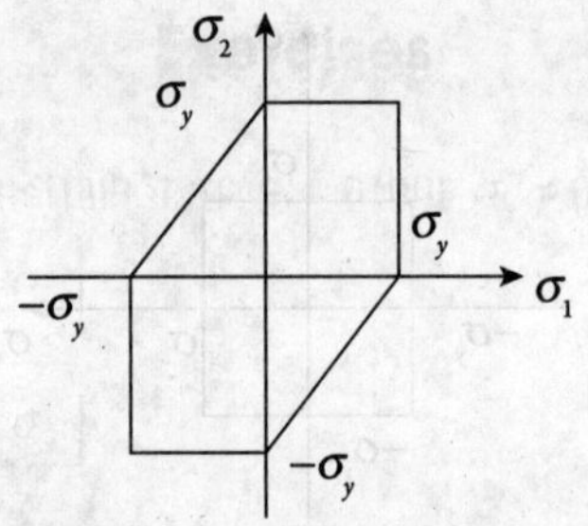

Figure 6.13 Yield surface corresponding to maximum shear stress theory

$$p_i\left(\frac{r_0^2+r_i^2}{r_0^2-r_i^2}\right)+p_i=\sigma_y.$$

This may be rearranged to give

$$\frac{t}{r_i}=\sqrt{\frac{1}{1-2\left(p_i/\sigma_y\right)}}-1. \tag{6.3.2}$$

Here again $t=r_0-r_i$.

6.3.3 Maximum Distortion Energy Theory

According to this theory, yielding would occur when total distortion energy per unit volume due to applied load exceeds the distortion energy at the tensile yield point. In 3D, we may write the yield criteria as

$$\frac{1+\nu}{6E}\left[\left(\sigma_1-\sigma_2\right)^2+\left(\sigma_2-\sigma_3\right)^2+\left(\sigma_3-\sigma_1\right)^2\right]=\frac{1+\nu}{3E}\sigma_y^2.$$

Substituting $\sigma_1=\sigma_\theta=p_i\left(\frac{r_0^2+r_i^2}{r_0^2-r_i^2}\right)$, $\quad\sigma_2=\sigma_r=\frac{p_ir_i^2}{\left(r_0^2-r_i^2\right)}\left(1-\frac{r_0^2}{r^2}\right)=-p_i$

and $\sigma_3=\sigma_z=\frac{p_ir_i^2}{\left(r_0^2-r_i^2\right)}$ in the above equation may be written as

$$x^2\left(1-3q\right)-2x+1=0,$$

where $q=\left(\frac{p_i}{\sigma_y}\right)^2$ and $x=\left(\frac{r_0}{r_i}\right)^2$.

Solving this quadratic equation, we have

$$1+\frac{t}{r_i}=\pm\frac{1}{\sqrt{1-\sqrt{3}\left(p_i/\sigma_y\right)}}. \tag{6.3.3}$$

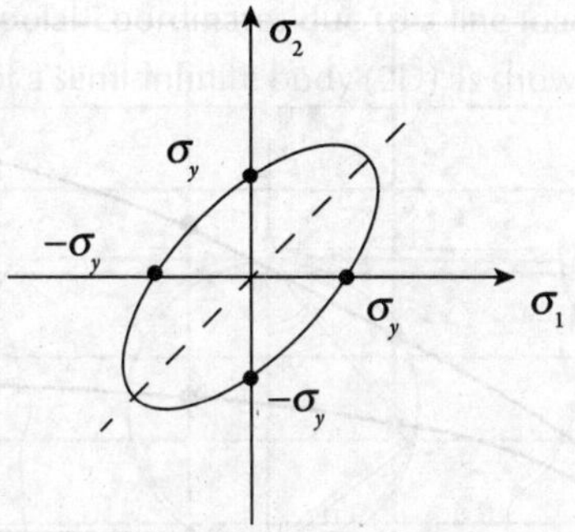

Figure 6.14 Yield surface corresponding to maximum distortion energy theory

The yield surface in this case in 2D is an ellipse, as shown in Figure 6.14, since the yield criterion in 2D reduces to

$$\sigma_1^2 + \sigma_2^2 - \sigma_1 \sigma_2 = \sigma_y^2.$$

This theory agrees well with the experimental results and is widely used for failure analysis of ductile materials.

Comparing equations (6.3.1), (6.3.2), and (6.3.3), it can be seen that for a given nondimensional internal pressure $\left(p_i / \sigma_y \right)$, nondimensional wall thickness $\left(t / r_i \right)$ is the smallest for maximum principal stresses theory and the largest for maximum shear stress theory, whereas wall thickness predicted by maximum distortion energy theory stands somewhere in between the above two theories.

It is now interesting to compare the results of autofrettage with maximum principal stress and shear stress theories. In order to do this, we write the maximum principal stress theory and maximum shear stress theory, and autofrettage in terms of radius ratio $\left(r_0 / r_i \right)$ and nondimensional pressure $\left(p_i / \sigma_y \right)$ and $\left(p_f / \sigma_y \right)$.

Maximum principal stress theory

$$\frac{r_0}{r_i} = \sqrt{\frac{1 + p_i / \sigma_y}{1 - p_i / \sigma_y}}.$$

Maximum shear stress theory

$$\frac{r_0}{r_i} = \sqrt{\frac{1}{1 - 2\left(p_i / \sigma_y \right)}}.$$

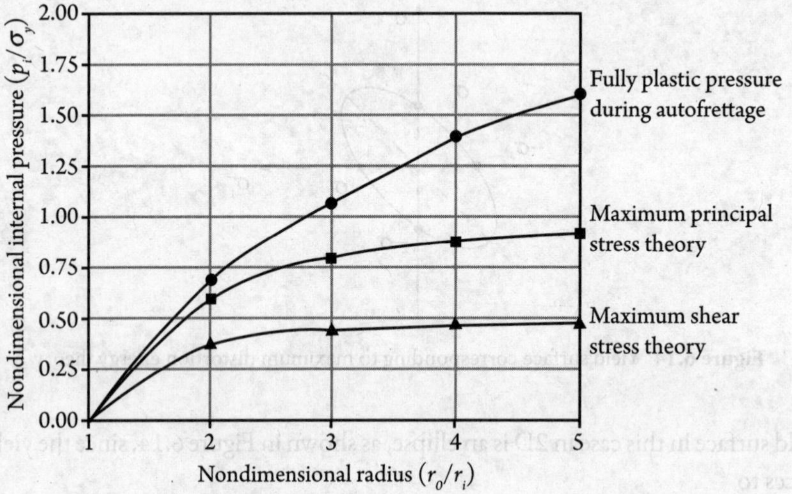

Figure 6.15 Comparison of maximum principal stress theory, maximum shear stress theory, and autofrettage pressure

Autofrettage

$$\ln\left(\frac{r_0}{r_i}\right) = -\frac{p_f}{\sigma_y}.$$

The last equation is written in a different form and positive values of the ratio $\frac{p_f}{\sigma_y}$ are plotted.

The comparison of the three theories is shown in Figure 6.15 and it can be seen that for a given radius ratio, autofrettage can allow much higher internal fluid pressure than the other two failure modes.

Example 6.1

A composite cylinder is formed by shrinking a jacket of 500 mm external diameter and 400 mm internal diameter, onto a main cylinder of external diameter 400 mm and internal diameter 300 mm. Find the stress distribution for the composite cylinder and the single cylinder of internal diameter 300 mm and external diameter 500 mm, both cylinders being made of the same material and subjected to the same internal pressure 70 MPa. Consider the material to be steel with $E = 200$ GPa and interference of 25 μm for the composite cylinder shrinkage.

Solution:

Although, based on Lame's equation, separate equations have been developed for different conditions, it is convenient to split the problems for three different effects:

(a) Effect of shrinkage pressure on the inside cylinder,
(b) Effect of shrinkage pressure on the jacket, and
(c) Effect of internal pressure on the composite and single cylinder.

Since the stress equations are linear, we may superimpose them in the case of composite cylinder.

Now, let us first consider the main cylinder, as shown in Figure 6.16(a).

From equation (6.2.15), the contact pressure is given by

$$p_s = \frac{E\delta}{r_s\left[\dfrac{r_0^2 + r_s^2}{r_0^2 - r_s^2} + \dfrac{r_s^2 + r_i^2}{r_s^2 - r_i^2}\right]}.$$

Substituting $r_i = 0.15\,\text{m}$, $r_s = 0.2\,\text{m}$, $r_0 = 0.25\,\text{m}$, $E = 200\times10^9\,\text{Pa}$, $\delta = 25\times10^{-6}\,\text{m}$.

Contact pressure $p_s = 3.076\,\text{MPa}$.

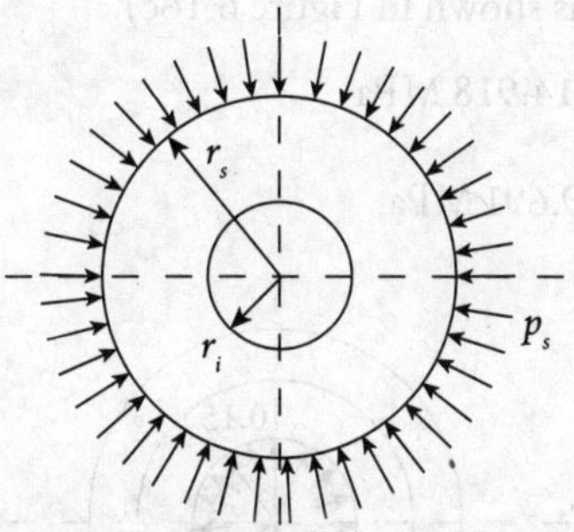

Figure 6.16a Main cylinder loading

Now, for the main cylinder, we use Lame's equation (6.2.3)

$$\sigma_r = C_1 + \frac{C_2}{r^2}$$

$$\sigma_\theta = C_1 - \frac{C_2}{r^2}.$$

At $r = 0.15\,\text{m}$ $\quad \sigma_r = 0$

$r = 0.2\,\text{m}$ $\quad \sigma_r = -3.076\,\text{MPa}.$

From the σ_r equation above we get $C_1 = -7.0308$, $C_2 = 0.1582$ and we have from the σ_θ equation above

$$\sigma_\theta\Big|_{r=0.15} = -14.062\,\text{MPa} \qquad \sigma_\theta\Big|_{r=0.2} = -10.986\,\text{MPa} \tag{a}$$

Now for the jacket

At $r = 0.2\,\text{m}$ $\quad \sigma_r = -3.076\,\text{MPa}$

$r = 0.25\,\text{m}$ $\quad \sigma_r = 0.$

Again from σ_r, equation above, we get $C_1 = 5.468$, $C_2 = -0.342$, and this gives

$$\sigma_\theta\Big|_{r=0.2} = 14.018\ \text{MPa} \qquad \sigma_\theta\Big|_{r=0.25} = 10.94\ \text{MPa} \tag{b}$$

Now, we consider the effect of internal pressure on the composite cylinder as shown in Figure 6.16(b).

Here at $r = 0.15$ m $\qquad \sigma_r = -70$ MPa

$r = 0.25$ m $\qquad \sigma_r = 0$.

Again from σ_r, equation above, we get $C_1 = 39.375$, $C_2 = -2.461$, and this gives

$$\sigma_\theta\Big|_{r=0.15} = 148.753\ \text{MPa} \qquad \sigma_\theta\Big|_{r=0.2} = 100.90\ \text{MPa} \qquad \sigma_\theta\Big|_{r=0.25} = 78.751\ \text{MPa} \tag{c}$$

Composite cylinder: Jacket (as shown in Figure 6.16c)

$$\sigma_\theta\Big|_{r=0.2} = 100.90 + 14.018 = 114.918\ \text{MPa}$$

$$\sigma_\theta\Big|_{r=0.25} = 78.751 + 10.94 = 89.691\ \text{MPa}$$

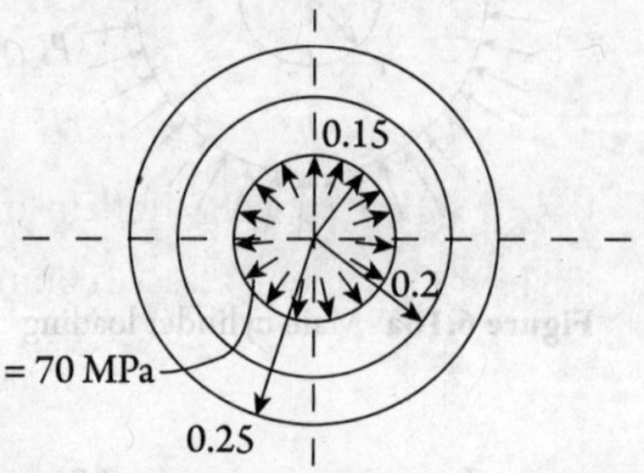

Figure 6.16b Composite cylinder with an internal fluid pressure of 70 MPa

Composite cylinder: Main cylinder

$$\sigma_\theta\Big|_{r=0.2} = 100.90 - 10.986 = 89.914\ \text{MPa}$$

$$\sigma_\theta\Big|_{r=0.15} = 148.753 - 14.062 = 134.691\ \text{MPa}$$

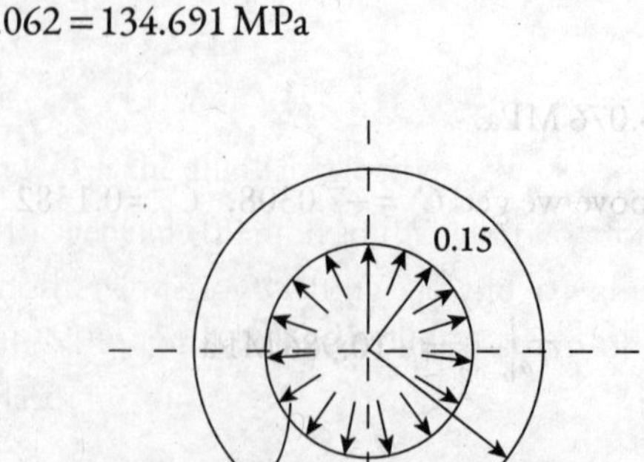

Figure 6.16c Single cylinder with an internal fluid pressure of 70 MPa

Single cylinder:

$$\sigma_\theta\Big|_{r=0.15} = 148.753\,\text{MPa}$$

$$\sigma_\theta\Big|_{r=0.25} = 78.751\,\text{MPa}$$

The hoop stress distribution in composite and single cylinder is shown in Figure 6.16(d).

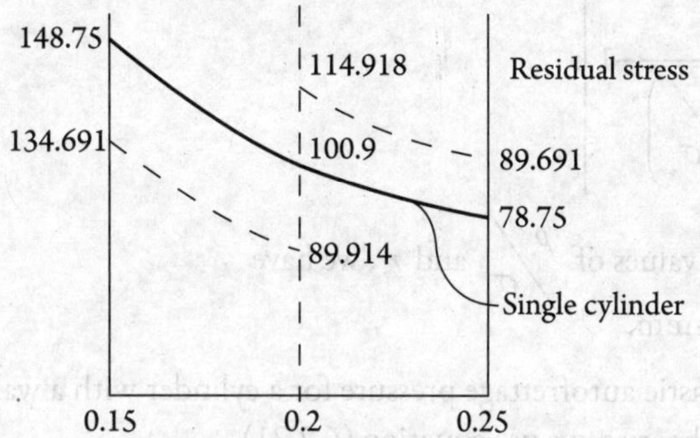

Figure 6.16d Hoop stress distribution in composite and single cylinder

Example 6.2

A pipe made of steel of allowable tensile yield stress of 280 MPa has an internal radius of 50 mm and it is subjected to an internal fluid pressure of 70 MPa. Find the wall thickness according to the maximum principal stress theory, maximum shear stress theory, and maximum distortion energy theory.

Find also the fully plastic autofrettage pressure corresponding to the least wall thickness according to the three theories.

Solution:

From equation (6.32) for the maximum principal stress theory

$$t = r_i\left[\sqrt{\frac{1+p_i/\sigma_y}{1-p_i/\sigma_y}}-1\right].$$

Substituting $p_i/\sigma_y = 0.25$, $r_i = 0.05\,\text{m}$, $t = 0.0145\,\text{m} = 14.5\,\text{mm}$.

From equation (6.33) for the maximum shear stress theory

$$t = r_i\left[\frac{1}{\sqrt{1-2\left(p_i/\sigma_y\right)}}-1\right].$$

Again, substituting the values of ${p_i}/{\sigma_y}$ and r_i, we have

$t = 0.02071\,\text{m} = 20.71\,\text{mm}$.

From equation (6.3.3) for the maximum distortion energy theory

$$t = r_i\left[\frac{1}{\sqrt{1-\sqrt{3}\left({p_i}/{\sigma_y}\right)}} - 1\right].$$

Again, substituting values of ${p_i}/{\sigma_y}$ and r_i, we have

$t = 0.0164\,\text{m} = 16.4\,\text{mm}$.

To find the fully plastic autofrettage pressure for a cylinder with a wall thickness of 14.5 mm, least among the above theories, we use equation (6.2.21).

$$\ln\frac{r_0}{r_i} = -\frac{p_f}{\sigma_y}.$$

Substituting $\sigma_y = 280$ MPa and writing $\frac{r_0}{r_i} = 1 + \frac{t}{r_i}$, we have

$p_f = 71.3\,\text{MPa}$,

which is higher than the inter fluid pressure. Here only the positive value of p_f is considered.

Example 6.3

A composite cylinder is made by shrinking a cylinder with an external diameter of 250 mm and internal diameter 200 mm on another cylinder with an external diameter of 200 mm and internal diameter of 150 mm. The composite cylinder is 100 mm long and requires an axial push of 100 kN to separate the two cylinders axially. If the coefficient of friction is 0.2 and elastic modulus $E = 200$ GPa, find the contact pressure at the interface, the circumferential stress of the two cylinders at the interface and also the interference.

Solution:

If the contact pressure is p_s. Substituting $r = 0.1\,\text{m},\ l = 0.1$.

Contact force, $N = p_s 2\pi r l = 0.0628 p_s$ N.

Axial force, $F_a = \mu N = 0.2 \times 0.0628\, p_s = 100 \times 10^3$ N.

This gives $p_s = 7.958$ MPa.

Now for the external cylinder

$\sigma_r = -7.958\,\text{MPa}$ at $r = 0.1\,\text{m}$.

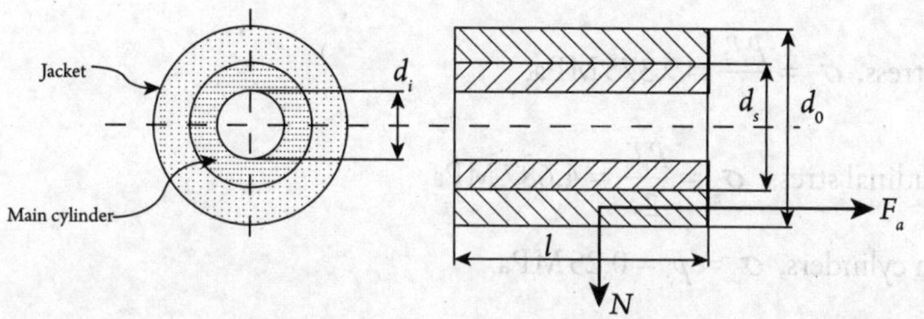

Figure 6.17 Contact force developed due to axial push during shrinking

$\sigma_r = 0$ at $r = 0.125\,\text{m}$.

Substituting this in $\sigma_r = C_1 + \frac{C_2}{r^2}$, we have $C_1 = 14.147$, $C_2 = -0.221$. This gives

$$\sigma_\theta\big|_{r=0.1} = 36.247\,\text{MPa}.$$

For the internal cylinder

$\sigma_r = -7.958\,\text{MPa}$ at $r = 0.1\,\text{m}$

$\sigma_r = 0$ at $r = 0.075\,\text{m}$.

This gives $C_1 = -18.190$, $C_2 = 0.1024$.

$$\sigma_\theta\big|_{r=0.1} = -18.22 - \frac{0.1024}{0.1^2} = -28.43\,\text{MPa}.$$

From equation (6.2.15), the total interference

$$\delta = \frac{p_s r_s}{E}\left[\frac{r_0^2 + r_s^2}{r_0^2 - r_s^2} + \frac{r_s^2 + r_i^2}{r_s^2 - r_i^2}\right]$$

$$= \frac{7.958\times10^6 \times 0.1}{200\times10^9}\left[\frac{0.125^2 + 0.1^2}{0.125^2 - 0.1^2} + \frac{0.1^2 + 0.075^2}{0.1^2 - 0.075^2}\right]\text{m} = 32.34\,\mu\text{m}.$$

Example 6.4

A thin cylindrical pressure vessel of internal diameter $1.5\,\text{m}$ and $5\,\text{m}$ long carried a fluid under $250\,\text{kPa}$ pressure. Determine the hoop and longitudinal stress developed and the change in internal diameter of the vessel. Take the wall thickness to be $20\,\text{mm}$ and $E = 200\,\text{GPa}$, $\nu = 0.3$.

Solution:

$D_i = 1.5\,\text{m}$, $l = 5\,\text{m}$, $p_i = 250\,\text{kPa}$, $t = 20\,\text{mm} = 0.02\,\text{m}$,

$r_i = 0.75\,\text{m}$, $r_0 = r_i + t = 0.77\,\text{m}$, $E = 200\,\text{GPa}$, $\nu = 0.3$.

Using thin cylinder approximation $(D_i \gg t)$,

(a) Hoop stress, $\sigma_\theta = \frac{p_i r_i}{t} = 9.375\,\text{MPa}.$

(b) Longitudinal stress, $\sigma_z = \frac{p_i r_i}{2t} = 4.687\,\text{MPa}.$

For thin cylinders, $\sigma_r = p_i = 0.25\,\text{MPa}$

(c) Circumferential strain, $\in_\theta = \frac{\Delta D}{2r_i} = \frac{1}{E}\left\{\sigma_\theta - \nu\left(\sigma_r + \sigma_z\right)\right\}.$

Change in the internal diameter of the vessel is thus obtained as:

$$\Delta D = \frac{2\times 0.75}{200\times 10^9}\times\left\{9.375 - 0.3\left(0.25 + 4.687\right)\right\}\times 10^6$$
$$= 5.92\times 10^{-5}\,\text{m} = 59.2\,\mu\text{m}. \qquad \text{... (Ans.)}$$

Example 6.5

If the internal pressure in the pressure vessel in Example 6.4 is increased to $2.5\,\text{MPa}$ and a torque of $50\,\text{kNm}$ is applied to the cylinder, check if the cylinder would yield according to the maximum principal stress theory, considering that the tensile yield stress of the cylinder material is $150\,\text{MPa}$.

Solution:

$p_i = 2.5\,\text{MPa}, \quad T = 50\,\text{kNm}, \quad \sigma_y = 150\,\text{MPa}$ (max. principal stress theory)

Considering the expression for shear stress due to torque on a hollow cylinder, the value for shear stress due to torque = 0.698 MPa.

$$\sigma_\theta\Big|_{pressure} = \frac{p_i r_i}{t} = 93.75\,\text{MPa}$$

$$\sigma_z\Big|_{pressure} = \frac{p_i r_i}{2t} = 46.875\,\text{MPa}.$$

Thus, the maximum principal stress is 93.75 MPa, which is less than the yield stress of 150 MPa. Therefore, the cylinder would not yield.

Example 6.6

A thick cylinder of internal diameter $150\,\text{mm}$ and external diameter $300\,\text{mm}$ is subjected to an external pressure of $15\,\text{MPa}$. If the hoop stress at the inside wall of the cylinder is limited to $25\,\text{MPa}$. Find the maximum internal pressure that can be applied. If the ends of the cylinder are closed, then find the change in outside diameter. Take $E = 200\,\text{GPa}$ and $\nu = 0.3$.

Solution:

$D_i = 150\,\text{mm}, \quad r_i = 0.075\,\text{m}, \quad D_0 = 300\,\text{mm}, \quad r_0 = 0.15\,\text{m},$

$p_o = 15\,\text{MPa}; \quad \sigma_\theta\Big|_{(r=r_i)} = 25\,\text{MPa}; \quad E = 200\,\text{GPa}; \quad \nu = 0.3$

$$\sigma_\theta\Big|_{(r=r_i)} = \sigma_{\theta max} = \frac{p_i r_i^2 - p_o r_o^2}{r_o^2 - r_i^2} - \frac{r_o^2 (p_o - p_i)}{r_o^2 - r_i^2} \quad \text{equation (6.2.5)}$$

Solving this, we get $p_i = 39$ MPa

$$\sigma_r\Big|_{(r=r_i)} = \frac{p_i r_i^2 - p_o r_o^2}{r_o^2 - r_i^2} + \frac{r_o^2 (p_o - p_i)}{r_o^2 - r_i^2}.$$

Solving, we obtain $\sigma_r\Big|_{(r=r_i)} = -39$ MPa

$$\sigma_z = \frac{p_i r_i^2 - p_o r_o^2}{r_o^2 - r_i^2} = -7 \text{ MPa}$$

$$\epsilon_\theta = \frac{\Delta D}{2r_o} = \frac{1}{E}\left(\sigma_\theta - \nu\sigma_r - \nu\sigma_z\right)$$

Therefore, $\Delta D = 58.2$ µm. ... (Ans.)

Example 6.7

A thick cylinder of internal diameter 160 mm and external diameter 200 mm is subjected to only internal pressure.

(a) If the maximum hoop stress at the external wall is 50 MPa, find the allowable internal fluid pressure.

(b) If the cylinder is autofrettaged before putting it in service, find the maximum autofrettaged pressure if the tensile yield stress of the cylinder material is 250 MPa.

Solution:

$D_i = 160$ mm; $r_i = 0.08$ m; $D_o = 200$ mm; $r_o = 0.1$ m

$\sigma_{\theta \max at\ r=r_o} = 50$ MPa;

(a) $$\sigma_r\Big|_{(r=r_o)} = \frac{p_i r_i^2}{r_o^2 - r_i^2}\left(1 + \frac{r_o^2}{r_o^2}\right) = \frac{2 p_i r_i^2}{r_o^2 - r_i^2} \quad \text{equation (6.2.6)}$$

Solving this, we get $p_i = 14.0625$ MPa.

(b) $\sigma_y = 250$ MPa

$$\frac{p_f}{\sigma_y} = \ln\left(\frac{r_i}{r_o}\right) \quad \text{equation (6.2.21)}$$

Solving, we get $p_f = -55.785$ MPa. ... (Ans.).

Example 6.8

At a point on the external surface of a thick cylinder of internal diameter 500 mm and external diameter 1 m, two strain gauges were attached: one axially and the other circumferentially. With an internal fluid pressure of 100 MPa, only axial and circumferential strain measured were 100×10^{-6} and 250×10^{-6} respectively. Find the axial and hoop stresses on the external surface of the cylinder. Compare these stresses with the theoretical values if the cylinder ends are closed. Take $E=200\,\text{GPa}$ and $\nu=0.3$.

Solution:

$D_i = 500\,\text{mm}$; $r_i = 0.25\,\text{m}$; $D_o = 1\,\text{m}$; $r_o = 0.5\,\text{m}$.

$\epsilon_z = 100\times10^{-6}$ (Axial strain) at $r=r_o$

$\epsilon_\theta = 250\times10^{-6}$ (Circumferential strain) at $r=r_o$

$E = 200\,\text{GPa}$; $\nu=0.3$.

(a) Experimental;

$$\epsilon_\theta = \frac{1}{E}\left(\sigma_\theta - \nu\sigma_z\right); \qquad \epsilon_z = \frac{1}{E}\left(\sigma_z - \nu\sigma_\theta\right);$$

Solving, we obtain:

$$\sigma_\theta = 61.538\,\text{MPa}; \qquad \sigma_z = 38.46\,\text{MPa}.$$

(b) Theoretical;

$$\sigma_\theta = \frac{p_i r_i^2}{r_o^2 - r_i^2}\left(1+\frac{r_o^2}{r^2}\right)$$

$$\sigma_z = \frac{p_i r_i^2 - p_o r_o^2}{r_o^2 - r_i^2}, \text{ here } p_o = 0$$

$$\sigma_\theta = \frac{2p_i r_i^2}{r_o^2 - r_i^2} = 66.67\,\text{MPa}.$$

$$\sigma_z = \frac{p_i r_i^2}{r_o^2 - r_i^2} = 33.33\,\text{MPa}.$$

Therefore,

$$\sigma_\theta\Big|_{experimental} = 61.538\,\text{MPa}, \qquad \sigma_\theta\Big|_{theoritical} = 66.67\,\text{MPa}$$

$$\sigma_z\Big|_{experimental} = 38.46\,\text{MPa}, \qquad \sigma_z\Big|_{theoritical} = 33.33\,\text{MPa}. \qquad \text{... (Ans.)}$$

Exercises

1. Write the general equation for determining stress distribution in axisymmetric problems and thereby derive Lame's equation for a thick cylinder.
   A thick cylinder of internal diameter 120 mm and external diameter 200 mm is made of steel with a tensile yield strength of 350 MPa.
   (a) Find the maximum internal pressure that the cylinder may be subjected to such that the maximum circumferential stress at the internal radius does not exceed the tensile yield strength of the material, assuming that there is no applied external pressure.
   (b) Find the maximum external pressure that the cylinder may be subjected to such that the maximum circumferential stress at the internal radius does not exceed the tensile yield strength of the material, assuming that there is no applied internal pressure.
   [164.7 MPa, –112 MPa]
2. A thick cylinder of internal diameter 800 mm is subjected to an internal fluid pressure of 5 MPa. Tensile yield strength of the cylinder material is 250 MPa. Find the thickness of the cylinder based on a factor of safety of 2.
   [16.667 mm]
3. A compound cylinder is made by shrinking a tube of 150 mm internal diameter and 200 mm outside diameter onto another tube of 100 mm internal diameter and 150 mm outside diameter. Based on a shrinkage allowance of 1 μm. Find the pressure developed at the interface between the two tubes and Hoop's stresses developed at the inner and outer radii of both tubes. Take E for both tube materials to be 200 GPa.
   [0.432 MPa, –1.555 MPa, –1.123 MPa. 1.543 MPa, 1.111 MPa]
4. A solid cylinder of 50 mm diameter is forced into another hollow cylinder of 50 mm internal diameter, 100 mm external diameter, and 50 mm long. It was found from the strain gauge readings on the external surface of the hollow cylinder that the circumferential stress on the outside surface to be 20 MPa. Find the force required to push the solid cylinder out if the coefficient of friction between the mating cylinder is 0.3.
   [70.65 kN]
5. (a) Discuss the need and methods of making thick-walled cylinders resisting elastically large internal pressure.
   (b) Show that in autofrettage or self-hooping in thick cylinders, the fully plastic pressure p_f is given by $p_f = \sigma_y \ln\frac{r_i}{r_0}$, where σ_y is the tensile yield strength and r_i, r_0 are the internal and external radii of the cylinder.
6. A thick cylinder with an internal diameter of 200 mm and external diameter 400 mm is subjected to increasing internal pressure. Take σ_y of the cylinder material to be 350 MPa.
   (a) Find the internal pressure at which yielding would just start at the insider surface of the cylinder.
   (b) Find the internal pressure at which the yield front would reach a radius of 150 mm.
   [131.25 MPa, –218.45 MPa]

7

Rotating Disks

Learning Objectives

After careful study of this chapter, students should be able to do the following:

LO1: Describe stresses and displacements for a rotating disk.

LO2: Compare the stress distribution in a flat disk with and without a central hole.

LO3: Illustrate the stress distribution in a disk of variable thickness.

LO4: Design the rotating disk of uniform stress.

7.1 INTRODUCTION [LO1]

The problems of stresses and deformations in disks rotating at high speeds are important in the design of both gas and steam turbines, generators and many such rotating machinery in industry. As discussed in earlier chapters, this is another example of axisymmetric problems in polar coordinates. Although the theoretical treatment of a flat disk is simpler, in many industrial applications, disks are tapered. They are usually thicker near the hub, and their theoretical analysis is slightly more involved. We shall first take up the analysis for flat disks.

In the case of rotating disks with centrifugal force as body force, the equation of equilibrium reduces to as in equation (6.1.3).

$$\frac{d\sigma_r}{dr}+\frac{\sigma_r-\sigma_\theta}{r}+\rho\omega^2 r=0. \tag{7.1.1}$$

Combining this with displacement equations, we have, as in equation (6.1.5), a general equation for determining the stress distribution in axisymmetric problems. This is given as

$$r\frac{d^2\sigma_r}{dr^2}+3\frac{d\sigma_r}{dr}=-(3+\nu)\rho\omega^2 r^2. \tag{7.1.2}$$

This is a nonhomogeneous differential equation. The associated homogeneous equation (complementary equation) is

$$r\frac{d^2\sigma_r}{dr^2}+3\frac{d\sigma_r}{dr}=0.$$

The solution of this equation is Lame's equation as discussed in Chapter 6, equation (6.2.3), and taking into consideration the particular solution, the solution to equation (7.1.2) turns out to be

$$\sigma_r = C_1 + \frac{C_2}{r^2} - \frac{(3+\nu)}{8}\rho\omega^2 r^2.$$

Combining this with equation (7.1.1), we have

$$\sigma_\theta = C_1 - \frac{C_2}{r^2} - \frac{(1+3\nu)}{8}\rho\omega^2 r^2.$$

We may also determine the radial displacement from equation (6.2.11), and this is given as

$$u_r = \frac{r}{E}(\sigma_\theta - \nu\sigma_r)$$

$$= \frac{r}{E}\left[C_1(1-\nu) - \frac{C_2}{r^2}(1+\nu) - \rho\omega^2 r^2\left(\frac{1-\nu^2}{8}\right)\right].$$

We may therefore write the stresses and displacement for the rotating disk under one bracket as

$$\begin{aligned} \sigma_r &= C_1 + \frac{C_2}{r^2} - \frac{(3+\nu)}{8}\rho\omega^2 r^2 \\ \sigma_\theta &= C_1 - \frac{C_2}{r^2} - \frac{(1+3\nu)}{8}\rho\omega^2 r^2 \\ u_r &= \frac{r}{E}\left[C_1(1-\nu) - \frac{C_2}{r^2}(1+\nu) - \rho\omega^2 r^2\left(\frac{1-\nu^2}{8}\right)\right]. \end{aligned} \tag{7.1.3}$$

With these introductory basic equations, we shall now set out to discuss the stress distribution and displacement in rotating disks.

7.2 STRESS DISTRIBUTION IN FLAT DISKS [LO2]

The stress distribution we wish to discuss here applies mainly to high-speed rotating disks in machinery, and clearly there the disks are mounted on shafts, indicating that the disks cannot be solid, i.e., inside radius $r_i \neq 0$ as shown in Figure 7.1. This also means that both σ_r and σ_θ are infinite at the center, if $C_2 \neq 0$.

We therefore consider the problem in two steps: (a) disks with no central hole, and (b) disks with a central hole.

(a) ***Disks with no central hole***

To write the stress and displacement equation in this case, we essentially set $C_2 = 0$ and consider a compressive pressure p_0 at the outside radius because the disk which is, in this case, a shaft that needs to be shrink-fitted inside another disk with a central hole.

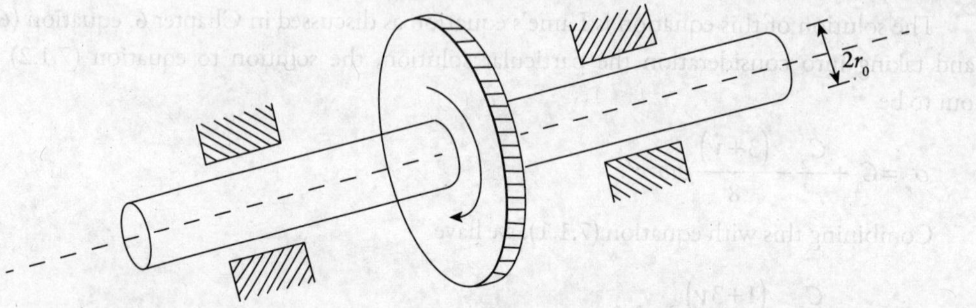

Figure 7.1 A rotor disk and shaft assembly: $2r_0$ is the outside diameter of the solid disk (shaft in this case)

Substituting in equation (7.1.3)

$C_2 = 0$ and

at $r = r_0$, $\sigma_r = -p_0$,

we get

$$C_1 = \frac{(3+\nu)}{8}\rho\omega^2 r_0^2 - p_0.$$

Substituting this value of C_1 in equation (7.1.3), we write the stress and displacement equations for a solid disk as

$$\sigma_r = \frac{(3+\nu)}{8}\rho\omega^2\left(r_0^2 - r^2\right) - p_0$$
$$\sigma_\theta = \frac{(3+\nu)}{8}\rho\omega^2\left(r_0^2 - \frac{1+3\nu}{3+\nu}r^2\right) - p_0 \quad (7.2.1)$$
$$u_r = \frac{1-\nu}{8E}\rho\omega^2 r\left[(3+\nu)r_0^2 - (1+\nu)r^2\right] - p_0 r\frac{1-\nu}{E}.$$

There are two important points to note here. Both stresses σ_r and σ_θ are the same and largest at $r = 0$, i.e., at the center of the solid disk, and in the absence of p_0, they are given as

$$\sigma_r\Big|_{r=0} = \sigma_\theta\Big|_{r=0} = \frac{3+\nu}{8}\rho\omega^2 r_0^2. \quad (7.2.2)$$

At the periphery, i.e., at $r = r_0$, $\sigma_r = 0$ but σ_θ has a finite value given as follows:

$$\sigma_r\Big|_{r=r_0} = 0$$
$$\sigma_\theta\Big|_{r=r_0} = \frac{1-\nu}{4}\rho\omega^2 r_0^2. \quad (7.2.3)$$

The nondimensional stress distributions $\overline{\sigma_r}$ and $\overline{\sigma_\theta}$ for a solid disk in the absence of external pressure are shown in Figure 7.2.

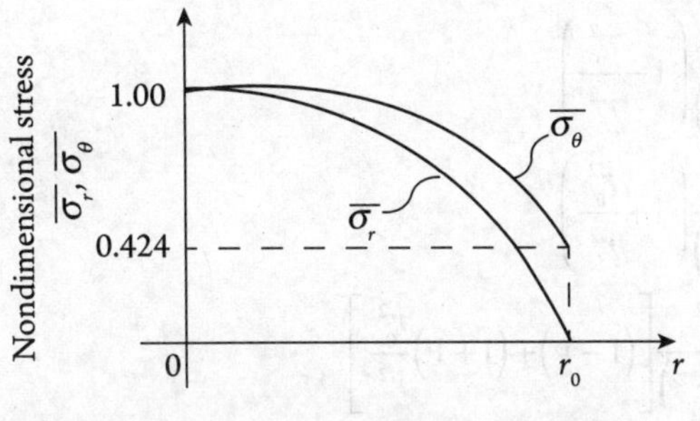

Figure 7.2 Stress distribution in a solid disk

Here,

$$\overline{\sigma_r} = \frac{\sigma_r}{\sigma_r\big|_{r=0}} \text{ and } \overline{\sigma_\theta} = \frac{\sigma_\theta}{\sigma_\theta\big|_{r=0}}. \tag{7.2.4}$$

This gives, at $r = 0$, $\overline{\sigma_r} = 1$ and at $r = r_0$, $\overline{\sigma_r} = 0$

at $r = 0$, $\overline{\sigma_\theta} = 1$ and at $r = r_0$, $\overline{\sigma_\theta} = \dfrac{2(1-\nu)}{3+\nu}$

with $\nu = 0.3\, \overline{\sigma_\theta}\big|_{r=r_0} = 0.424$.

The curves are parabolic due to the second term in equation (7.2.1).

The second point to note is that the stresses in a solid disk consist of two terms as can be seen in equation (7.2.1). The first term is proportional to $\rho\omega^2$ whereas the second term due to external pressure is independent of centrifugal stresses.

(b) ***Disk with a central hole***

There are two important issues to be considered before we proceed to develop equations for the radial and tangential stresses and radial displacement for the disk with a central hole. Since the disk is likely to be shrink-fitted onto a shaft, a contact pressure will be developed at the interface. Secondly, as the assembly rotates with increasing speed, the difference between the radial displacement at the inner radius of the disk and the radial displacement of the shaft at the contact must be less than the interference between the shaft and disk due to shrink fit for the assembly to rotate as one unit.

The problem may now be split into the following two cases:

(i) A nonrotating disk with a central hole and with an internal compressive pressure p_i at $r = r_i$.

(ii) A rotating disk with no pressure either at the internal radius $r = r_i$ or at the external radius $r = r_0$.

The first problem can be treated using the theory of thick cylinders as given in Section 6.2, equations (6.2.6), and we have

$$\sigma_r = \frac{p_i r_i^2}{\left(r_0^2 - r_i^2\right)}\left(1 - \frac{r_0^2}{r^2}\right)$$

$$\sigma_\theta = \frac{p_i r_i^2}{\left(r_0^2 - r_i^2\right)}\left(1 + \frac{r_0^2}{r^2}\right) \tag{7.2.5}$$

$$u_r = \frac{r}{E}\frac{p_i r_i^2}{\left(r_0^2 - r_i^2\right)}\left[(1-\nu) + (1+\nu)\frac{r_0^2}{r^2}\right].$$

Here, p_i represents the constant pressure developed between the shaft and disk due to interference.

The second problem can be treated by solving the general equation for a rotating disk discussed in Section 7.1, equation (7.1.3).

To determine the constants C_1 and C_2 in equation (7.1.3), we consider the following boundary conditions:

At $r = r_i$ and $r = r_0$, $\sigma_r = 0$. The constants C_1 and C_2 work out to be

$$C_1 = \frac{3+\nu}{8}\left(r_i^2 + r_0^2\right)\rho\omega^2$$

$$C_2 = -\frac{3+\nu}{8}\left(r_i^2 r_0^2\right)\rho\omega^2.$$

This gives the stresses and radial displacement in a rotating disk with a central hole with no boundary loadings.

$$\sigma_r = \frac{3+\nu}{8}\rho\omega^2\left[r_i^2 + r_0^2 - \left(\frac{r_i r_0}{r}\right)^2 - r^2\right]$$

$$\sigma_\theta = \frac{3+\nu}{8}\rho\omega^2\left[r_i^2 + r_0^2 + \left(\frac{r_i r_0}{r}\right)^2 - \frac{1+3\nu}{3+\nu}r^2\right] \tag{7.2.6}$$

$$u_r = \frac{r}{E}\left(\sigma_\theta - \nu\sigma_r\right) = \rho\omega^2\frac{r}{E}\frac{(3+\nu)(1-\nu)}{8}\left[r_i^2 + r_0^2 + \frac{1+\nu}{1-\nu}\left(\frac{r_i r_0}{r}\right)^2 - \frac{1+\nu}{3+\nu}r^2\right].$$

It now remains to combine equations (7.2.6) and (7.2.5) to solve problems of rotors fitted onto a shaft.

A typical question of practical importance on the above discussion is: at what rotating speed would the rotor shaft assembly loosen up? Clearly, this would occur when

$$u_{r(disc)}\Big|_{(r=r_i)} - u_{r(shaft)}\Big|_{(r=r_0)} = \delta, \tag{7.2.7}$$

where δ is the interference due to shrinkage.

The radial displacement at the inner radius $u_{r(disc)}$ is given by the third equation in (7.2.6) by substituting $r = r_i$. The radial displacement of the shaft at the outside surface is given by the third equation in (7.2.1). It must be taken into consideration that as the assembly loosens up, there will be no contact pressure, i.e., $p_0 = 0$, and also we substitute $r = r_0$. Therefore, equation (7.2.1) must be used with $p_0 = 0$ and $r = r_0$.

Another interesting point that relates to the present discussion is the stress concentration due to a small hole at the center of a solid disk. As stated earlier in equation (7.2.2), the maximum stresses occur at the center of a solid disk, and these are given by

$$\sigma_r = \sigma_\theta = \frac{3+\nu}{8}\rho\omega^2 r_0^2. \tag{7.2.8}$$

If we now consider a disk with a central hole (as shown in Figure 7.3), maximum σ_θ occurs at $r = r_i$, and from equation (7.2.6), this is given by

$$\sigma_{\theta(\max)} = \frac{\rho\omega^2}{4}\left[(3+\nu)r_0^2 + (1-\nu)r_i^2\right].$$

For a small hole, we may set $r_i \to 0$ and this gives

$$\sigma_{\theta(\max)} = \frac{\rho\omega^2}{4}\left[(3+\nu)r_0^2\right]. \tag{7.2.9}$$

Comparing equations (7.2.8) and (7.2.9), we can conclude that the presence of a small hole at the center of a flat disk raises the tangential stress by a factor of two as shown in Figure 7.4. This is an example of stress concentration due to the presence of a small hole. However, the radial stress in the center of a small-holed disk is zero.

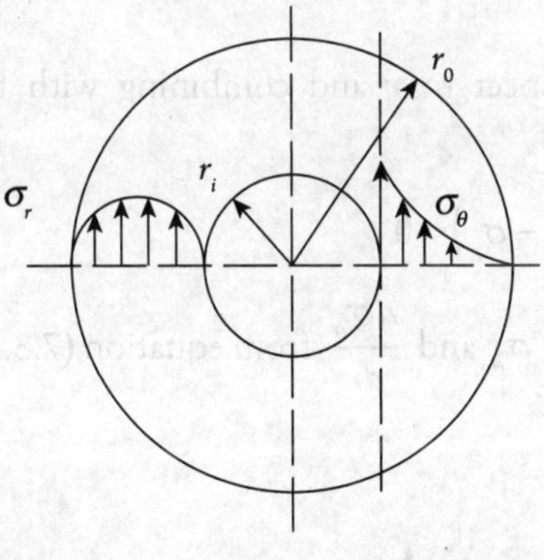

Figure 7.3 Schematic stress distribution in a rotating disk with a central hole. Both σ_r and σ_θ are tensile, but σ_r is zero at $r = r_i$, whereas σ_θ is maximum at $r = r_i$

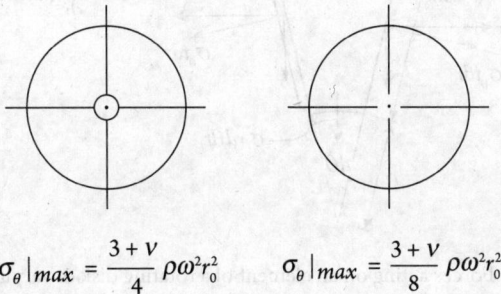

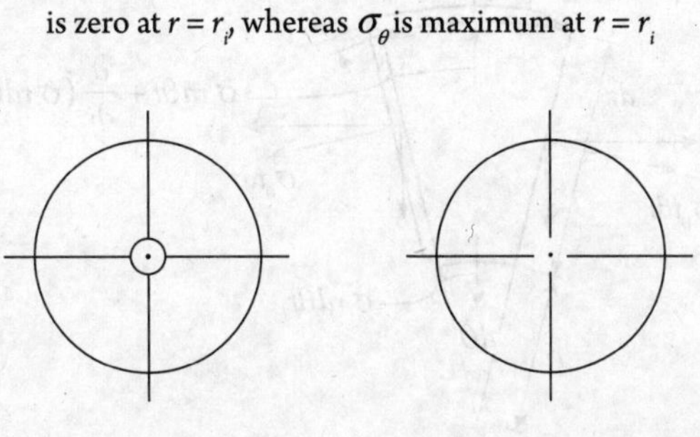

$$\sigma_\theta\big|_{max} = \frac{3+\nu}{4}\rho\omega^2 r_0^2 \qquad \sigma_\theta\big|_{max} = \frac{3+\nu}{8}\rho\omega^2 r_0^2$$

Figure 7.4 Stress concentration occurs at the center of a small-holed flat disk. Circumferential stress rises by a factor of two due to a small hole

7.3 DISK OF VARIABLE THICKNESS [LO3]

Rotating disks such as turbine blades are not always flat. Turbine blades are normally thick near the hub and thin at the periphery. This is because, as seen in Section 7.2, tensile tangential stress is largest at the inside radius of a rotating disk with a central hole. In analyzing the stress distribution in a flat disk with a central hole, the standard equation of equilibrium with centrifugal force as the body force was convenient. But for a disk with variable thickness, we consider the force equilibrium of a small element of a rotating disk acted upon by radial and circumferential stresses and also the centrifugal force as shown in Figure 7.5. In such an analysis, we can easily introduce the dependence of radius on variable thickness.

Considering the equilibrium of forces in the radial direction, we have

$$\frac{d}{dr}\left(tr\sigma_r\right)-t\sigma_\theta+t\rho\omega^2r^2=0. \tag{7.3.1}$$

Now we use the displacement equation in the same manner shown in Section 6.2, and we rewrite the radial and circumferential strains as

$$\begin{aligned}\epsilon_r&=\frac{du_r}{dr}=\frac{1}{E}\left(\sigma_r-\nu\sigma_\theta\right)\\ \epsilon_\theta&=\frac{u_r}{r}=\frac{1}{E}\left(\sigma_\theta-\nu\sigma_r\right).\end{aligned} \tag{7.3.2}$$

Differentiating u_r with respect to r and combining with the first of equations (7.3.2), we have

$$r\frac{d\sigma_\theta}{dr}-\nu r\frac{d\sigma_r}{dr}+\left(1+\nu\right)\left(\sigma_\theta-\sigma_r\right)=0. \tag{7.3.3}$$

To proceed further, we find σ_θ and $\frac{d\sigma_\theta}{dr}$ from equation (7.3.1), as shown below.

$$\sigma_\theta=\frac{1}{t}\frac{d\left(tr\sigma_r\right)}{dr}+\rho\omega^2r^2$$

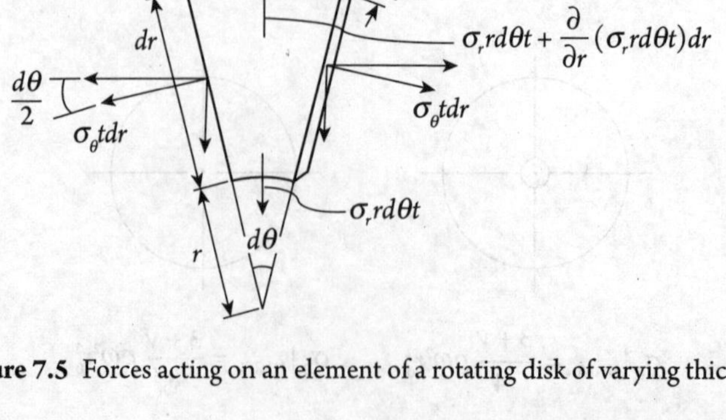

Figure 7.5 Forces acting on an element of a rotating disk of varying thickness

$$\frac{d\sigma_\theta}{dr}=\frac{1}{t}\frac{d^2\left(tr\sigma_r\right)}{dr^2}+2\rho\omega^2 r.$$

We now substitute these in equation (7.3.3), keeping $\left(tr\sigma_r\right)$ as a new variable σ_r' in which "t" also depends on r. This gives

$$r^2\frac{d^2\sigma_r'}{dr^2}+r\frac{d\sigma_r'}{dr}-\sigma_r'-\frac{r}{t}\frac{dt}{dr}\left[r\frac{d\sigma_r'}{dr}-\nu\sigma_r'\right]+\left(3+\nu\right)\rho\omega^2 r^3 t=0. \tag{7.3.4}$$

Following Aruel Stadola (1945) and DenHartog (1952), another thickness function that gives the hyperbolic profiles of different order for the disk (shown in Figure 7.6) is introduced, and that is given as

$$t=\frac{t_1}{r^q},$$

where t_1 is the thickness at the bore, and q is a constant that gives the order of the hyperbola. This allows us to choose the profile of the disk and its thickness at the root and at the periphery. For $q=0$, thickness is constant, and for $q=1$, the disk profile is that of an ordinary hyperbola.

Substituting $\frac{dt}{dr}=-\frac{qt}{r}$ in equation (7.3.4), we have

$$r^2\frac{d^2\sigma_r'}{dr^2}+r\frac{d\sigma_r'}{dr}\left(1+q\right)-\left(1+\nu q\right)\sigma_r'+\left(3+\nu\right)\rho\omega^2 r^3 t=0. \tag{7.3.5}$$

The equation is similar to equation (7.1.2), and if we let $\rho\omega^2=0$ the solution is of the form

$$\sigma_r'=Cr^n,$$

where C and n are constants, as discussed in Section 6.2.

Substituting this in equation (7.3.5) with $\rho\omega^2=0$, we get the following two values of n

$$n_1,\ n_2=-\frac{q}{2}\pm\sqrt{\frac{q^2}{4}+\nu q+1}. \tag{7.3.6}$$

We may write

$$\sigma_r'=C_1 r^{n_1}+C_2 r^{n_2}.$$

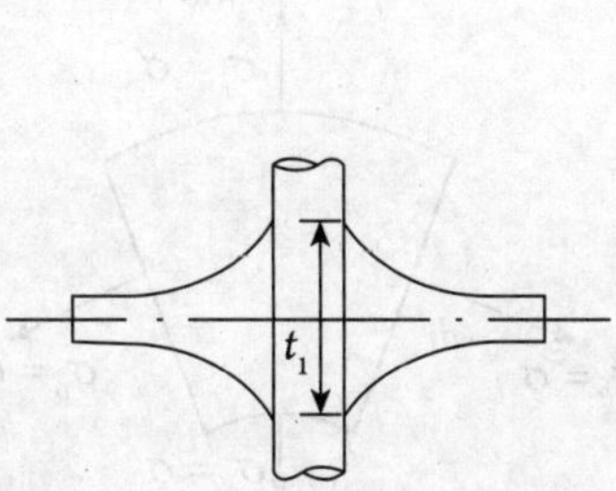

Figure 7.6 Rotating disk with a hyperbolic profile of any chosen order

In addition, we need to add a particular solution. If we proceed with a solution of the form $C_3 r^3 t$, then by substitution in equation (7.3.5) C_3 works out to be

$$C_3 = -\frac{(3+\nu)\rho\omega^2}{8-(3+\nu)q}.$$

Therefore, we may write the general solution for σ'_r as

$$\sigma'_r = C_1 r^{n_1} + C_2 r^{n_2} - \frac{(3+\nu)\rho\omega^2}{8-(3+\nu)q} r^3 t. \tag{7.3.7}$$

It is convenient to determine C_1 and C_2 for individual cases numerically. This will be demonstrated in Example 7.2.

7.4 ROTATING DISK OF UNIFORM STRESS [LO4]

In Section 7.3, we discussed the need for rotating disks with variable thickness from the practical utility point of view. However, there is enormous complication in designing such disks with hyperbolic profiles. There exists a simple solution, and that is to design these high speed disks for constant stress conditions under the action of large centrifugal forces to which they are subjected.

It is convenient to start the analysis using a small element of the disk subjected to equal hoop and radial stresses, i.e., $\sigma_r = \sigma_\theta = \sigma$ (say). Let the element be of radial length dr and included angle $d\theta$, as shown in Figure 7.7. Let the thickness at radius r and $(r+dr)$ be t and $(t+dt)$ respectively.

Considering the force equilibrium in the radial direction, we may write

$$(\rho t r d\theta dt)\omega^2 r + \sigma(r+dr)d\theta(t+dt) = \sigma r t d\theta + 2\sigma t dr \sin\frac{d\theta}{2}.$$

This gives $\sigma r \frac{dt}{dr} = -\rho t \omega^2 r^2$.

We now have $\ln t = -\frac{\rho\omega^2 r^2}{2\sigma} + C.$

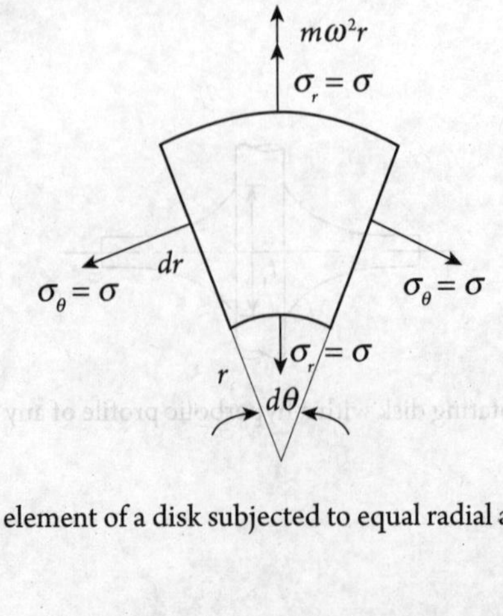

Figure 7.7 An element of a disk subjected to equal radial and hoop stresses

The constant may be evaluated using the boundary condition that at $r = 0$, $t = t_0$. This gives $C = \ln t_0$. We then have a general form of thickness variation equation as

$$t = t_0 e^{-\frac{\rho\omega^2 r^2}{2\sigma}}. \tag{7.4.1}$$

This applies to the constant stress condition over the entire disk without a central hole, commonly used in small single-wheel steam turbines. If $k = \sqrt{\frac{2\sigma}{\rho\omega^2}}$, then we may write a nondimensional form of equation (7.4.1) as:

$$\frac{t}{t_0} = e^{-\left(\frac{r}{k}\right)^2}. \tag{7.4.2}$$

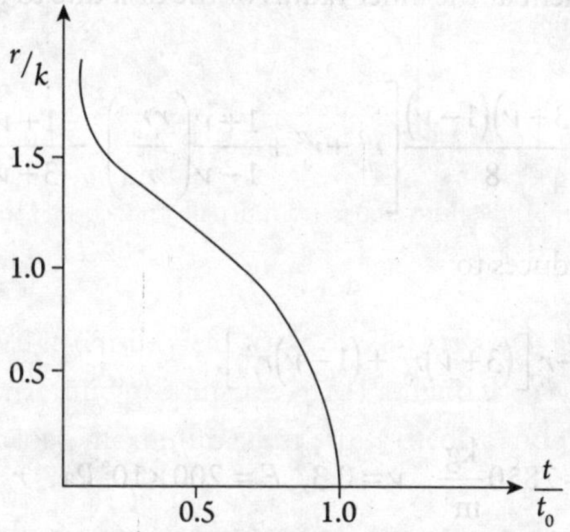

Figure 7.8 Plot of nondimensional radius versus nondimensional thickness for a solid rotating disk subjected to equal radial and hoop stress

A plot of nondimensional radius versus nondimensional thickness is shown in Figure 7.8. This gives the disk shape roughly. Inspecting equation (7.4.2), we find that there is a point of inflection at $\frac{r}{k} = 0.707$, and by choosing suitable values of k and t_0 we may design the disk shape of our choice, which would be thicker in the middle and asymptotically thin down the periphery. This would give constant stress all over the disk. This analysis, however, is not useful for a disk with a central hole. For a disk with a central hole, at the hole boundary, $\sigma_r = 0$, and this means, according to this analysis, zero constant stress would occur all over the disk.

Example 7.1

A disk c outside diameter of 300 mm and a central hole of 50 mm diameter is press-fitted to a shaft with a diameter 50.01 mm. If both the disk and the shaft are made of steel with the following properties:

Modulus of elasticity, $E = 200\,\text{GPa}$, Poisson's ratio, $\nu = 0.3$, and density, $\rho = 7850\,\text{kg/m}^3$. Find

(a) The speed at which the shaft and the disk loosen up.
(b) Stresses in the disk at the speed found in (a).
(c) Stress at the inside radius of the disk at half the speed found in (a).

Solution:

(a) Stresses developed in a rotating disk mounted on a shaft have two components:
   i. Stresses due to contact pressure developed by shrinkage or interference
   ii. Stresses due to rotation

Since in the first problem, the shaft and the disk loosen up, stresses due to interference need not be considered.

Radial displacement at the inner radius of the disk due to rotation is given by the third of equation (7.2.6).

$$u_{r(disk)} = \rho\omega^2 \frac{r}{E}\frac{(3+\nu)(1-\nu)}{8}\left[r_i^2 + r_0^2 + \frac{1+\nu}{1-\nu}\left(\frac{r_i r_0}{r}\right)^2 - \frac{1+\nu}{3+\nu}r^2\right].$$

At $r = r_i$, this reduces to

$$u_{r(disk)}\Big|_{r=r_i} = \frac{\rho\omega^2}{4E} r_i\left[(3+\nu)r_0^2 + (1-\nu)r_i^2\right].$$

Substituting $\rho = 7850\frac{\text{kg}}{\text{m}^3}$, $\nu = 0.3$, $E = 200\times10^9\,\text{Pa}$, $r_0 = 0.15\,\text{m}$, and $r_i = 0.025\,\text{m}$, we have

$$u_{r(disk)} = 1.83217\times10^{-11}\omega^2.$$

The radial displacement at the outer radius of the shaft from equation (7.2.1) is

$$u_{r(shaft)} = \frac{1-\nu}{8E}\rho\omega^2 r\left[(3+\nu)r_0^2 - (1+\nu)r^2\right].$$

Here we take $r = r_{os} = 0.025\,\text{m}$ and neglect the influence of interference pressure. Substituting the values of ρ, ν, and E as before, we have

$$u_{r(shaft)} = 1.0732\times10^{-13}\omega^2.$$

Interference, $\delta = (r_{os} - r_i) = \left[u_{r(disk)} - u_{r(shaft)}\right]_{r=r_i}$

$$\omega^2 = 274508.38\left(\frac{\text{rad}}{\text{s}}\right)^2;\ \omega = 523.93\frac{\text{rad}}{\text{s}}$$

This gives the required speed, $N = 5003.22\,\text{rpm}$.

(b) Disk stresses only due to rotation are given by equation (7.2.6).

$$\sigma_r = \frac{3+\nu}{8}\rho\omega^2\left[r_i^2 + r_0^2 - \left(\frac{r_i r_0}{r}\right)^2 - r^2\right]$$

$$\sigma_\theta = \frac{3+\nu}{8}\rho\omega^2\left[r_i^2 + r_0^2 + \left(\frac{r_i r_0}{r}\right)^2 - \frac{1+3\nu}{3+\nu}r^2\right].$$

σ_r is zero at both $r = r_i$ and $r = r_0$

σ_θ is maximum at $r = r_i$ and given by

$$\sigma_\theta\Big|_{(r=r_i)} = \frac{\rho\omega^2}{4}\left[(1-\nu)r_i^2 + (3+\nu)r_0^2\right] = 146.574\omega^2 = 40.235\,\text{MPa}.$$

(c) Here we need to consider stresses due to both shrinkage pressure and rotation.

The displacements in the disk and the shaft in part (a) need to be corrected for half the speed. This gives $u_{r(disk)}\Big|_{r=r_i} = 1.83217\times10^{-11}\times\left(\frac{274503.56}{4}\right) = 1.26\,\mu\text{m}$

and $u_{r(shaft)} = 1.07\times10^{-13}\times\left(\frac{274503.56}{4}\right) = 0.00735\,\mu\text{m}.$

Therefore, the inner radius of the disk $= 25 + 0.00126 = 25.00126\,\text{mm}$.

The outer radius of the shaft $= 25.005 + 0.00000735 = 25.00500735\,\text{mm}$.

This gives interference $\delta = 25.00500735 - 25.00126 = 3.75\,\mu\text{m}$.

Contact pressure p_s is given by equation (6.2.15), which needs to be modified in this case.

Considering solid shaft, $r_i = 0$, which gives:

$$p_s = \frac{E\delta}{r_s\left[\frac{r_0^2 + r_s^2}{r_0^2 - r_s^2} + 1\right]},$$

where r_0 = disk outside radius = 0.15 m

r_s = contact surface radius = 0.025 m

$E = 200\times10^9$ Pa

Substituting values in the above equation, we have

$p_s = 14.583\,\text{MPa}.$

The radial stress σ_r at the inside radius of the disk is

$\sigma_{r_i} = -14.583\,\text{MPa}.$

Here i stands for shrinkage. Therefore, σ_{r_i} is the radial stress due to interference. The tangential stress at the inside radius of the disk is given by equation (7.2.5).

$$\sigma_\theta = \frac{p_i r_i^2}{r_0^2 - r_i^2}\left(1 + \frac{r_0^2}{r_i^2}\right).$$

Substituting $p_i = p_s = 14.583\,\text{MPa}$.

$r_0 = 0.15\,\text{m},\ r_i = 0.025\,\text{m}.$

We have $\sigma_{\theta_i} = 15.41\,\text{MPa}$.

Here σ_{θ_i} gives the tangential stress due to interference.

Now we proceed to find the stresses due to rotation. This can be easily found from the results in part (b) by substituting new values of ω, which is simply $\frac{\omega}{2}$.

This gives the stresses due to rotation as

$$\sigma_{rr} = 0$$

$$\sigma_{\theta r} = \frac{40.235}{4} = 10.058\,\text{MPa}.$$

Here σ_{rr} and $\sigma_{\theta r}$ denote radial and tangential stresses due to rotation.

Therefore, the final stresses at the inside radius of the disk are

$$\sigma_r = \sigma_{r_i} + \sigma_{rr} = -14.583\,\text{MPa}$$

$$\sigma_\theta = \sigma_{\theta_i} + \sigma_{\theta r} = 15.41 + 10.058 = 25.468\ \text{MPa}.$$

Example 7.2

For the same disk-rotor assembly as in example 7.1, we consider a disk of hyperbolic shape with 120 mm thickness at the shaft end and 20 mm at the periphery. Find

(a) The speed at which the disk and the shaft loosen up and
(b) Stresses in the disk at the speed found in (a).

Solution:

(a) The hyperbolic profile of different orders is given by

$$t = \frac{t_1}{r^q}.$$

This gives $\frac{dt}{dr} = -\frac{qt}{r}$.

q gives the order of the hyperbola. We may therefore find q considering that at

$r_i = 0.025\,\text{m},\ t_i = 0.12\,\text{m}$ and at $r_0 = 0.15\,\text{m},\ t_i = 0.02\,\text{m}.$

This gives $\frac{t_i}{t_0} = \left(\frac{r_0}{r_i}\right)^q$ and $q = 1$.

We then find the exponent of r from equation (7.3.6)

$$n_1, n_2 = -\frac{q}{2} \pm \sqrt{\frac{q^2}{4} + \nu q + 1}.$$

Substituting $q = 1$ and $\nu = 0.3$, we have $n_1 = -1.745$, $n_2 = 0.745$.

To find the stresses in the disk, we may use equations (7.3.7) and (7.3.1). At this stage, we may write equation (7.3.7) with *tr* as a constant and divide the left- and right-hand terms by *tr*. This gives

$$\sigma_r = C_3 r^{n_1} + C_4 r^{n_2} - \frac{(3+\nu)\rho\omega^2}{8-(3+\nu)q} r^2,$$

where C_3 and C_4 are new constants after dividing C_1 and C_2 by *tr*.

Now, we use the boundary conditions when the disk and the shaft loosen up. The boundary conditions are:

$\sigma_r = 0$ at $r = 0.025\,\text{m}$ and $r = 0.15\,\text{m}$.

This gives $C_3 = -6.02\times10^{-6}\,\rho\omega^2$ and $C_4 = 0.06562\,\rho\omega^2$.

From equation (7.3.1), $\sigma_\theta = \dfrac{1}{t}\dfrac{d(tr\sigma_r)}{dr} + \rho\omega^2 r^2$

or, $\sigma_\theta = \dfrac{1}{t}\dfrac{dt}{dr} r\sigma_r + \dfrac{1}{t} t \dfrac{d(r\sigma_r)}{dr} + \rho\omega^2 r^2$

or, $\sigma_\theta = -\sigma_r + \dfrac{d(r\sigma_r)}{dr} + \rho\omega^2 r^2.$

Simplifying, $\sigma_\theta = n_1 C_3 r^{n_1} + n_2 C_4 r^{n_2} - 2\dfrac{(3+\nu)\rho\omega^2}{8-(3+\nu)q} r^2 + \rho\omega^2 r^2.$

We then have

$$\sigma_r = \left(-6.02\times10^{-6} r^{-1.745} + 0.06562 r^{0.745} - 0.702 r^2\right)\rho\omega^2. \qquad \text{(I)}$$

$$\sigma_\theta = \left(10.5049\times10^{-6} r^{-1.745} + 0.04888 r^{0.745} - 0.404 r^2\right)\rho\omega^2. \qquad \text{(II)}$$

The radial displacement of the disk at the inner radius is given by

$$u_r\Big|_{(r=r_i)} = \frac{r_i}{E}\left[\sigma_\theta - \nu\sigma_r\right].$$

Substituting σ_r and σ_θ from above and

$r_i = 0.025\,\text{m}$, $E = 200\times10^9$ Pa, $\nu = 0.3$, and $\rho = 7850\,\text{kg/m}^3$,

we have $u_r\Big|_{(r=0.025\,\text{m})} = 9.2612\times10^{-12}\,\omega^2.$

Since the radial displacement of the shaft is small compared to disk displacement, we may equate $u_{r(disk)}$ to the interference of 0.005×10^{-3} m. This gives

$$9.2612\times10^{-12}\,\omega^2 = 5\times10^{-6}.$$

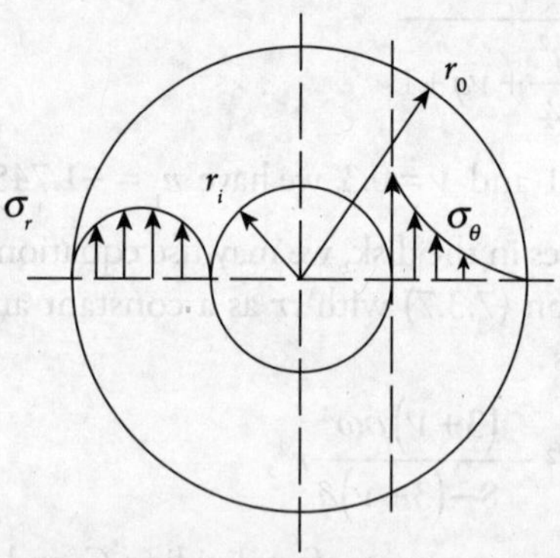

Figure 7.9 Schematic variation of radial and circumferential stresses in a rotating disk of varying thickness

This gives loosening speed to be $N = 7019.53\,\text{rpm}$.

This is nearly 1.4 times more than the loosening up speed for constant thickness disk. This is an advantage from a design point of view.

(b) The disk stresses evaluated in part (a) are given in equations marked (I) and (II), and their variations within the disk thickness are shown in Figure 7.9.

As we can see in the above figure, radial stresses are zero at both inner and outer radii. The circumferential stress is largest at the inner radius and we proceed to calculate this stress at $r = r_i$ from equation (II).

$$\sigma_\theta\Big|_{(r=0.025)} = \left[10.5049\times10^{-6}(0.025)^{-1.745} + 0.04888(0.025)^{0.745} - 0.404(0.025)^2\right] \times 7850\times0.5398\times10^6$$

$$= 40\,\text{MPa}.$$

Example 7.3

A turbine rotor with an outer diameter of 350 mm and inner diameter of 70 mm is to be designed for uniform strength under rotational conditions. If the disk thickness is 15 mm at the outer diameter and design stress and speed are 300 MPa and 20000 rpm respectively, find the maximum thickness at the inner diameter. Take the density of the steel disk to be 7850 kg/m^3.

Solution:

Thickness variation for uniform strength is given by equation (7.4.1).

$$t = t_0 e^{-\frac{\rho\omega^2 r^2}{2\sigma}}, \tag{7.4.1}$$

where t is the thickness at any radius r and t_0 is the thickness at $r = 0$.

Substituting $t = 0.015\,\text{m}$, $\rho = 7850\,\dfrac{\text{kg}}{\text{m}^3}$, $\omega = 2094.35$ rad/s, $r = 0.175\,\text{m}$, and

$\sigma = 300 \times 10^6$ Pa.

We have $t_0 = 0.0869$ m.

With this t_0 value using equation (7.4.1) for $r = 0.035$ m, maximum thickness $t = 0.07820$ m, i.e., 78.20 mm.

Example 7.4

A steel disk with an outside diameter of 625 mm is shrunk on a solid steel shaft with diameter of 125 mm. If the radial interference is 0.01 mm, find

(a) radial and circumferential stresses in the disk at standstill.
(b) Speed at which the disk and the shaft loosen up.
(c) Radial and circumferential stresses in the disk at half the loosening up speed.

Take $E = 200$ GPa and $\rho = 7850$ kg/m^3.

Solution:

Given $D_o = 625\,\text{mm};\ r_o = 0.3125\,\text{m}$

$D_s = 125\,\text{mm};\ r_s = 0.0625\,\text{m}.$

$\delta_i = 0.01\,\text{mm} = 10 \times 10^{-6}\ \text{m (initial)}$

$E = 200 \times 10^9\,\text{N/m}^2;\ \rho = 7850\,\text{kg/m}^3.$

(a) At the standstill condition: Interference causes a contact pressure given by the relation;

$$p_s = \frac{E\delta}{r_s\left[\dfrac{r_0^2 + r_s^2}{r_0^2 - r_s^2} + \dfrac{r_s^2 + r_i^2}{r_s^2 - r_i^2}\right]}.$$

Considering $r_s = r_i$ and $r_i = 0$

$$p_s = \frac{E\delta}{r_i\left[\dfrac{r_0^2 + r_i^2}{r_0^2 - r_i^2} + 1\right]} = \frac{200 \times 10^9 \times 10 \times 10^{-6}}{0.0625\left[\dfrac{0.3125^2 + 0.0625^2}{0.3125^2 - 0.0625^2} + 1\right]} = 15.36\,\text{MPa}$$

$$\sigma_r = -p_s = -15.36\,\text{MPa}$$

$$\sigma_\theta = \frac{p_i r_i^2}{r_0^2 - r_i^2}\left(1 + \frac{r_0^2}{r_i^2}\right) = \frac{15.36 \times 0.0625^2}{0.3125^2 - 0.0625^2}\left(1 + \frac{0.3125^2}{0.0625^2}\right) = 16.64\,\text{MPa}. \quad \text{...(Ans.)}$$

(b) At the point of loosening up: The interference pressure = 0.

Therefore, stresses are contributed due to rotation only.

$$u_{r(disk)} = u_{r(disk)}\Big|_{r=r_i} = \frac{\rho\omega^2}{4E} r_i\left[(3+\nu)r_o^2 + (1-\nu)r_i^2\right] = 1.993 \times 10^{-10} \times \omega^2$$

$r_{os} = r_i$ (neglecting interference); $r_o = r_{os}$ (for shaft)

$$u_r\Big|_{r=r_{os}} = u_{r(shaft)} = \frac{1-\nu}{8E}\rho\omega^2 r\left[(3+\nu)r_0^2 - (1+\nu)r^2\right]$$

$$= \frac{(1-0.3)\times 7850\times\omega^2\times(0.0625)^3}{8\times 200\times 10^9}\left[(3+0.3)-(1+0.3)\right]$$

$$= 1.677\times 10^{-12}\,\omega^2.$$

Interference, $\delta = u_{r(disk)} - u_{r(shaft)}$

$$\Rightarrow 10\times 10^{-6} = \left(1.993\times 10^{-10} - 1.677\times 10^{-12}\right)\omega^2$$

$$\Rightarrow \omega^2 = 50598.84;\ \omega = 224.54\ \text{rad/s}$$

Therefore, $N = \dfrac{60\omega}{2\pi} = 2144.24\ \text{rpm}.$ ...(Ans.)

(c) At half the loosening speed: An interference exists; hence, both rotational and interference stresses are to be accounted for.

At, $\omega = \omega_{max}/2$;

$$u_{r(disk)} = 1.993\times 10^{-10}\left(\frac{\omega_{max}}{2}\right)^2 = 1.993\times 10^{-10}\left(\frac{50598.838}{4}\right) = 2.52\ \mu\text{m}$$

$$u_{r(shaft)} = 1.677\times 10^{-12}\left(\frac{\omega_{max}}{2}\right)^2 = 0.0212\ \mu\text{m}$$

$$\delta\Big|_{\omega=\omega_{max}/2} = u_{r(disk)} - u_{r(shaft)} = 2.4988\ \mu\text{m}.$$

(Interference) Contact pressure = $p_s = \dfrac{E\delta}{r_i\left[\dfrac{r_0^2+r_i^2}{r_0^2-r_i^2}+1\right]} = 3.84\ \text{MPa}$

$$\sigma_r = -p_s = -3.84\ \text{MPa}$$

$$\sigma_\theta = \frac{p_s r_i^2}{\left(r_0^2-r_i^2\right)}\left(1+\frac{r_0^2}{r_i^2}\right) = 4.16\ \text{MPa}.$$

Due to rotation,

$$\sigma_{r_{rot}} = 0;$$

$$\sigma_{\theta_{rot}} = \frac{\rho\omega^2}{4}\left[(3+\nu)r_0^2 + (1-\nu)r_i^2\right] = 32.27\ \text{MPa}$$

$$\sigma_{r_{total}} = \sigma_{r_{interference}} + \sigma_{r_{rotation}} = -3.84\ \text{MPa}$$

$$\sigma_{\theta_{total}} = \sigma_{\theta_{interference}} + \sigma_{\theta_{rotation}} = 36.43\ \text{MPa}.$$ ...(Ans.)

Example 7.5

A steel disk with an outside diameter of 250 mm and inside diameter of 50 mm is shrunk on a shaft of diameter 50.01 mm. If the disk is of a hyperbolic shape of 125 mm thickness at the shaft end and 25 mm thickness at the periphery. Find the rpm at which the shaft loosens up and also the stress in the disk at the loosening up speed. Take $\nu = 0.3$, $\rho = 7850\ \text{kg/m}^3$, and $E = 200$ GPa.

Solution:

Given, $D_o = 250\,\text{mm};\ r_o = 0.125\,\text{m}$

$D_i = 50\,\text{mm};\ r_i = 0.025\,\text{m}$

$D_{os} = 50.01\,\text{mm};\ r_{os} = 0.025005\,\text{m}$

$t_{shaft} = t_s = 125\,\text{mm} = 0.125\,\text{m}$

$t_{periphery} = t_p = 25\,\text{mm} = 0.025\,\text{m}.$

Hyperbolic rotor-disk: $t = t_0/r^q$.

At $r = r_i;\ t = t_s;\ 0.125 = t_0/0.025^q$

At $r = r_o;\ t = t_p;\ 0.025 = t_0/0.125^q$

Solving the above, we get

$q = 1$

$$\{n_1, n_2\} = \frac{-q}{2} \pm \sqrt{\frac{q^2}{4} + \nu q + 1} = \{-1.745,\ 0.745,\}$$

$$\sigma_r = C_3 r^{n_1} + C_4 r^{n_2} - \frac{(3+\nu)\rho\omega^2}{8-(3+\nu)q} r^2.$$

Applying boundary conditions: $\sigma_r = 0$, at $r = r_i$ and $r = r_o$.

At $r = r_i$;

$$0 = C_3 r_i^{n_1} + C_4 r_i^{n_2} - \frac{(3+\nu)\rho\omega^2}{8-(3+\nu)q}(0.025)^2$$

$$\Rightarrow 0 = C_3 (0.025)^{-1.745}\ 4 + C_4\, 0.025^{0.745} - 0.702\rho\omega^2 (0.025)^2 \qquad \text{(I)}$$

At $r = r_0$;

$$0 = C_3 (0.125)^{-1.745} + C_4\, 0.125^{0.745} - 0.702\rho\omega^2 (0.125)^2 \qquad \text{(II)}$$

Solving (I) and (II), we get

$C_3 = -4.688 \times 10^{-6} \rho\omega^2;\ C_4 = 0.052\rho\omega^2.$

From equation (7.3.1), $\sigma_\theta = \dfrac{1}{t}\dfrac{d(tr\sigma_r)}{dr} + \rho\omega^2 r^2$

$$\text{or, } \sigma_\theta = \frac{1}{t}\frac{dt}{dr} r\sigma_r + \frac{1}{t} t \frac{d(r\sigma_r)}{dr} + \rho\omega^2 r^2$$

$$\text{or, } \sigma_\theta = -\sigma_r + \frac{d(r\sigma_r)}{dr} + \rho\omega^2 r^2.$$

$$\text{Simplifying, } \sigma_\theta = n_1 C_3 r^{n_1} + n_2 C_4 r^{n_2} - 2\frac{(3+\nu)\rho\omega^2}{8-(3+\nu)q} r^2 + \rho\omega^2 r^2.$$

We then have

$$\sigma_r = \left[-4.688\times10^{-6} r^{-1.745} + 0.052 r^{0.745} - 0.702 r^2\right]\rho\omega^2. \quad \text{(I)}$$

$$\sigma_\theta = \left(8.8105\times10^{-6} r^{-1.745} + 0.0387 r^{0.745} - 0.404 r^2\right)\rho\omega^2. \quad \text{(II)}$$

The radial displacement of the disk is given by

$$u_r = \frac{r}{E}\left(\sigma_\theta - \nu\sigma_r\right).$$

At $r = r_i = 0.025\,\text{m}$;

$$u_r\Big|_{(r=r_i)} = \frac{r_i}{E}\left[10.2169\times10^{-6} r^{-1.745} + 0.0231 r^{0.745} - 0.1934 r^2\right]\rho\omega^2$$

$$u_r\Big|_{(r=r_i)} = 6.5276\times10^{-12}\,\omega^2 = u_{r(disk)}$$

Interference: $\delta = u_{r(disc)} - u_{r(shaft)}$.

Since $(u_{r(shaft)} = 0)$

$$\Rightarrow r_{os} - r_i = 5\times10^{-6} = 6.5276\times10^{-12}\,\omega^2.$$

Hence, $\omega^2 = 765978.3$; $\omega = 875.2\,\text{rad/s}$

and $N = \dfrac{60\omega}{2\pi} = 8361.78\,\text{rpm}.$ ...(Ans.)

Stress at loosening speed; at $r = r_i$

$$\sigma_\theta\Big|_{(r=0.025)} = \left[8.8105\times10^{-6}(0.025)^{-1.745} + 0.0387(0.025)^{0.745} - 0.404(0.025)^2\right]\times7850\times765978.3$$

$$= 46.467\,\text{MPa}.$$ ...(Ans.)

Example 7.6

A steel disk with an outer diameter of 300 mm and inner diameter of 100 mm is shrunk onto a solid steel shaft. Interference is arranged such that the interference pressure does not fall below 25 MPa during rotation of the assembly. If the maximum hoop stress at the maximum speed at the inner surface of the disk is 250 MPa. Find the maximum speed at which the assembly can rotate without any relative slip between the disk and the shaft. For steel, take $E = 200\,\text{GPa}$, $\nu = 0.3$, and $\rho = 7850\,\text{kg/m}^3$.

Solution:

Given: $D_o = 300\,\text{mm};\ r_o = 0.15\,\text{m}$

$D_i = 100\,\text{mm};\ r_i = 0.05\,\text{m}$

$p_s = 25\,\text{MPa}$ (Interference pressure)

$(\sigma_\theta)_{max} = 250\,\text{MPa}$ (at inner surface)

$E = 200\times10^9\,\text{N/m}^2;\ \rho = 7850\,\text{kg/m}^3;\ \nu = 0.3$

To find the point where the disengagement starts

$p_s = 25\,\text{MPa};\ \sigma_r\Big|_{(r=r_o)} = 0;\ \sigma_r\Big|_{(r=r_i)} = -25\,\text{MPa}.$

Also, $\sigma_r = C_1 + \dfrac{C_2}{r^2} - \dfrac{(3+\nu)}{8}\rho\omega^2 r^2.$

Hence for $r = r_o$ and $r = r_i$, we get

$C_1 + 400C_2 - 8.0953125\omega^2 = -25\times10^6.$

And, $C_1 + 44.44C_2 - 72.8578\omega^2 = 0$ respectively.

Also, $\sigma_\theta = C_1 - \dfrac{C_2}{r^2} - \dfrac{(1+3\nu)}{8}\rho\omega^2 r^2.$

And, $(\sigma_\theta)_{max} = \sigma_\theta\Big|_{(r=r_i)}$; from this, we get

$C_1 - 400C_2 - 4.661\omega^2 = 250\times10^6.$

$$\text{Hence solving}\begin{cases} C_1 + 400C_2 - 8.0953125\omega^2 = -25\times10^6 \\ C_1 + 44.44C_2 - 72.8578\omega^2 = 0 \\ C_1 - 400C_2 - 4.661\omega^2 = 250\times10^6 \end{cases}$$

We get, $C_1 = 121854636.6;\ C_2 = -337453.75;\ \omega = 1211.06\,\text{rad/s}.$

Hence, $N = \dfrac{60\omega}{2\pi} = 11564.78\,\text{rpm}.$ ...(Ans.)

Example 7.7

A steel disk of a turbine assembly with an outer diameter of 500 mm and inner diameter of 100 mm has a uniform thickness of 30 mm. 100 blades, each of 0.5 kg mass, are evenly pitched around the periphery of the disk at an effective radius of 300 mm. Find the rotational speed at which the yielding would be initiated at the inside surface of the disk according to the maximum shear stress failure criterion. Take $E = 200\,\text{GPa}$, $\nu = 0.3$, $\rho = 7850\,\text{kg/m}^3$, and the shear yield stress is 300 MPa for the rotor steel.

Solution:

Given, $D_o = 500\,\text{mm};\ r_o = 0.25\,\text{m}$

$D_i = 100\,\text{mm};\ r_i = 0.05\,\text{m}$

$D_{eff} = 300\,\text{mm};\ r_{eff} = 0.15\,\text{m}$

$t = 30\,\text{mm} = 0.03\,\text{m};\ m = 0.5\dfrac{\text{kg}}{\text{blade}};\ \eta_{blades} = 100\,\text{blades}$

$E = 200\times10^9\,\text{N/m}^2;\ \rho = 7850\dfrac{\text{kg}}{\text{m}^3};\ \nu = 0.3;\ \tau_y = 300\,\text{MPa}$

Centrifugal force of rotating blades, F_C (acting radially outwards)

$$F_C = Mr\omega^2 = \eta_{blades} \times m \times r_{eff} \times \omega^2 = 7.5\,\omega^2$$

$$p_s = \frac{F_C}{2\pi r_i t} = 795.775\,\omega^2$$

$$\sigma_r\Big|_{(r=r_i)} = -p_s = -795.775\,\omega^2$$

From equation (7.1.3), we have

$$\sigma_r = C_1 + \frac{C_2}{r^2} - \frac{(3+\nu)}{8}\rho\omega^2 r^2$$

$$\sigma_\theta = C_1 - \frac{C_2}{r^2} - \frac{(1+3\nu)}{8}\rho\omega^2 r^2.$$

Boundary condition: $\sigma_r\Big|_{(r=r_o)} = 0$ and $\sigma_r\Big|_{(r=r_i)} = -p_s$.

Hence for $r = r_o$ and $r = r_i$, we get

$C_1 + 16C_2 = 202.383\omega^2$ and $C_1 + 400\,C_2 = -787.68\,\omega^2$ respectively.

Solving, we get

$C_1 = 243.635\,\omega^2;\ C_2 = -2.5783\,\omega^2$.

By the maximum shear stress theory:

$\tau_y = (\sigma_1 - \sigma_2)/2$, where σ_1 and σ_2 are principal stresses.

Or, $\tau_y = \sigma_{eff};\ \sigma_{eff} = \sigma_\theta - \nu\sigma_r$

$$\sigma_{eff} = (1-\nu)C_1 - (1+\nu)\frac{C_2}{r^2} - \left[\frac{(1+3\nu) - \nu(3+\nu)}{8}\right]\rho\omega^2 r^2$$

At $= r_i$; $\sigma_{eff} = 1509.03\omega^2$.

Since $\tau_y = \sigma_{eff}$.

Therefore, $\omega^2 = 198803.205$ and $\omega = 445.87\dfrac{\text{rad}}{\text{s}}$.

Hence, $N = \dfrac{60\omega}{2\pi} = 4257.75$ rpm. ...(Ans.)

Example 7.8

A turbine rotor disk is designed for uniform stress during rotation. The outer diameter of the rotor is 600 mm and its inner diameter is 100 mm. The rotor thickness at the shaft end is 50 mm and at the periphery is 10 mm. Find the design speed if the design stress is to be 300 MPa. Take the density of the rotor steel to be $7850\,\text{kg/m}^3$.

Solution:

Given, $D_o = 600\,\text{mm}$; $r_o = 0.3\,\text{m}$

$D_i = 100\,\text{mm}$; $r_i = 0.05\,\text{m}$

$t_{shaft} = t_s = 50\,\text{mm} = 0.05\,\text{m}$

$t_{periphery} = t_p = 10\,\text{mm} = 0.01\,\text{m}$

$\sigma_{design} = 300\,\text{MPa}$; $\rho = 7850\,\text{kg/m}^3$.

In order to design for uniform strength, we have

$$t = t_0 e^{-\frac{\rho\omega^2 r^2}{2\sigma}}$$

at, $r = r_i$; $t = t_s$

$$0.05 = t_0 e^{-\frac{\left(7850\times\omega^2\times0.05^2\right)}{2\times300\times10^6}} \quad \text{(a)}$$

at, $r = r_o$; $t = t_p$

$$0.01 = t_0 e^{-\frac{\left(7850\times\omega^2\times0.3^2\right)}{2\times300\times10^6}} \quad \text{(b)}$$

Solving (a) and (b), we get

$\omega^2 = 1405880.5$

$\omega = 1185.7\,\text{rad/s}$

Hence, $N = \dfrac{60\omega}{2\pi} = 11322.58$ rpm.

Exercises

1. Derive equations for radial and circumferential stresses in a high-speed rotating disk with a central hole from the basic considerations. Show how they vary across the disk.

   State also the condition that determines the speed at which the rotor-shaft assembly loosens up.

2. A steel disk of 750 mm outside diameter is shrunk onto a 75 mm diameter shaft with an interference of 20 μm.

   Find the speed at which the contact pressure is zero. Find also the maximum circumferential stress at this speed. Take $E = 200\,\text{GPa}$, $\rho = 7850\dfrac{\text{kg}}{\text{m}^3}$ and $\nu = 0.3$ for the steel disk and shaft.

   [3276.75 rpm, 107 MPa]

3. Show that the maximum circumferential and radial stresses in a rotating flat disk of outside radius r_0 and inside radius r_i with no loading at the boundary are

   $$\sigma_{\theta(max)} = \frac{\rho\omega^2}{4}\left[(3+\nu)r_0^2 + (1-\nu)r_i^2\right]$$

   $$\sigma_{r(max)} = \frac{\rho\omega^2}{8}(3+\nu)\left(r_0 - r_i\right)^2.$$

   Find where these maximum stresses occur in the disk.

4. Discuss the effect of the presence of a small hole at the center of a flat solid disk.

   Find the circumferential stresses at the center of a flat rotating disk with and without a small hole, and show that the presence of a small hole raises the stress by a factor of 2.

5. Find the speed in rpm of a steel disk with an outer radius of 200 mm and inner radius of 100 mm, if the radial displacement of the outer radius of the disk is not to exceed 10 μm. Take $E = 200\,\text{GPa}$, $\rho = 7850\dfrac{\text{kg}}{\text{m}^3}$, and $\nu = 0.3$ for steel.

   [2761.38 rpm]

6. A steel disk with an external diameter of 800 mm is shrink fitted onto a steel shaft of 80 mm diameter with an interference of 20 μm. Take $E = 200\,\text{GPa}$, $\rho = 7850\dfrac{\text{kg}}{\text{m}^3}$, and $\nu = 0.3$ for steel.

   (a) Find the maximum circumferential stress in the disk in its standstill position.

   (b) Find the maximum circumferential stress at the speed when the contact pressure reaches zero.

   [5.5 MPa, 100.247 MPa]

8

Torsion

Learning Objectives

After careful study of this chapter, students should be able to do the following:

LO1: Identify torsion members.

LO2: Describe the torsion formula for a circular member.

LO3: Apply the torsion formula for a noncircular cross-section.

LO4: Apply Prandtl's stress function approach.

LO5: Analyze Prandtl's membrane analogy.

LO6: Assess the torsion of hollow sections.

LO7: Design a thin-walled hollow section of torsion members.

8.1 INTRODUCTION [LO1]

In simple words, the application of a torque on a prismatic member causes twisting or torsion. This causes shear stress if a torque alone is applied. However, this is rarely true in practical cases. A circular bar, used to transmit torque between a prime mover and a machine, is a typical example of a torsion member. However, in many applications, a torque along with a bending moment and axial loading are applied, and there we need to combine these effects and find the principal stresses. A typical example of such combined stresses is a propeller shaft. Torsional problems are important in many applications both in industry and in our daily life. Therefore, we consider torsion alone in this chapter in some detail.

Torsional problems for circular members are generally solved assuming that plane sections normal to the axis of the bar remain plane even after twisting. This assumption was first made by Coulomb intuitively in 1784, and he came up with a correct usable equation for members with circular sections. However, this assumption does not apply to bars with a noncircular cross-section. Navier attempted to solve torsional problems with noncircular sections using Coulomb's assumption and came up with an erroneous solution. The correct solution was provided by St. Venant in 1853 using a warping function. Much later, in 1903, Prandtl came up with a membrane analogy method that could solve problems with any complicated cross-section. First, we shall consider torsional problems with circular cross-sections.

8.2 TORSION OF MEMBERS WITH CIRCULAR CROSS-SECTION [LO2]

The torsion analysis of members with a circular cross-section starts with simplified assumptions made by Coulomb. In order to establish a relation between the applied torque and shear stress developed and the angle of twist in such cases, the following assumptions are made:

1. Material is homogeneous and isotropic.
2. Plane sections perpendicular to the axis of a circular member remain plane after twisting. No warping or distortion of the parallel planes occurs.
3. Shear strain varies linearly from the central axis.
4. Material obeys Hooke's Law.

Now we shall consider the assumptions in detail. The first assumption is necessary to develop a simplified model. Non-homogeneity and anisotropy would introduce complications in the model. To demonstrate the second assumption, consider a circular member, as shown in Figure 8.1, fixed at end *ab*, and a torque *T* is applied at end *gh*.

According to the assumption, parallel sections, for example, *cd* and *ef* remain plane even after the application of torque. Therefore, due to the application of torque, parallel planes rotate such that points c, e, and g move to their new positions c', e', and g'. This is because an imaginary plane o_1o_2ga moves to the new position $o_1o_2g'a$. This is demonstrated in the third assumption as shown in Figure 8.2. If radius o_1a is fixed, similar radii o_3c and o_2g rotate to o_3c' and o_2g' respectively and they remain straight. Here the angle γ_{max} between the original and the rotated planes represents the maximum shear strain at the surface, and $d\phi$ represents the angle of twist.

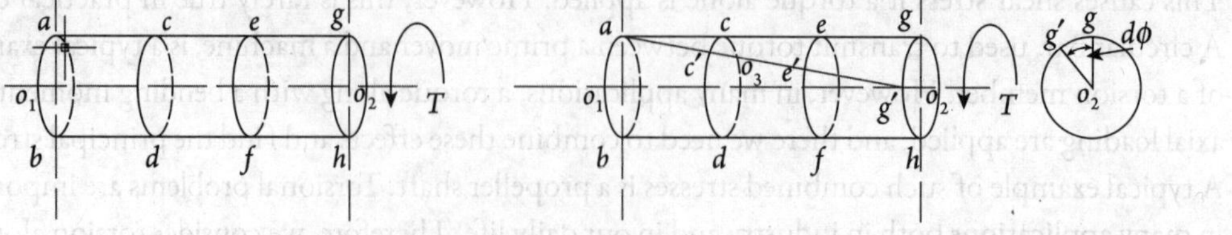

Figure 8.1 Plane sections before and after torsion

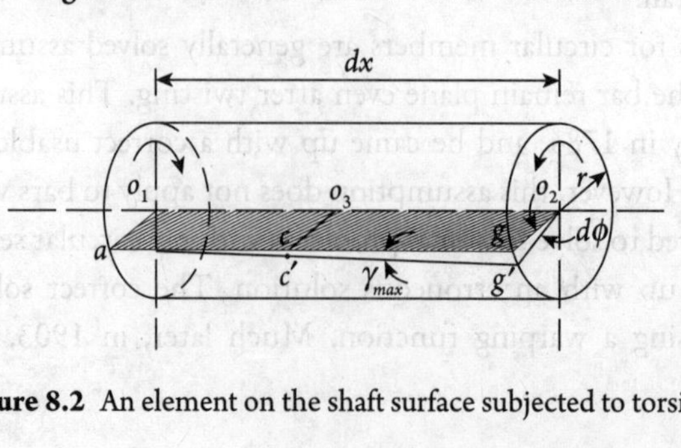

Figure 8.2 An element on the shaft surface subjected to torsion

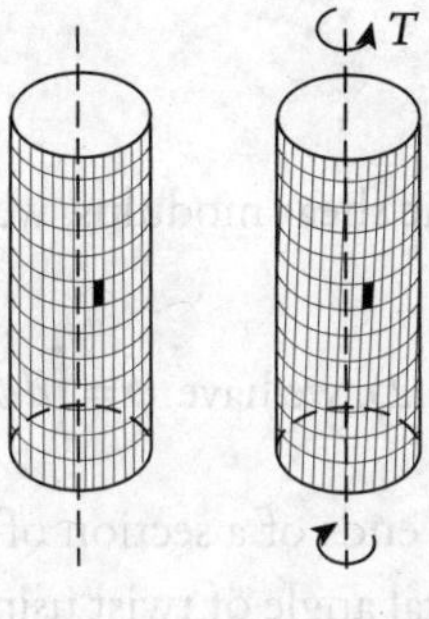

Figure 8.3 A circular shaft with a grid before and after the application of torque

This may be further demonstrated by an experiment with a circular cylinder of plasticine on which square grids are formed on the surface, as shown in Figure 8.3.

On the application of torque *T*, grids are deformed with no warping of the parallel section. The shaded elements show deformation without warping. A typical small section of length *dx* and radius *r* from the shaft is shown exaggerated in Figure 8.2. Here, the arc gg' is given by

$$\gamma_{max} dx = r d\phi.$$

This gives a relation between the maximum shear strain and the angle of twist, and for any radius *r*, we may write

$$\gamma dx = r d\phi. \tag{8.2.1}$$

Now, we shall consider the relation between the applied torque *T* and shear stress τ developed in a circular member.

According to the assumption, the shear strain varies linearly from the central axis. Since in linear elastic situations, shear stress is proportional to shear strain, we may conclude that the shear stress also varies linearly from the central axis of the circular shaft. This is shown in Figure 8.4, which shows the cross-section of a circular shaft of radius *R* and shear stress varies linearly from the central axis to a maximum value of τ_{max}.

We now consider a small element of area dA at a radius *r*. We now equate the integrated torque to the applied torque, and this gives

$$\int_A \tau dAr = T$$

From Figure 8.4, we may write $\tau = \tau_{max} \dfrac{r}{R}$ and this gives $T = \dfrac{\tau_{max}}{R} \int_A r^2 dA.$

Since $\int_A r^2 dA = J$, the polar moment of inertia of the cross-section, we have $\dfrac{T}{J} = \dfrac{\tau_{max}}{R}.$

Therefore, for any radius *r*, the equation reduces to

$$\frac{T}{J} = \frac{\tau}{r}. \tag{8.2.2}$$

We shall now combine equations (8.2.1) and (8.2.2). In order to do this, we need to manipulate equation (8.2.1), which can be written as

$$\frac{d\phi}{dx}=\frac{\gamma}{r}.$$

Now writing $\gamma=\frac{\tau}{G}$, where G is the shear modulus, we have $\frac{d\phi}{dx}=\frac{\tau}{Gr}$.

Substituting τ from equation (8.2.2), we have $\phi=\int d\phi=\int_A^B \frac{Tdx}{GJ}$.

Here A and B represent the two ends of a section of a shaft. In a shaft with varying torque and cross-sections, we may find the total angle of twist using the following equation.

$$\phi=\int_A^B \frac{T(x)dx}{GJ}. \tag{8.2.3}$$

We may take an example, as shown in Figure 8.5, to demonstrate equation (8.2.3).

The shaft carries five pulleys A,B,C,D, and E. The pulleys carry toques 30 Nm clockwise, 800 Nm clockwise, 250 Nm anticlockwise, 500 Nm anticlockwise, and 100 Nm clockwise respectively.

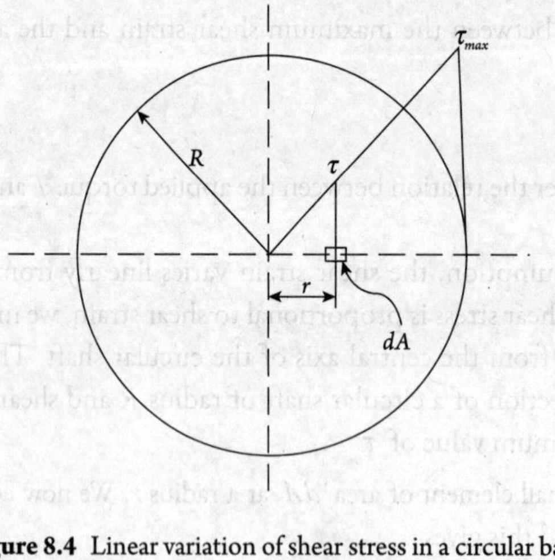

Figure 8.4 Linear variation of shear stress in a circular bar

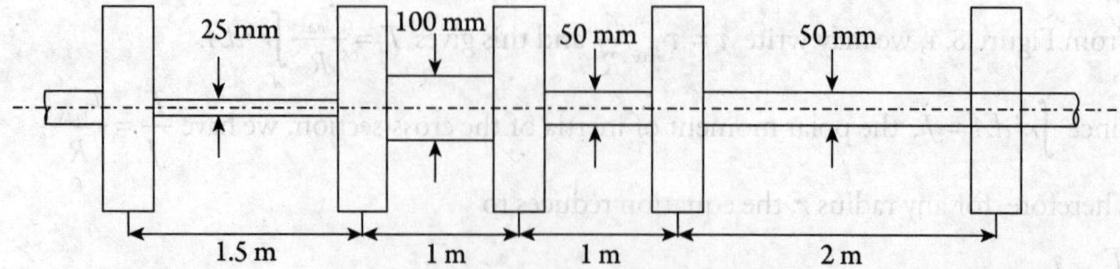

Figure 8.5 A shaft with varying cross-sections carrying pulleys transmitting varying torques

To find the total rotation, we may start from end A and write

$$\phi=\int_{A}^{B}\frac{T_{AB}dx}{GJ_{AB}}+\int_{B}^{C}\frac{T_{BC}dx}{GJ_{BC}}+\int_{C}^{D}\frac{T_{CD}dx}{GJ_{CD}}+\int_{D}^{E}\frac{T_{DE}dx}{GJ_{DE}}.$$

Substituting the values, we may evaluate each section and then add up to find the total effect.

$$T_{AB}=+30\,\text{Nm}\,; \qquad J_{AB}=\frac{\pi(0.025)^4}{32}=3.83\times10^{-8}\ \text{m}^4$$

$$T_{BC}=+(800+30)\,\text{Nm}\,; \qquad J_{BC}=\frac{\pi(0.1)^4}{32}=9.81\times10^{-6}\ \text{m}^4$$

$$T_{CD}=+(830-250)\,\text{Nm}; \qquad J_{CD}=\frac{\pi(0.05)^4}{32}=6.136\times10^{-7}\ \text{m}^4$$

$$T_{DE}=+(580-500)\,\text{Nm}\,; \qquad J_{DE}=\frac{\pi(0.05)^4}{32}=6.136\times10^{-7}\ \text{m}^4$$

Taking $G=80\,\text{GPa}$ and $\phi=1.7°$.

We may simplify equation (8.2.3) for a straight circular cylinder of length l to

$$\phi=\frac{Tl}{GJ}\ \text{or}\ \frac{T}{J}=\frac{G\phi}{l}. \tag{8.2.4}$$

Combining this with equation (8.2.2), we have a simple torsion formula for a circular member of radius r and length l as

$$\frac{T}{J}=\frac{\tau}{r}=\frac{G\phi}{l}. \tag{8.2.5}$$

Here ϕ represents the angle of twist. For most problems with the design of a circular shaft in engineering practice, equation (8.2.5) provides sufficient information. Here J is the polar moment of inertia for solid circular bars of diameter d and is taken as $\frac{\pi}{32}d^4$, and for tubular bars of internal diameter d_i and external diameter d_0, J is taken as $\frac{\pi}{32}\left(d_0^4-d_i^4\right)$. The design for strength requires the yield shear stress of the shaft material. In many cases, the angle of twist of the shaft is specified, and there we need to specify the value of ϕ in radians in equation (8.2.5).

Another important point that needs to be considered in designing circular shafts is that, for an element on the shaft surface, the shear stresses would act in planes perpendicular to the shaft axis. However, these stresses cannot exist alone, and for equilibrium, complimentary shear stresses must act, as shown in Figure 8.6. This means that the element is under pure shear, where all shear stresses are of equal magnitude.

This does not pose any problem for isotropic materials, but for anisotropic materials such as wood, where shear on planes parallel to the grains is much less than that perpendicular to the grains, the complementary shear stress is an important issue.

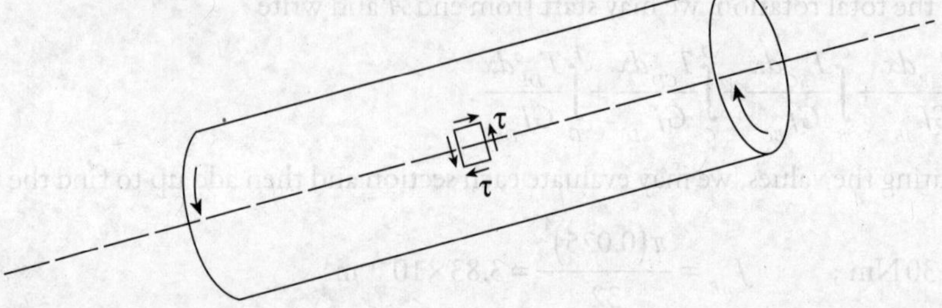

Figure 8.6 State of stress for an element on the surface of a shaft subjected to pure torsion

8.3 TORSION OF PRISMATIC BARS WITH NONCIRCULAR CROSS-SECTION [LO3]

As discussed, earlier plane sections do not remain plane after twisting for bars with noncircular cross-sections. This can be demonstrated clearly by an experiment with a rectangular bar of plasticine on whose four surfaces rectangular grids are formed. On application of twisting moment at two ends, distortion and warping of the parallel sections can be noticed clearly. This is shown in Figure 8.7.

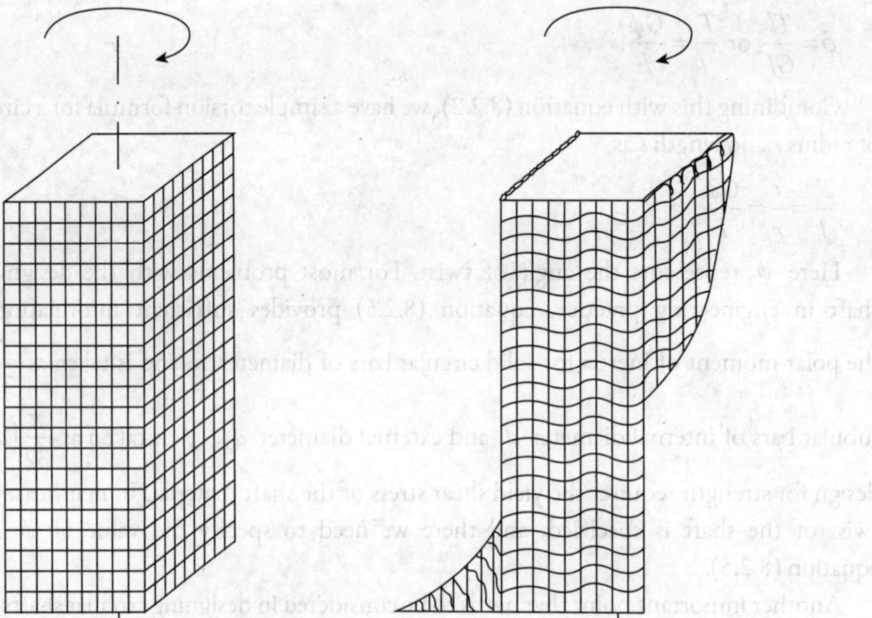

Figure 8.7 Experiment showing warping of parallel planes after twisting in prismatic bars of rectangular cross-section

Unlike Navier, St. Venant proposed that the deformation of a bar of noncircular cross-section consists of rotation of the cross-section and warping. We now consider a prismatic bar of any cross-section as shown in Figure 8.8.

Displacements of a point $P(x,y)$ on a plane z-distance away from the origin are given by (as shown in Figure 8.8(b)

$$x = r\cos\phi \qquad y = r\sin\phi$$

$$dx = -r\sin\phi d\phi = -y d\phi \qquad dy = r\cos\phi d\phi = x d\phi.$$

Here $d\phi$ represents rotation, and if we define rotation per unit length in the z-direction as θ, we may then write the displacements u and v in the x and y directions respectively as

$$u = -\theta zy \text{ and } v = \theta zx. \tag{8.3.1}$$

We also need to define warping of the cross-sections, and this is given by a warping function $\psi(x,y)$.

$$w = \theta\psi(x,y). \tag{8.3.2}$$

Here w is the displacement in the z-direction.

It is now necessary that these assumptions satisfy compatibility, constitutive, and equilibrium equations. Now the strain field derived from the displacement components in equations (8.3.1) and (8.3.2) is

$$\begin{aligned}
&\in_x = \frac{\partial u}{\partial x} = 0 \qquad \gamma_{xy} = \frac{\partial u}{\partial y} + \frac{\partial v}{\partial x} = -\theta z + \theta z = 0\\
&\in_y = \frac{\partial v}{\partial y} = 0 \qquad \gamma_{yz} = \frac{\partial v}{\partial z} + \frac{\partial w}{\partial y} = \theta x + \theta\frac{\partial \psi}{\partial y} = \theta\left(x + \frac{\partial \psi}{\partial y}\right)\\
&\in_z = \frac{\partial w}{\partial z} = 0 \qquad \gamma_{zx} = \frac{\partial w}{\partial x} + \frac{\partial u}{\partial z} = -\theta y + \theta\frac{\partial \psi}{\partial x} = \theta\left(-y + \frac{\partial \psi}{\partial x}\right).
\end{aligned} \tag{8.3.3}$$

It can be shown that these results satisfy the compatibility equations. Using the constitutive equations, stress fields may be obtained from equations (8.3.3), and these are given in equations (8.3.4):

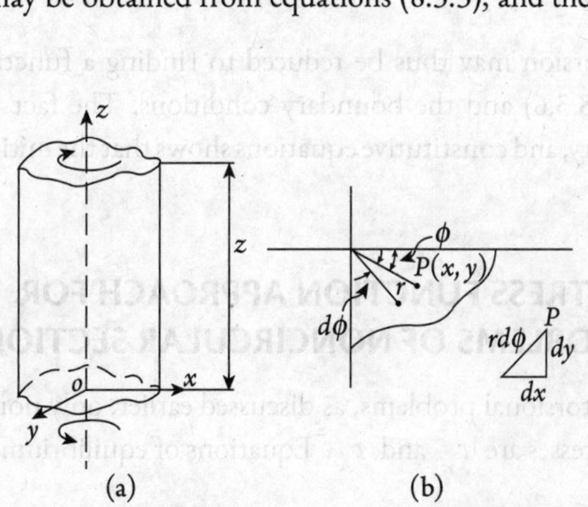

Figure 8.8 Displacements corresponding to the rotation of a cross-section

$$\sigma_x = \sigma_y = \sigma_z = \tau_{xy} = 0 \text{ and also } \gamma_{xy} = 0$$

$$\tau_{yz} = G\theta\left(\frac{\partial \psi}{\partial y} + x\right) \tag{8.3.4}$$

$$\tau_{xz} = G\theta\left(\frac{\partial \psi}{\partial x} - y\right).$$

Now we consider the equilibrium equations in the absence of body forces reproduced for convenience, as given below:

$$\frac{\partial \sigma_x}{\partial x} + \frac{\partial \tau_{xy}}{\partial y} + \frac{\partial \tau_{xz}}{\partial z} = 0$$

$$\frac{\partial \tau_{yx}}{\partial x} + \frac{\partial \sigma_y}{\partial y} + \frac{\partial \tau_{yz}}{\partial z} = 0 \tag{8.3.5}$$

$$\frac{\partial \tau_{zx}}{\partial x} + \frac{\partial \tau_{zy}}{\partial y} + \frac{\partial \sigma_z}{\partial z} = 0.$$

Substituting the stress field from equation (8.3.4) in equilibrium equations shown in equation (8.3.5), we find that the only equation that is not automatically satisfied is

$\frac{\partial \tau_{xz}}{\partial x} + \frac{\partial \tau_{yz}}{\partial y} + \frac{\partial \sigma_z}{\partial z} = 0.$ Since $\sigma_z = 0$, this reduces to

$\frac{\partial \tau_{xz}}{\partial x} + \frac{\partial \tau_{yz}}{\partial y} = 0$, and substituting τ_{xz} and τ_{yz} from equation (8.3.4), we have

$$\frac{\partial}{\partial x}\left[G\theta\left(\frac{\partial \psi}{\partial x} - y\right)\right] + \frac{\partial}{\partial y}\left[G\theta\left(\frac{\partial \psi}{\partial y} + x\right)\right] = 0.$$

This gives $\frac{\partial^2 \psi}{\partial x^2} + \frac{\partial^2 \psi}{\partial y^2} = 0$, i.e., $\nabla^2 \psi = 0$. (8.3.6)

Every problem on torsion may thus be reduced to finding a function ψ that satisfies the harmonic equation in (8.3.6) and the boundary conditions. The fact that these results satisfy equilibrium, compatibility, and constitutive equations shows that the initial assumptions regarding deformation are true.

8.4 PRANDTL'S STRESS FUNCTION APPROACH FOR TORSION PROBLEMS OF NONCIRCULAR SECTIONS [LO4]

Prandtl noticed that for torsional problems, as discussed earlier, only nonzero strains are γ_{yz} and γ_{xz}, and only nonzero stresses are τ_{yz} and τ_{xz}. Equations of equilibrium thus reduce to

$$\frac{\partial \tau_{xz}}{\partial x} + \frac{\partial \tau_{yz}}{\partial y} = 0. \tag{8.4.1}$$

Therefore, a suitable stress function ϕ must be so chosen that it automatically satisfies equation (8.4.1).

We may therefore define a stress function as shown below:

$$\tau_{xz} = \frac{\partial \phi}{\partial y} \qquad \tau_{yz} = -\frac{\partial \phi}{\partial x}. \tag{8.4.2}$$

This gives $\gamma_{xz} = \dfrac{1}{G}\dfrac{\partial \phi}{\partial y} \qquad \gamma_{yz} = -\dfrac{1}{G}\dfrac{\partial \phi}{\partial x}.$

It can be shown that under these conditions, only compatibility equations that are not automatically satisfied are

$$\frac{\partial}{\partial x}\left(\frac{\partial \gamma_{xy}}{\partial z} - \frac{\partial \gamma_{yz}}{\partial x} + \frac{\partial \gamma_{zx}}{\partial y}\right) = 2\frac{\partial^2 \epsilon_x}{\partial y \partial z}$$

$$\frac{\partial}{\partial y}\left(\frac{\partial \gamma_{yz}}{\partial x} - \frac{\partial \gamma_{zx}}{\partial y} + \frac{\partial \gamma_{xy}}{\partial z}\right) = 2\frac{\partial^2 \epsilon_y}{\partial x \partial y}.$$

Since $\gamma_{xy} = \epsilon_x = \epsilon_y = 0$, the equations reduce to

$$\frac{\partial}{\partial x}\left(\frac{\partial \gamma_{zx}}{\partial y} - \frac{\partial \gamma_{yz}}{\partial x}\right) = 0$$

$$\frac{\partial}{\partial y}\left(\frac{\partial \gamma_{zx}}{\partial y} - \frac{\partial \gamma_{yz}}{\partial x}\right) = 0.$$

This means $\dfrac{\partial \gamma_{xz}}{\partial y} - \dfrac{\partial \gamma_{yz}}{\partial x} =$ constant.

Substituting the warping function expressions from equation (8.3.3), it can be shown that

$$\frac{\partial \gamma_{xz}}{\partial y} - \frac{\partial \gamma_{yz}}{\partial x} = -2\theta.$$

Having obtained the value of the constant, we revert to the stress function formulation, and we have

$$\frac{\partial}{\partial y}\left(\frac{1}{G}\frac{\partial \phi}{\partial y}\right) + \frac{\partial}{\partial x}\left(\frac{1}{G}\frac{\partial \phi}{\partial x}\right) = -2\theta.$$

This gives a harmonic equation of the stress function

$$\nabla^2 \phi = -2G\theta. \tag{8.4.3}$$

θ being the angle of rotation per unit length.

Now we shall consider the boundary conditions for a torsional problem, and this indicates that the perimeter must be free from shear stress.

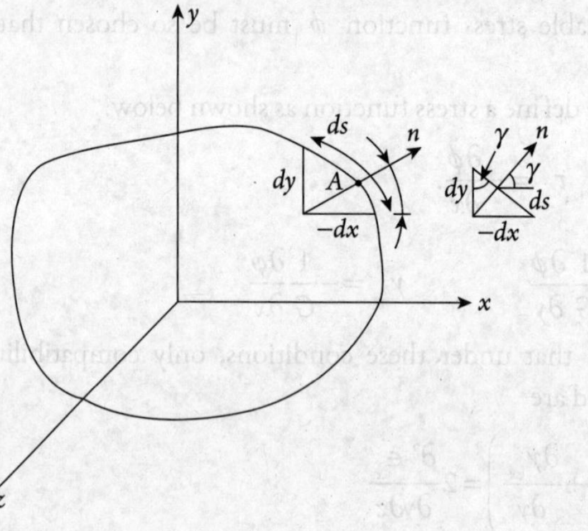

Figure 8.9 Shear stress at the boundary is directed along the tangent to the boundary

We consider an arbitrary cross-section, and let n be the normal to the perimeter at a point A (Figure 8.9), making an angle γ with the x-axis while ds is the perimeter. Therefore, the direction cosines of the normal are in general

$l = \cos\gamma \quad m = \sin\gamma$, and $n = 0$,

since the cross-section is perpendicular to the z-axis. Using the general expression for shear stress in terms of direction cosines of two orthogonal directions, we have

$$\tau_{nz} = \sigma_x l_1 l_2 + \sigma_y m_1 m_2 + \sigma_z n_1 n_2 + \tau_{xy}\left(l_1 m_2 + m_1 l_2\right) + \tau_{yz}\left(m_1 n_2 + n_1 m_2\right) + \tau_{zx}\left(n_1 l_2 + l_1 n_2\right).$$

Substituting $l_1 = \cos\gamma,\quad m_1 = \sin\gamma,\quad n_1 = 0,\quad l_2 = 0,\quad m_2 = 0$, and $n_2 = 1$, we have

$$\tau_{nz} = \tau_{yz}\sin\gamma + \tau_{xz}\cos\gamma.$$

Substituting τ_{yz} and τ_{xz} from equation (8.4.2) and taking $\sin\gamma = -\dfrac{dx}{ds}$ and $\cos\gamma = \dfrac{dy}{ds}$, we have

$$\tau_{nz} = \frac{\partial\phi}{\partial x}\frac{dx}{ds} + \frac{\partial\phi}{\partial y}\frac{dy}{ds} = \frac{d\phi}{ds} = 0,$$

since shear stress along the perimeter needs to be zero. This means that ϕ is constant along the boundary and we may set the boundary value for a solid bar of any cross-section to be zero, i.e.,

$$\phi = 0, \text{ along the boundary.} \tag{8.4.4}$$

The two essential equations for solving torsion problems of noncircular cross-sections are (8.4.3) and (8.4.4).

8.4.1 Torque

The magnitude of the couple required to twist the bar, as shown in Figure 8.10, is given by

$$T = \iint\left(x\tau_{yz} - y\tau_{xz}\right)dxdy.$$

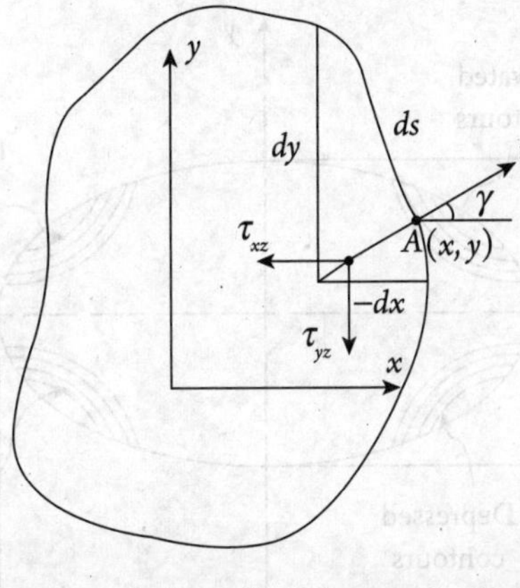

Figure 8.10 Shear stresses at the boundary of a cross-section of a solid prismatic bar

Substituting τ_{xz} and τ_{yz} in terms of stress function, we have

$$T=-\iint\left(x\frac{\partial\phi}{\partial x}+y\frac{\partial\phi}{\partial y}\right)dxdy$$

and this may be written as

$$T=-\iint\left[\frac{\partial}{\partial x}(x\phi)+\frac{\partial}{\partial y}(y\phi)\right]dxdy+2\iint\phi dxdy.$$

Applying Green's theorem, the double integral may be reduced to the line integral and we have

$$T=-\oint_s(x\phi dy-y\phi dx)+2\iint\phi dxdy.$$

Since the line integral is performed around the boundary of the cross-section and ϕ is zero around the boundary, the line integral vanishes. We therefore have

$$T=2\iint\phi dxdy. \tag{8.4.5}$$

Equation (8.4.5) applies to any cross-section. We shall first examine an elliptical cross-section. The boundary of the elliptical section is given by

$$\frac{x^2}{a^2}+\frac{y^2}{b^2}=1$$

and a stress function of the following form satisfies the boundary condition $\phi=0$.

$$\phi=m\left(\frac{x^2}{a^2}+\frac{y^2}{b^2}-1\right),\ \text{where } m \text{ is a constant.}$$

Substituting this in the governing differential equation $\nabla^2\phi=-2G\theta$, we have

$$\frac{\partial^2}{\partial x^2}\left[m\left(\frac{x^2}{a^2}+\frac{y^2}{b^2}-1\right)\right]+\frac{\partial^2}{\partial y^2}\left[m\left(\frac{x^2}{a^2}+\frac{y^2}{b^2}-1\right)\right]=-2G\theta.$$

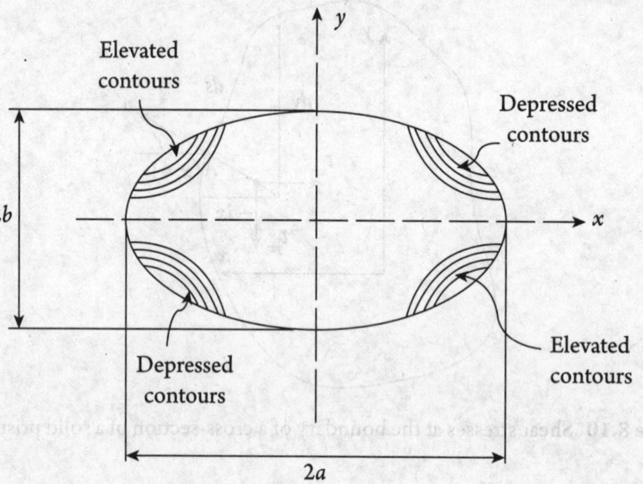

Figure 8.11 Displacement field in an elliptical cross-section of a bar subjected to torsion

This gives $m=-\dfrac{G\theta a^2b^2}{a^2+b^2}$, and we thus have $\phi=-\dfrac{G\theta a^2b^2}{a^2+b^2}\left(\dfrac{x^2}{a^2}+\dfrac{y^2}{b^2}-1\right)$.

We now find the torque requirement of such a section using equation (8.4.5).

$$T=-\frac{G\theta a^2b^2}{a^2+b^2}\iint\left(\frac{x^2}{a^2}+\frac{y^2}{b^2}-1\right)dxdy.$$

Now, we note

$$\iint\frac{x^2}{a^2}dxdy=\frac{I_y}{a^2}$$

$$\iint\frac{y^2}{b^2}dxdy=\frac{I_x}{b^2}$$

and $\iint dxdy=A.$

Now, since for an elliptical section shown in Figure 8.11

$I_y=\dfrac{\pi ba^3}{4}$, $I_x=\dfrac{\pi ab^3}{4}$ and $A=\pi ab$, we have $T=\dfrac{G\theta\pi a^3b^3}{a^2+b^2}$.

We may now write this in our usual form of torsion equation as

$$\frac{T}{J'}=G\theta,$$

where J', the effective polar moment of inertia is given as $J'=\dfrac{\pi a^3b^3}{a^2+b^2}$,

whereas the true polar moment of inertia for an ellipse, shown in Figure 8.11, is given by

$$J=\frac{\pi ab}{4}\left(a^2+b^2\right).$$

If $a=b$, the ellipse reduces to a circle and the polar moment of inertia reduces to that of a circle

$$J=\frac{\pi a^4}{2} \text{ and } \frac{T}{J}=G\theta.$$

It is now interesting to investigate why the simple formula for torsion $\frac{T}{J}=\frac{\tau}{a}=G\theta$ works well for members with a circular cross-section.

In order to do this, we look back at the shear stress τ_{yz} in terms of warping and stress functions in equations (8.3.4) and (8.4.2). This is given by

$$\tau_{yz}=G\theta\left(\frac{\partial\psi}{\partial y}+x\right)=-\frac{\partial\phi}{\partial x}.$$

This gives $\frac{\partial\psi}{\partial y}=-\left(\frac{1}{G\theta}\frac{\partial\phi}{\partial x}\right)-x.$

Substituting the stress function for an elliptical section, we have

$$\frac{\partial\psi}{\partial y}=-\frac{1}{G\theta}\frac{\partial}{\partial x}\left[-\frac{G\theta a^2b^2}{a^2+b^2}\left(\frac{x^2}{a^2}+\frac{y^2}{b^2}-1\right)\right]-x.$$

This reduces to $\frac{\partial\psi}{\partial y}=\frac{a^2b^2}{a^2+b^2}\cdot\frac{2x}{a^2}-x.$

On integration $\psi=\frac{2b^2xy}{a^2+b^2}-xy+C.$

Since $\psi=0$ at the axis of rotation, $C=0$.

We then have $\psi=\left(\frac{b^2-a^2}{a^2+b^2}\right)xy.$

This means that when $a=b$, i.e., for circular sections $\psi=0$.

This shows that the simple torsion formula derived for members with a circular section without warping is valid.

This may be further clarified by solving the torsion problem for a circular cross-section using the stress function approach.

Since the boundary of a circle is given by $x^2+y^2=a^2$, a being the radius, the stress function may be of the following form with a boundary condition of $\phi=0$.

$$\phi=m\left(x^2+y^2-a^2\right).$$

Substituting this in the governing equation

$$\frac{\partial^2\phi}{\partial x^2}+\frac{\partial^2\phi}{\partial y^2}=-2G\theta.$$

We have $m=-\frac{G\theta}{2}$ and $\phi=-\frac{G\theta}{2}\left(x^2+y^2-a^2\right).$

Substituting ϕ in equation (8.4.5), we have the required torque T as

$$T = 2\iint -\frac{G\theta}{2}\left(x^2 + y^2 - a^2\right)dxdy = -G\theta\left[I_y + I_x - a^2 A\right]$$

$$= -G\theta\left[\frac{\pi a^4}{2} - \pi a^4\right] = G\theta\left[\frac{\pi a^4}{2}\right] = G\theta J$$

and since the polar moment of inertia for a circular section of radius a is given by

$$J = \frac{\pi a^4}{2},$$

we have $T = G\theta J$, i.e., $\frac{T}{J} = G\theta$.

If we now write the two nonzero shear stresses τ_{xz} and τ_{yz} in terms of stress functions, we have

$$\tau_{xz} = \frac{\partial \phi}{\partial y} = -G\theta y, \qquad \tau_{yz} = -\frac{\partial \phi}{\partial x} = \frac{G\theta}{2}.2x = G\theta x.$$

Therefore, the resultant shear stress τ is given as

$$\tau = G\theta\sqrt{x^2 + y^2} = G\theta a.$$

Combining this with the torque equation, we have

$$\frac{T}{J} = \frac{\tau}{a} = G\theta,$$

which is our familiar torque formula for members of a circular cross-section.

8.5 PRANDTL'S MEMBRANE ANALOGY [LO5]

St. Venant in 1855 first showed the correct approach for solving the torsional problems of members with noncircular cross-sections, such as rectangular, elliptical, and triangular. However, there remains a large variety of noncircular sections that are used in common engineering practice. Mathematical approaches to torsional problems for channels or I-sections are difficult analytically. With the progress in present-day computational techniques, any such sections can be treated easily. Prandtl's membrane analogy was proposed in 1903, when no such computational technique was known, and the analogy technique was then considered to be a simple and novel technique to solve the torsional problem for almost any cross-section.

The development of this thin membrane analogy technique evolved with the observation that there is an almost exact similarity between the harmonic torsion equation (8.4.3) and the deflection equation of a stretched membrane.

We now consider a thin elastic membrane of an arbitrary shape initially stretched flat with a rigid boundary, such that the tension s per unit length in the membrane is large and equal in all directions. The membrane is then blown up at one end with a small pressure p. This is shown in Figure 8.12. The significance of large tension s and small pressure p is that the large tension in the membrane remains almost unaffected even after blowing up. The rigid boundary indicates

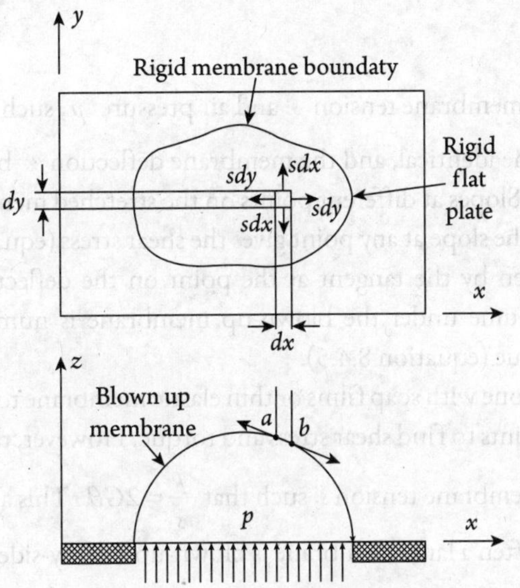

Figure 8.12 A flat membrane with a rigid boundary and large tension *s* blown up by a small pressure *p*. The arbitrary shape of the membrane is cut in a rigid flat plate

zero deflection at the boundary, and deflection of the rest of the membrane is small due to small but finite pressure. Now we consider a small element of sides dx and dy in the x–y plane. The forces on the two sides of the element are sdy and sdx in the x–y plane. There is also another vertical force $pdxdy$ acting on the element in the z-direction.

Equilibrium in the x- and y-directions is automatically satisfied. For equilibrium in the z-direction, we consider the component of sdx and sdy at faces a and b, and sum them up to give the deflection equation of the blown-up membrane. The z component of sdy on face a is $sdy\left(\frac{\partial z}{\partial x}\right)$ downward. The z component of sdy on face b would have an increment due to a change in slope, and this is given by

$$sdy\left(\frac{\partial z}{\partial x}+\frac{\partial}{\partial x}\frac{\partial z}{\partial x}dx\right)=s\frac{\partial z}{\partial x}\;dy+s\frac{\partial^2 z}{\partial x^2}dxdy.$$

Net upward tension on membrane at a and b is thus $s\frac{\partial^2 z}{\partial x^2}dxdy$. This is proportional to $\frac{\partial^2 z}{\partial x^2}$, which is the curvature for the small slope. Similarly, we can find the net upward tension on the membrane resulting from the two dx faces to be $s\frac{\partial^2 z}{\partial y^2}dxdy$. These two tensions are balanced by the air pressure given by $p\,dxdy$. This gives

$$\frac{\partial^2 z}{\partial x^2}+\frac{\partial^2 z}{\partial y^2}=-\frac{p}{s}. \tag{8.5.1}$$

The equation is very similar to the harmonic torsion equation (8.4.3) in terms of stress function ϕ reproduced here for clarity

$$\frac{\partial^2\phi}{\partial x^2}+\frac{\partial^2\phi}{\partial y^2}=-2G\theta.$$

If we now adjust the membrane tension s and air pressure p, such that $\frac{p}{s}=2G\theta$, equations (8.5.1) and (8.4.3) become identical, and the membrane deflection z becomes numerically equal to the stress function ϕ. Slopes at different points on the stretched membrane are measured using a traveling microscope. The slope at any point gives the shear stress (equation 8.4.2). The direction of the shear stress is given by the tangent at the point on the deflection contour as shown in Figure 8.13, and the volume under the blown-up membrane is numerically equal to half the twisting moment or torque (equation 8.4.5).

Experiments can be done with soap films or thin elastic membrane to determine the deflections and slopes at different points to find shear stress and torque. However, this requires an adjustment of fluid pressure p and membrane tension s, such that $\frac{p}{s}=2G\theta$. This adjustment is cumbersome in reality. To avoid this, often a large membrane is blown-up side-by-side, one for the desired cross-section and the other one is a circular section, as shown in Figure 8.14. Since $\frac{p}{s}$ is the same for both cases, $2G\theta$ is also the same.

We may now measure the volume under the bulged portion in both cases, as well as the slopes. Then, we may write

$$\frac{v}{\alpha}=k\frac{T}{\tau}. \tag{8.5.2}$$

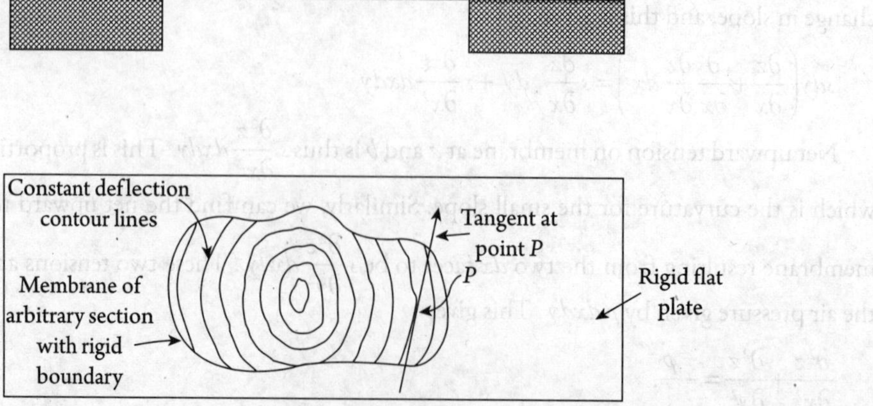

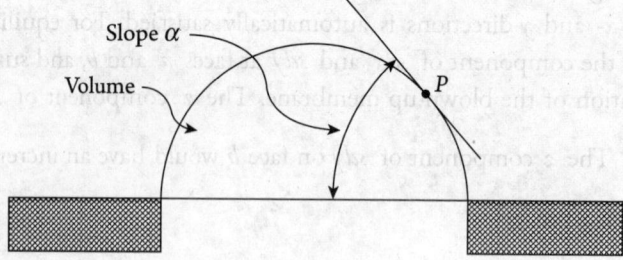

Figure 8.13 Slope and volume of a bulged membrane give the shear stress and twisting moment respectively

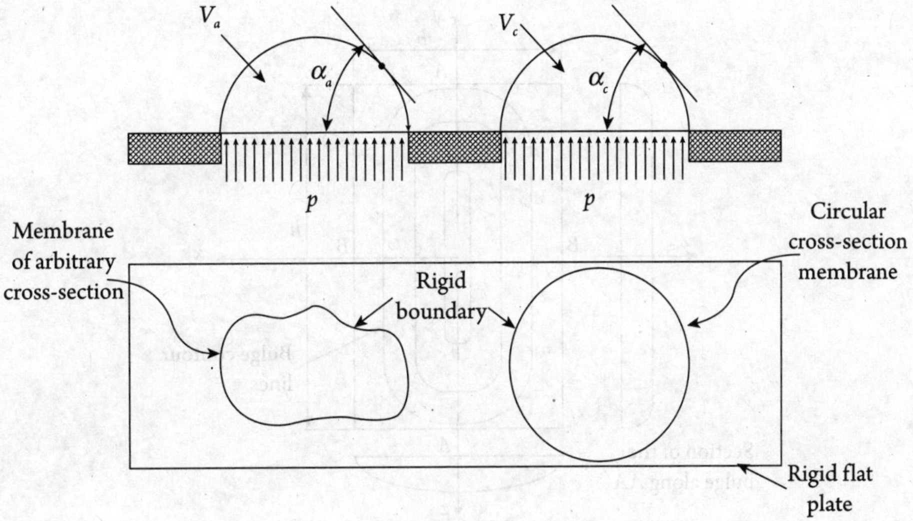

Figure 8.14 A large membrane blown-up side-by-side, one for an arbitrary cross-section and another for a circular cross-section. Here α_a and α_c are slopes for the arbitrary cross-section and circular cross-section respectively

where v is the volume under the bulged portion, α is the slope, T is the torque, and τ is the shear stress. Here k is the constant of proportionality. It is not difficult to find the constant k for circle, and since the constant being the same for all cross-sections, we may use this for any arbitrary section. This gives us the torque–stress relation for any cross-section with a known angle of twist.

We may also compare the ratios of the volumes under the blown-up membrane for the two cases with the torques per unit angle of twist as follows:

$$\frac{V_a}{V_c} = \frac{T_a/\theta}{T_c/\theta}, \tag{8.5.3}$$

where V_a and V_c are volumes under the blown-up membrane of the arbitrary and circular sections respectively and T_a and T_c are the torques for the arbitrary and circular sections respectively.

We shall now take up a simple case of a narrow rectangular cross-section of height h and width b to demonstrate the use of the membrane analogy technique. This is shown in Figure 8.15. Due to symmetry, there is no slope of the membrane bulge along the y-direction except at the edges. We consider the slope of the bulge in the $x-z$ plane. If we consider the equilibrium of a small portion of the membrane of length l along the y-axis and width x along the x-axis, the upward force due to fluid pressure p would balance the component of the membrane tension force in the z-direction, and we may write

$$-sl\frac{dz}{dx} = pxl.$$

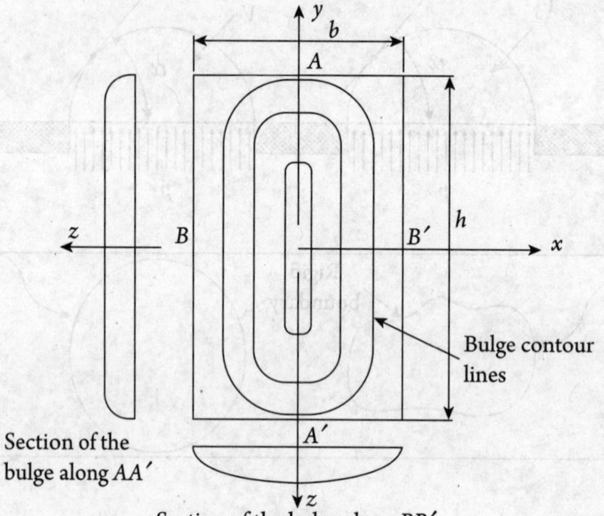

Figure 8.15 Membrane bulge and contour lines for a rectangular cross-section

We may now write this as $\frac{dz}{dx} = -\frac{px}{s}$, and on integration, we get $z = -\frac{px^2}{2s} + Constant.$

Here the boundary condition is $z = 0$ at $x = \pm b/2$.

This gives $z = \frac{p}{2s}\left(\frac{b^2}{4} - x^2\right)$,

which indicates that the membrane bulge shape is parabolic.

This also gives the maximum shear stress as

$$\tau_{max} = \left(\frac{dz}{dx}\right)_{max} = \left(-\frac{px}{s}\right)_{x=\pm b/2} = \frac{pb}{2s}.$$

Since the membrane slope indicates shear stress when $\frac{p}{s} = 2G\theta$, we have

$$\tau_{max} = G\theta b. \tag{8.5.4}$$

Now we consider the volume of the bulge portion along section BB', which is given in this case as a parabola. This gives

$$V = \frac{2}{3} bhz_{max}.$$

z_{max} occurs at $x = 0$ and this is given by the second equation in this derivation as

$$\left(z_{max}\right)_{x=0} = \frac{p}{2s}\frac{b^2}{4}.$$

We then have $V = \frac{1}{12} b^3 h \frac{p}{s}$.

This means for the shaft under torsion

$$\frac{T}{\theta} = Gh\frac{b^3}{3}. \tag{8.5.5}$$

Combining equations (8.5.4) and (8.5.5), we have in general

$$\tau_{max} = \frac{3T}{hb^2}. \tag{8.5.6}$$

Equation (8.5.6) is true only when $h \gg b$, but for different ratios of h/b a numerical factor can be introduced, and we may write equations (8.5.5) and (8.5.6) in the following forms:

$$T = C_1 G\theta hb^3 \text{ and } \tau_{max} = \frac{T}{C_2 hb^2},$$

where the constants C_1 and C_2 are given in the Table 8.1.

Table 8.1 Values of C_1 and C_2

h/b	1.0	1.5	2.0	2.5	3.0	4.0	5.0	10.0
C_1	0.1406	0.196	0.229	0.249	0.263	0.281	0.291	0.312
C_2	0.208	0.231	0.246	0.258	0.267	0.282	0.291	0.312

8.6 TORSION OF HOLLOW SECTIONS [LO6]

Solid cross-sections, such as solid circular, rectangular, or triangular sections, are mathematically described as singly connected cross-sections, whereas hollow sections are described as multiply connected (Figure 8.16). In solving such problems, St. Venant's theory can still be applied. This means that the deformation due to torsion may be considered to consist of (a) rotation and (b) warping. In the stress function approach, the governing equation is

$$\nabla^2 \phi = -2G\theta$$

and the boundary condition is $\phi = constant$ along the boundary, where ϕ is the stress function and θ is the angle of twist per unit length.

For singly connected cross-sections, the constant can be chosen arbitrarily, and it is taken as zero. For multiply connected cross-sections, the constant cannot be chosen arbitrarily because the stress function along each boundary is also constant but different.

Let us start our discussion with a simple problem of a hollow circular shaft with the inner boundary coinciding with a stress line of a solid shaft with the same outside boundary. Stress lines in this case are concentric circles as shown in Figure 8.17.

$$x^2 + y^2 = (\lambda a)^2. \tag{8.6.1}$$

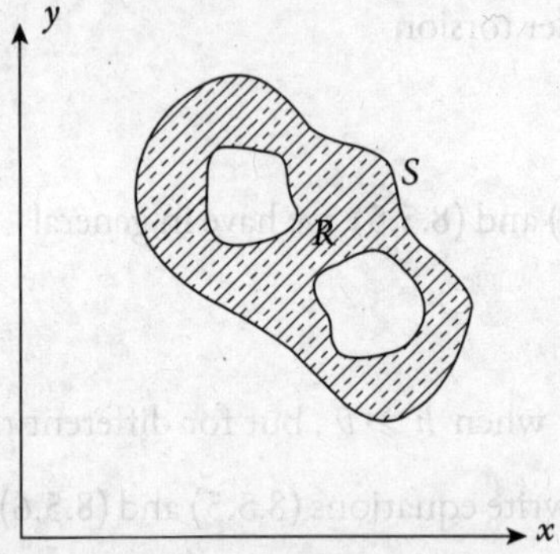

Figure 8.16 A typical multiply connected section. Here R is the closed bounded region and S is the boundary consisting of many smooth curves

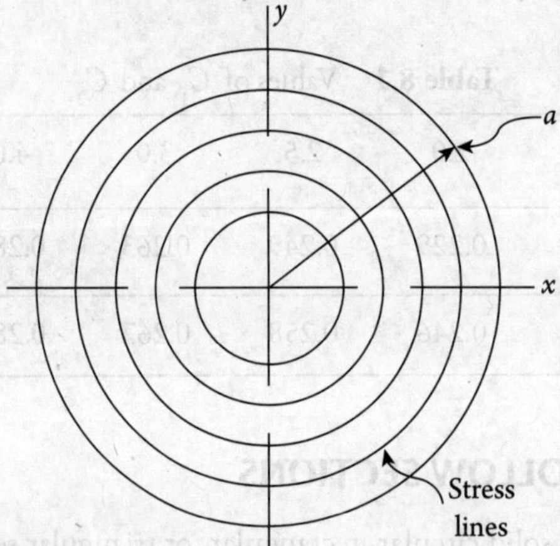

Figure 8.17 Stress lines in a solid circular cross-section

It is a curve that is geometrically similar to the outer boundary, and therefore, curves with any value of $\lambda < 1$ represent the stress lines for the solid section.

Considering the shear stress at any point, the stress lines are tangent to the line and there will be no shear stress across the cylinder surface. This means that if we remove the material along any stress line, the stress distribution of the remaining hollow shaft remains unaltered, but the torque transmitted would vary. Following examples in Section 8.4, we take the stress function of a hollow shaft as:

$\phi = m\left(x^2 + y^2 - \lambda^2 a^2\right).$

Satisfying the governing equation $\nabla^2\phi = -2G\theta$.

The stress function reduces to

$$\phi = -\frac{G\theta}{2}\left(x^2 + y^2 - \lambda^2 a^2\right). \tag{8.6.2}$$

Torque is given by equation (8.4.5) as

$$T = 2\iint \phi \, dx \, dy.$$

Substituting ϕ, we have

$$T = \frac{G\theta\pi a^4}{2}\left(1-\lambda^4\right). \tag{8.6.3}$$

We now consider the two nonzero shear stresses in the section τ_{yz} and τ_{xz} in terms of stress function.

$$\tau_{yz} = -\frac{\partial\phi}{\partial x} = G\theta x$$

$$\tau_{xz} = \frac{\partial\phi}{\partial y} = -G\theta y.$$

For a given angle of twist θ per unit length, the shear stress is the same as that in the corresponding solid shaft but the torque is less as shown above.

The maximum shear stress is thus

$$\tau_{max} = G\theta a. \tag{8.6.4}$$

Combining equations (8.6.3) and (8.6.4), we have

$$\tau_{max} = \frac{2T}{\pi a^3\left(1-\lambda^4\right)}. \tag{8.6.5}$$

$$\theta = \frac{2T}{G\pi a^4\left(1-\lambda^4\right)}. \tag{8.6.6}$$

For $\lambda = 0$, the expressions reduce to familiar formulae for solid cross-sections as discussed in Section 8.4.1, and for hollow cylinders (Figure 8.18) we may choose suitable values for λ.

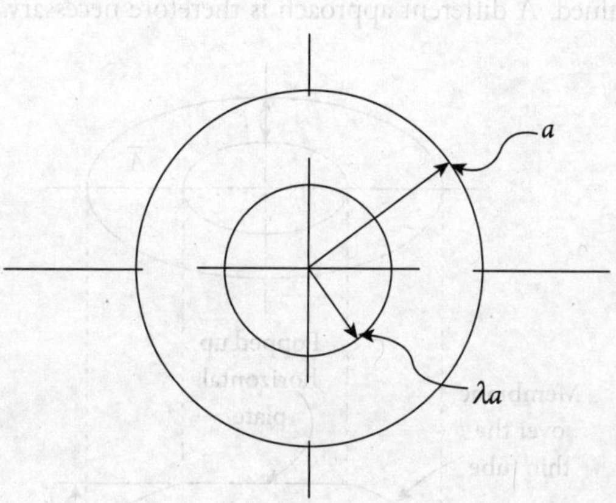

Figure 8.18 A concentric circle coinciding with a stress line removed from the solid section to generate a hollow section

In the membrane analogy, the central portion of the cross-section of a hollow shaft is replaced by a flat horizontal plate as shown in Figure 8.19. The uniform pressure distributed over the plate is balanced by the tensile force in the membrane along the edge of the plate. The analysis of membrane analogy with the measurement of slope and volume would apply here too.

Some approximate analysis of torsion of thin tubes may be effectively carried out by membrane analogy. Consider the cross-section in Figure 8.19, where a membrane or soap film is applied over the cross-section and air is blown from beneath. The middle plate pops up and it is maintained horizontally by some devices. Torque transmitted T may be given by the average volume enclosed by the membrane and the plate. Referring to Figure 8.19, we may write

$$T = 2\bar{A}h.$$

where $\bar{A}$ is the mean of the areas enclosed by the outer and inner boundaries and h is the difference in levels between the two boundaries.

For this tube, we assume that the membrane is straight over the tube thickness, which indicates that the shear stress is uniformly distributed over the thickness. Referring to Figure 8.19, we may write

$$\tau = \frac{h}{t}.$$

Therefore, we may write

$$T = 2\bar{A}t\tau \text{ and } \tau = \frac{T}{2\bar{A}t}.$$

It must be remembered that $t\tau$ is constant along the wall thickness of the cross-section and this is known as shear flow. Shear stress is maximum at the point where the thickness t is minimum. A more detailed discussion is given in Section 8.7.

Let us now consider a general case of a multiply connected section. The different constants of the stress function along the different boundaries must be determined in such a manner that the displacements are single valued. A different approach is therefore necessary. Again, as in Figure 8.12,

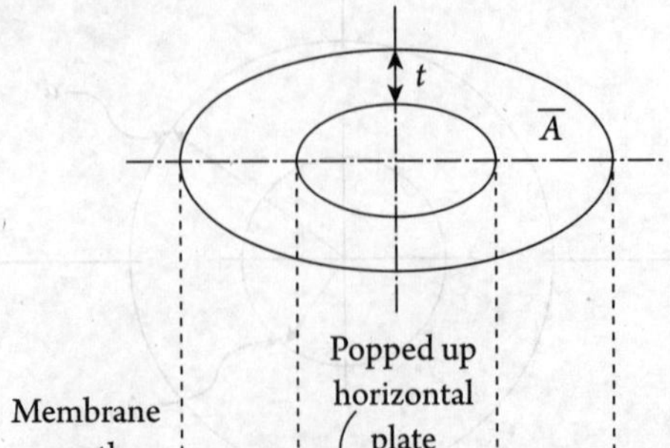

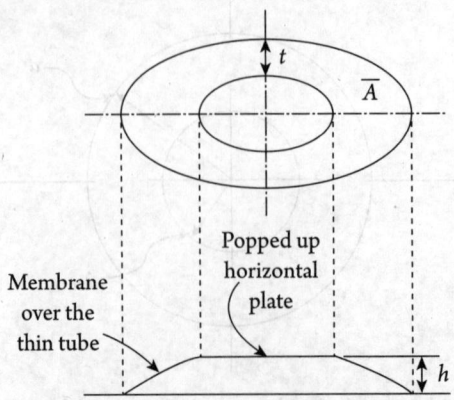

Figure 8.19 Membrane analogy for a thin tube

let s measure the distance along a boundary of the cross-section and the resultant shear stress at the boundary τ_{sz} can be obtained by resolving the nonzero shear stresses along the boundary.

$$\tau_{sz} = \tau_{zy} \cos\gamma - \tau_{zx} \sin\gamma.$$

We now substitute τ_{zy} and τ_{zx} in terms of the warping function ψ from equation (8.3.4) and put $\sin\gamma = -\frac{dx}{ds}$ and $\cos\gamma = \frac{dy}{ds}$ from Figure 8.9 and we have

$$\tau_{sz} = G\theta\left(\frac{\partial\psi}{\partial y} + x\right)\frac{dy}{ds} + G\theta\left(\frac{\partial\psi}{\partial x} - y\right)\frac{dx}{ds}.$$

Since displacement ω is given by the warping function as in equation (8.3.2),

$\omega = \theta\psi(x, y).$

θ being the angle of twist, we have

$$\tau_{sz} = G\left(\frac{\partial\omega}{\partial y} + \theta x\right)\frac{dy}{ds} + G\left(\frac{\partial\omega}{\partial x} - \theta y\right)\frac{dx}{ds}.$$

This may also be written as

$$\tau_{sz}\, ds = G\left(\frac{\partial\omega}{\partial y}dy + \frac{\partial\omega}{\partial x}dx\right) + G\theta(xdy - ydx).$$

The integral of this stress around the boundary is

$$\int \tau_{sz}\, ds = G\oint\left(\frac{\partial\omega}{\partial y}dy + \frac{\partial\omega}{\partial x}dx\right) + G\theta\oint(xdy - ydx).$$

Since the boundary is closed and the displacement ω is continuous,

$\oint\left(\frac{\partial\omega}{\partial y}dy + \frac{\partial\omega}{\partial x}dx\right) = \oint d\omega = 0$, and applying Green's theorem

$$\oint(xdy - ydx) = \iint\left[\frac{\partial}{\partial x}(x) + \frac{\partial}{\partial y}(y)\right]dxdy = 2\iint dxdy = 2A.$$

Here A is the area enclosed by the boundary (hatched portion in Figure 8.16).

This gives $\int \tau ds = 2G\theta A$. This indicates that in the membrane the vertical load on a plate such as the one shown in Figure 8.19 is balanced by the vertical component of the resultant of the tensile forces produced by the membrane.

If we now rewrite τ_{sz} in terms of stress function ϕ, and substitute $\sin\gamma = \frac{dy}{dn}$ and $\cos\gamma = \frac{dx}{dn}$ from Figure 8.20, we have

$$\tau_{sz} = \tau_{zy} \cos\gamma - \tau_{zx} \sin\gamma = -\frac{\partial\phi}{\partial n}.$$

And the line integral of this function must be $2G\theta A$.

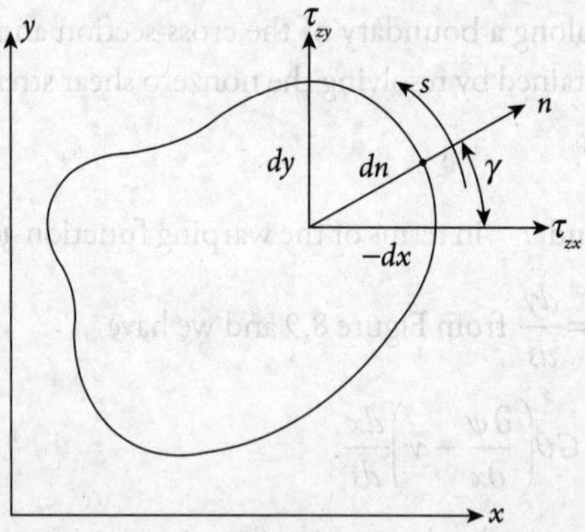

Figure 8.20 Resultant shear stress at the boundary

$$\oint\left(-\frac{\partial\phi}{\partial n}\right)ds = 2G\theta A. \tag{8.6.7}$$

In order to satisfy the condition in equation (8.6.7) for each boundary, a suitable constant value of ϕ must be assigned for each boundary. Analytical solutions for complicated cases are difficult and numerical method must be adopted there. But for simple cases, we may adopt the following procedure:

Let us consider i smooth curves within a bounded region. The governing equation $\nabla^2\phi = -2G\theta$ is solved i times with different boundary conditions. For example, let us consider in the first solution, $\phi = 0$ holds for all the boundaries except the first one where we take $\phi = 1$, and in the second solution, $\phi = 0$ holds for all the boundaries except the second one where we take $\phi = 1$, and so on.

For the last solution, $\phi = 0$ holds for all boundaries.

If the stress functions obtained from the above solutions are $\phi_1(x,y)$, $\phi_2(x,y)$, $\phi_3(x,y)$, etc. and $\phi_0(x,y)$ for the last solution, then we may write the stress functions for the different boundaries as $C_1(\phi_1 - \phi_0)$, $C_2(\phi_2 - \phi_0)$, and so on, then the required solution is

$$\phi = \phi_0 + C_1(\phi_1 - \phi_0) + C_2(\phi_2 - \phi_0) + \ldots\ldots\ldots\ldots\ldots + C_{i-1}(\phi_{i-1} - \phi_0).$$

Therefore, to satisfy the boundary condition, we have

$$\oint\left[\frac{\partial\phi_0}{\partial n} + C_1\left(\frac{\partial\phi_1}{\partial n} - \frac{\partial\phi_0}{\partial n}\right) + C_2\left(\frac{\partial\phi_2}{\partial n} - \frac{\partial\phi_0}{\partial n}\right) + \ldots + C_{i-1}\left(\frac{\partial\phi_{i-1}}{\partial n} - \frac{\partial\phi_0}{\partial n}\right)\right]ds = 2G\theta A_1, \tag{8.6.8}$$

where A_1 is the area of the first boundary.

Similarly, other $(i-2)$ equations can be written and C_1, C_2, C_3 may be obtained. Figure 8.21 shows a typical multiply connected section with different boundary constants.

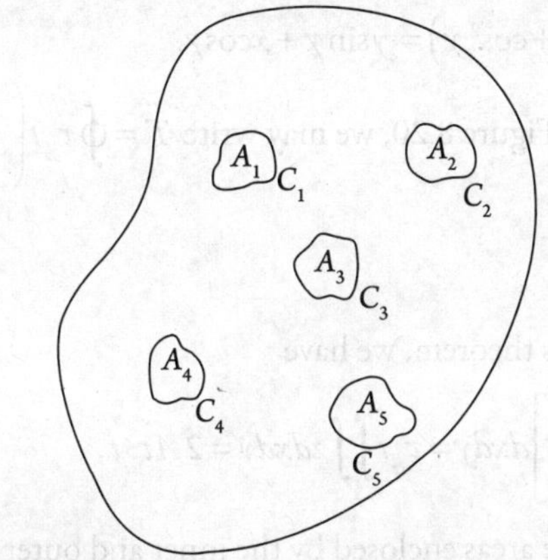

Figure 8.21 Multiply connected sections with different boundary constants

8.7 THIN-WALLED HOLLOW SECTIONS [LO7]

There are many practical applications of thin-walled hollow sections, such as closed tubes, box girders in building construction, and aerospace construction. The problem may be approached mathematically, following the methods discussed in the previous section. Consider a thin-walled section, as shown in Figure 8.22, with two boundary constant values of stress function, say C_1 and C_2, and wall thickness t. Since the wall thickness is small, shear stress may be treated as constant across the wall thickness. We may then write

$$\tau_{sz} = -\frac{\partial \phi}{\partial n} = \frac{C_2 - C_1}{t}.$$

Shear force on a small element ds on the boundary is $\tau_{sz}\, t ds$. Integrating the moment of this constant shear force about an arbitrary point O, r distance away, we get the torque T as follows:

$$T = \oint \tau_{sz}\, t\left(y \sin\gamma + x\cos\gamma\right) ds.$$

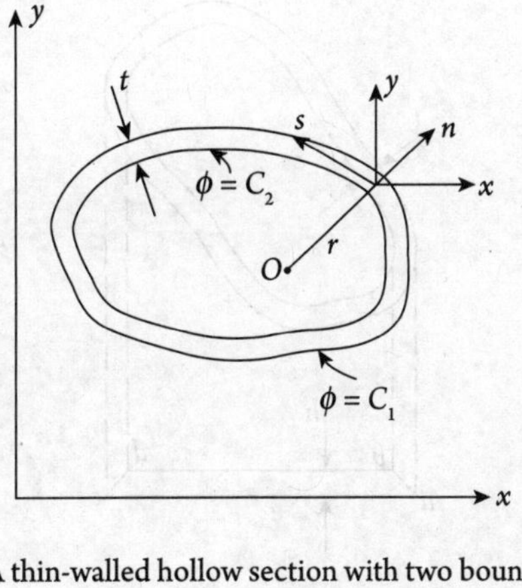

Figure 8.22 A thin-walled hollow section with two boundary constants

Here we write $r = r\left(\sin^2\gamma + \cos^2\gamma\right) = y\sin\gamma + x\cos\gamma$.

Following the details from Figure 8.20, we may write $T = \oint \tau_{sz} t\left(y\left(-\frac{dx}{ds}\right)ds + x\left(\frac{dy}{ds}\right)ds\right)$,

which reduces to

$$T = \oint \tau_{sz} t\left(xdy - ydx\right).$$

And then, applying Green's theorem, we have

$$T = \tau_{sz} t\iint\left[\frac{\partial}{\partial x}(x) + \frac{\partial}{\partial y}(y)\right]dxdy = \tau_{sz} t\iint zdxdy = 2\tilde{A}\tau_{sz} t,$$

where $\tilde{A}$ is the mean of the areas enclosed by the inner and outer boundaries. This gives

$$\tau_{sz} = \frac{T}{2\tilde{A}t}. \tag{8.7.1}$$

Again, we have, from equation (8.6.7), $\oint \tau_{sz} ds = 2G\theta\tilde{A}$.

We may write this as $\frac{T}{2\tilde{A}}\oint\frac{ds}{t} = \frac{Ts}{2\tilde{A}t}$,

where t is the constant thickness and s is the length of the perimeter.

From the above two equations, we have

$$\theta = \frac{Ts}{4\tilde{A}^2 Gt}. \tag{8.7.2}$$

We now look at the membrane analogy of a thin-walled hollow section. An arbitrary section with a constant wall thickness t with a membrane is shown in Figure 8.23. It is important to note that the central plate must be maintained in a horizontal position under the fluid pressure p and let the difference in height between the two boundaries be h. We assume that the membrane slope within the wall thickness remains constant, which means that the shear stress is also constant across the thickness. We may therefore write

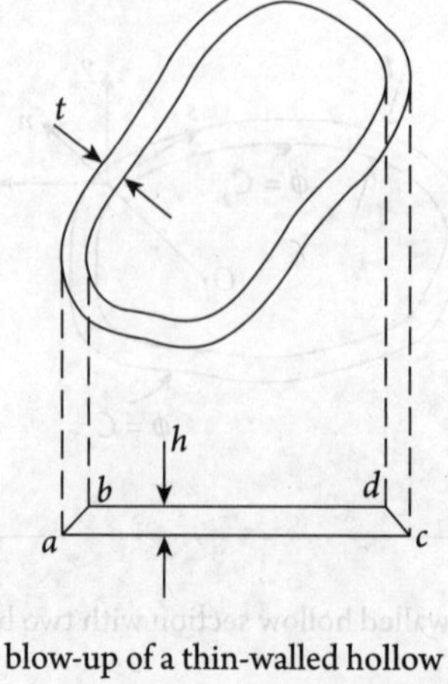

Figure 8.23 Membrane blow-up of a thin-walled hollow section with a central plate

$$\tau = {}^{h}\!/_{t}.$$

Here bd and ac are the inner and outer boundaries of the membrane respectively and therefore the blown-up volume is given by $2\tilde{A}h$, where $\tilde{A}$ is the mean of the areas enclosed by inner and outer boundaries. We therefore have

$$T = 2\tilde{A}h = 2\tilde{A}t\tau.$$

From this, we have a simple expression for the shear stress as

$$\tau = \frac{T}{2\tilde{A}t}. \tag{8.7.3}$$

Again since $\oint \tau ds = 2G\theta\tilde{A}$.

Substituting τ from equation (8.7.3), we have

$$\theta = \frac{T}{4\tilde{A}^2 G}\oint \frac{ds}{t}.$$

For constant wall thickness t, we have

$$\theta = \frac{Ts}{4\tilde{A}^2 Gt}. \tag{8.7.4}$$

This shows that the same expressions are obtained analytically and by membrane analogy.

Example 8.1

Find the maximum shear stress and torque for a given angle of twist θ for a member of a cross-section of the shape of an equilateral triangle.

Solution:

For torsional problems, the governing equation is $\nabla^2\phi = -2G\theta$, and the boundary condition is $\phi = 0$ around the boundary.

This requires finding the appropriate stress function. Stress functions for different cross-sections may be represented by polynomials of different degrees.

For convenience, let us consider the equilateral triangle as shown in the figure.

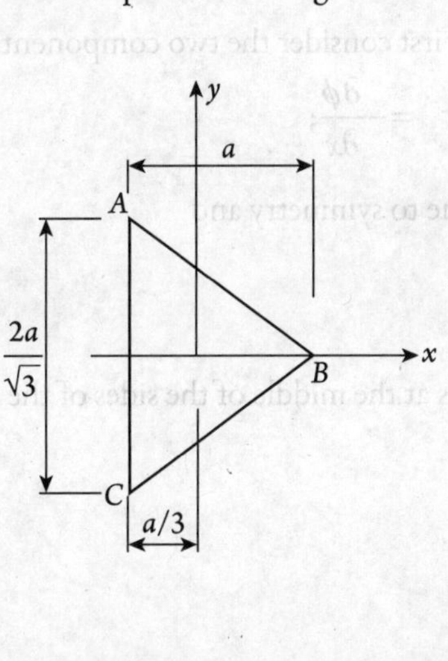

Equation of side AB

$$\frac{1}{\sqrt{3}}\left(\frac{2}{3}a-x\right)=y,\ \text{i.e.,}\ x+\sqrt{3}y-\frac{2}{3}a=0 \qquad \text{...(a)}$$

Equation of side BC

$$\frac{1}{\sqrt{3}}\left(\frac{2}{3}a-x\right)=-y,\ \text{i.e.,}\ x-\sqrt{3}y-\frac{2}{3}a=0 \qquad \text{...(b)}$$

Equation of side AC

$$x=-\frac{a}{3},\ \text{i.e.,}\ x+\frac{a}{3}=0 \qquad \text{...(c)}$$

The boundary of the cross-section is given by the product of equations (a), (b), and (c) as

$$\left(x^2+y^2\right)-\frac{1}{a}\left(x^3-3xy^2\right)-\frac{4}{27}a^2=0$$

We may now set the stress function as

$$\phi=m\left[\left(x^2+y^2\right)-\frac{1}{a}\left(x^3-3xy^2\right)-\frac{4}{27}a^2\right],$$

and we find the constant m by satisfying the governing equation.

This gives $\frac{\partial^2\phi}{\partial x^2}+\frac{\partial^2\phi}{\partial y^2}=m\left[2-\frac{6}{a}x\right]+m\left[2+\frac{6}{a}x\right]=-2G\theta.$

We then have $m=-G\theta/2.$

Substituting this in equation (8.4.5), we get the torque T as

$$T=2\iint\phi\,dxdy=-G\theta\int_{x=-\frac{a}{3}}^{\frac{2a}{3}}\int_{y=-\left(\frac{2a}{3}-x\right)/\sqrt{3}}^{\left(\frac{2a}{3}-x\right)/\sqrt{3}}\left\{\left(x^2+y^2\right)-\frac{1}{a}\left(x^3-3xy^2\right)-\frac{4}{27}a^2\right\}dxdy.$$

Solving $T=\frac{G\theta}{15\sqrt{3}}a^4.$

To find the shear stress, we first consider the two components τ_{xz} and τ_{yz}.

Shear stress, $\tau_{xz}=\frac{\partial\phi}{\partial y};\qquad \tau_{yz}=-\frac{\partial\phi}{\partial x};$

Along the x-axis; $\tau_{xz}=0,$ due to symmetry and

$$\tau_{yz}=\frac{G\theta}{2}\left[2x-\frac{3x^2}{a}\right].$$

Since the largest stress occurs at the middle of the sides of the triangle, we have

$$\tau_{max}=\frac{1}{2}G\theta a.$$

Example 8.2

An I section with the dimensions, as shown in the figure, is subjected to a torque of 450 Nm. Find the maximum shear stress and twist per unit length.

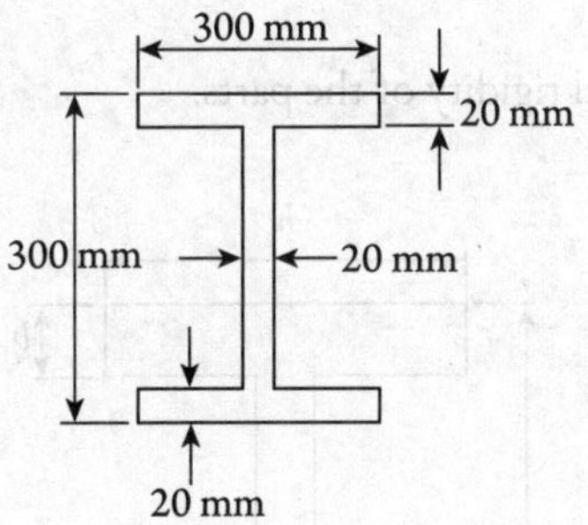

Solution:

It is important to note that equations (8.5.5) and (8.5.6) work equally well for sections that can be built up of rectangular sections, as shown below, because the membrane will not change its shape except for the local effects at the bends. Here, in applying equations (8.5.5) and (8.5.6), the total length of the sections must be taken as the total aggregate length of the wall. For example, for the *L*-section, h is equal to lengths $ac+gf$ (as shown in the figure).

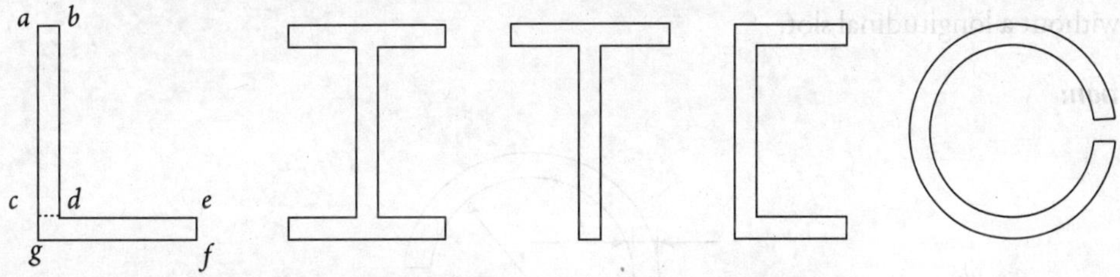

Another important point to note is that if the thicknesses of the two rectangles in the *L*-section, for example, are different, the above equation will still apply except that we need to find torques for the two sections separately and then add up the torques.

With these comments, we return to the main problem, where:

$h = 300 + 300 + 260 = 860\,\text{mm}$ and $b = 20\,\text{mm}$.

We then have

$$\tau_{max} = \frac{3T}{hb^2} = \frac{3 \times 450}{0.86 \times (0.02)^2} = 3.92\,\text{MPa}.$$

And the angle of twist θ is given by

$$\theta = \frac{3T}{Ghb^3} = \frac{3 \times 450}{200 \times 10^9 \times 0.86 \times (0.02)^3} = 0.001\,\text{radian/m}.$$

In the present problem, where the torque is specified, we find τ_{max} as follows for the section shown:

$$\tau_{max} = \frac{3T}{\left(h_1 - 2b_1\right)b_2^{\,2} + 2h_2 b_1^{\,2}}.$$

Here we sum up the torsional rigidity of the parts.

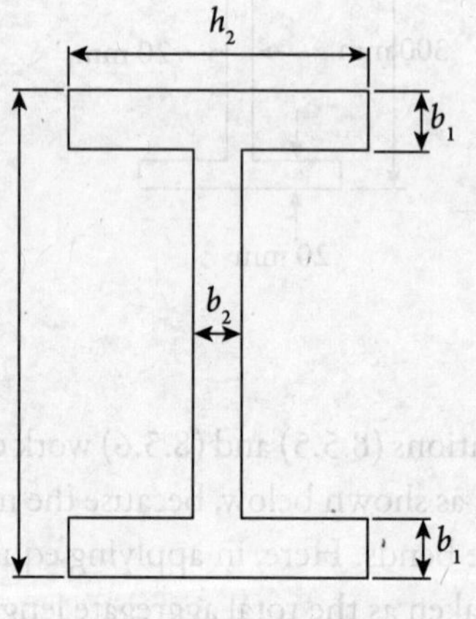

Example 8.3

Compare the shear stress and torsional rigidity of a thin-walled tube of circular cross-section with and without a longitudinal slot.

Solution:

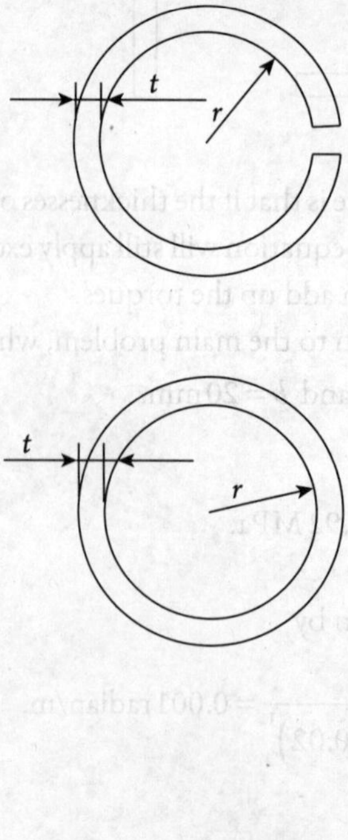

First, let us consider the circular section with a slot.

As discussed in Example 8.2, this case may be treated with equations developed for a narrow rectangular cross-section as long as the total length is taken suitably. From equation (8.5.6), we have

$$\tau_{max} = \frac{3T}{hb^2} = \frac{3T}{2\pi r t^2}. \qquad \text{(a)}$$

For the circular section without a slot, we need to consider the equations developed for thin-walled hollow sections. Here, we use equation (8.7.1)

$$\tau_{max} = \frac{T}{2\tilde{A}t} = \frac{T}{2\pi r^2 t}. \qquad \text{(b)}$$

Thus, the ratio of stresses for sections with and without a slot is $\frac{3r}{t}$.

Now we use equations (8.5.5) and (8.7.2) to find $\frac{T}{\theta}$.

For sections with a slot $\frac{T}{\theta} = \frac{Ghb^3}{3} = \frac{G2\pi r t^3}{3}$.

For sections without a slot $\frac{T}{\theta} = \frac{4\tilde{A}^2 Gt}{s} = \frac{4\pi^2 r^4 Gt}{2\pi r} = 2\pi r^3 Gt$.

Thus, the ratio of torsional rigidity for sections with and without a slot is: $\frac{t^2}{3r^2}$.

Example 8.4

Find the rotation at the end e. Take shear modulus $G = 80\,\text{GPa}$, length ac = length ce = 0.1 m.

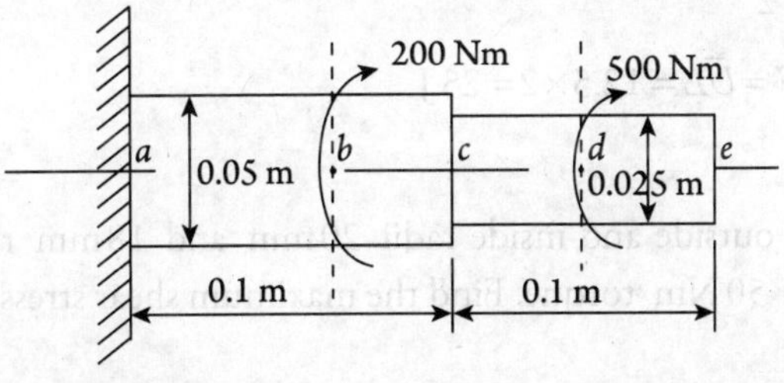

Solution:

The shafts are connected in series, hence,

$$T_2 = T_{ce} = 500\,\text{Nm}, \qquad T_1 = T_{ce} + T_{ac} = 200 + 500 = 700\,\text{Nm}.$$

$$L_{ac} = 0.1\,\text{m}, \qquad L_{ce} = 0.1\,\text{m}, \qquad d_{ac} = 0.05\,\text{m}, \qquad d_{ce} = 0.025\,\text{m}.$$

$$G_1 = G_2 = G = 80\,\text{GPa}.$$

The polar moment of inertia, $J_1 = \frac{\pi}{32} d_1^4 = \frac{\pi}{32} 0.05^4$, $J_2 = \frac{\pi}{32} d_2^4 = \frac{\pi}{32} 0.025^4$.

$$\theta_{ae} = \sum_{i=1}^{2} \frac{T_i L_i}{G_i J_i} = \frac{T_1 L_1}{G_1 J_1} + \frac{T_2 L_2}{G_2 J_2} = 0.01426 + 0.01629 = 0.0177\,\text{rad}. \qquad \text{... (Ans.)}$$

Example 8.5

A thin-walled steel tube of 200 mm diameter and 2 mm wall thickness is 2 m long. Find the total strain energy that can be stored if it is subjected to 5 kNm torque. Take $G=80\,\text{GPa}$.

Solution:

$r=0.1\,\text{m}$, $t=0.002\,\text{m}$, $G=80\,\text{GPa}$.

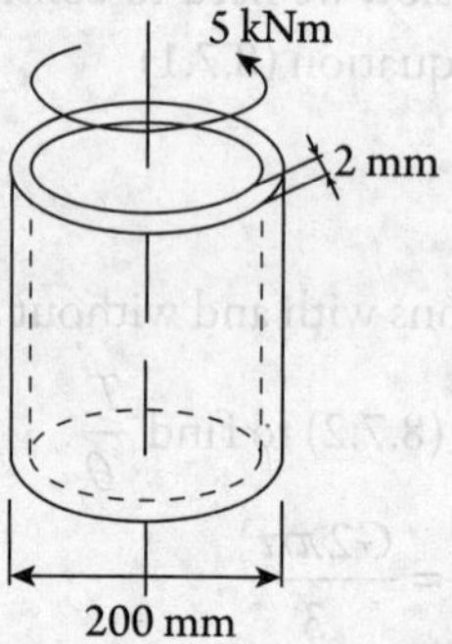

Torque transmitted is given by: $T=2\pi r^3 tG\theta$.

Hence, $\theta=\dfrac{5\times10^3}{2\times\pi\times0.1^3\times0.002\times80\times10^9}=4.97\times10^{-3}=0.005\,\text{rad/m}.$

Strain energy per unit length,

$\bar{U}=\dfrac{1}{2}\times Torque\times\theta=\dfrac{1}{2}\times5\times10^3\times0.005=12.5\ \text{J/m}.$

Total strain energy, $U=\bar{U}L=12.5\times2=25\,\text{J}.$ ... (Ans.)

Example 8.6

A thin-walled tube with outside and inside radii 20 mm and 18 mm respectively, as shown in Figure (a), is twisted with 50 Nm torque. Find the maximum shear stress and the angle of twist.

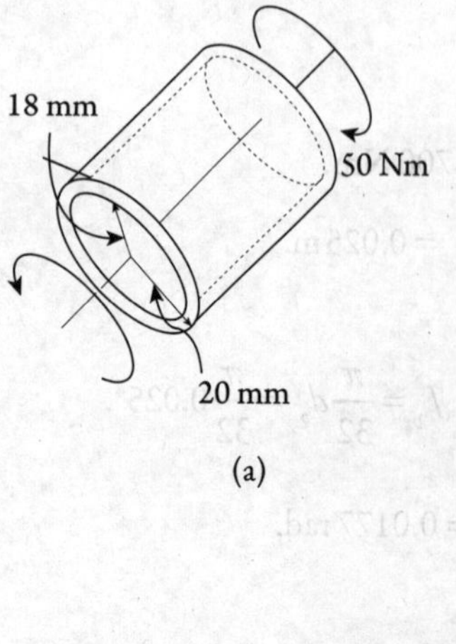

(a)

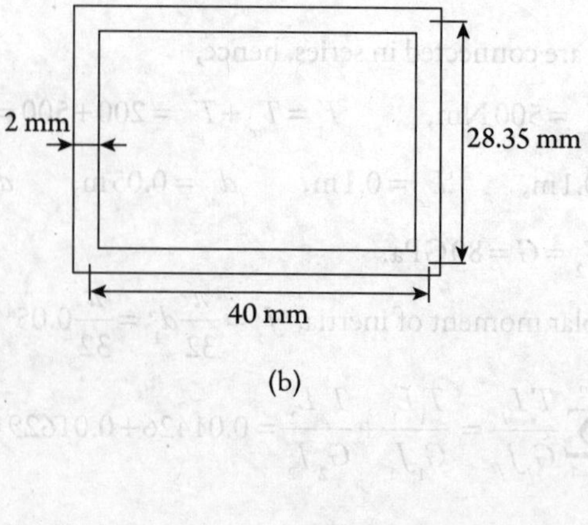

(b)

Also show that the maximum shear stress for the rectangular tube with dimensions, as shown in Figure (b), is the same as that in Figure (a) when subjected to the same torsion of 50 Nm but the angle of twist is different. State why it is so.

Solution:

(i) ***Thin-walled Circular Tube:***

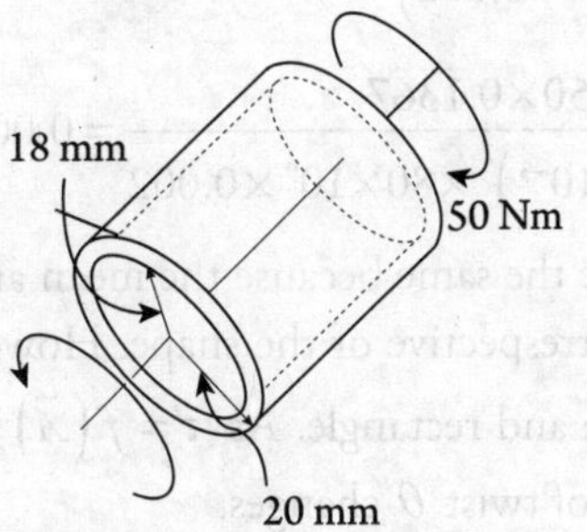

$r_i = 18\,\text{mm}, \quad r_0 = 20\,\text{mm}, \quad T = 50\,\text{Nm}, \quad G = 80\,\text{GPa}$

Maximum shear stress, $\tau_{max} = \dfrac{T}{2\bar{A}t}$,

where $\bar{A} = \dfrac{\pi}{2}\left(0.02^2 + 0.018^2\right) = 1.137 \times 10^{-3}\ \text{m}^2$

$t = \left(r_0 - r_i\right) = 2\,\text{mm}.$

Hence, $\tau_{max} = \dfrac{50}{2 \times 1.137 \times 10^{-3} \times 0.002} = 10.99 \times \dfrac{10^6\,N}{m^2} = 11\,\text{MPa}.$ ...(Ans.)

Maximum angle of twist, $[r = \left(r_0 + r_i\right)/2 = 0.019\ \text{m}]$

$$\theta_{max} = \frac{T}{2\pi r^3 Gt} = \frac{50}{2\pi \times (0.019)^3 \times 80 \times 10^9 \times 0.002} = 7.25 \times 10^{-3} = 0.0072\,\text{rad/m}. \quad \text{...(Ans.)}$$

(ii) ***Thin-walled Square Tube:***

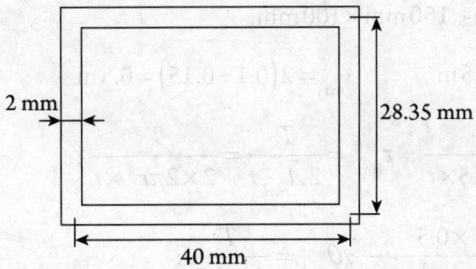

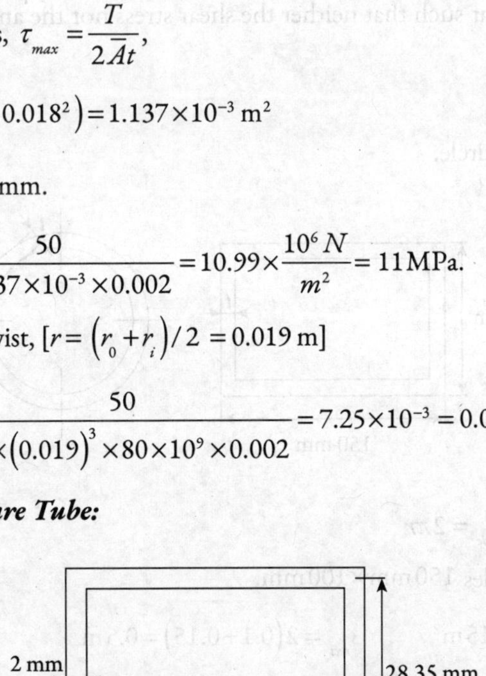

$\bar{A}$ is the mean area enclosed by the inner and outer boundaries, for the square tube:

$$\bar{A}=(40\times 28.35)\times 10^{-6}=1.134\times 10^{-3}\ \text{m}^2$$

$$s=2\times(40+28.35)\,\text{mm}=0.1367\,\text{m}$$

$$\tau_{max}=\frac{T}{2\bar{A}t}=\frac{50}{(2\times 1.134\times 10^{-3}\times 0.002)}=11\,\text{MPa} \qquad \text{... (Ans.)}$$

$$\theta_{max}=\frac{Ts}{4\bar{A}^2Gt}=\frac{50\times 0.1367}{4\times(1.134\times 10^{-3})^2\times 80\times 10^9\times 0.002}=0.0083\,\text{rad/m.} \qquad \text{... (Ans.)}$$

The stresses for both cases are the same because the mean area $\bar{A}$ enclosing the boundary of the hollow sections is the same irrespective of the shape. However, due to change in shape, the mean periphery s varies for circle and rectangle. As $\tau=f(\bar{A})$, $\theta=g(\bar{A},s)$, the shear stress τ remains constant while the angle of twist θ changes.

Example 8.7

A rectangular bar of 150 mm × 100 mm sides is to be replaced by a solid circular bar. Determine the diameter of the circular bar such that neither the shear stress nor the angle of twist exceeds those for the rectangular bar.

Solution:

Let r be the radius of the circle.

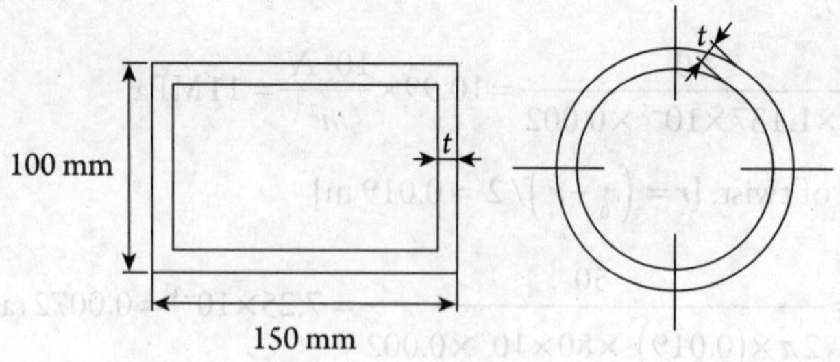

$$\bar{A}_{circle}=2\pi r^2, \qquad s_{circle}=2\pi r$$

For the rectangle of sides $150\,\text{mm}\times 100\,\text{mm}$,

$$\bar{A}_{rect}=0.1\times 0.15=0.015\,\text{m}^2, \qquad s_{rect}=2(0.1+0.15)=0.5\,\text{m}$$

$$\tau_{rect}=\frac{T}{2\bar{A}_{rect}t}=\frac{T}{2\times 0.015\times t};\ \tau_{circle}=\frac{T}{2\bar{A}_{circle}t}=\frac{T}{2\times 2\pi r^2\times t};$$

$$\theta_{rect}=\frac{Ts}{4\bar{A}^2Gt}=\frac{T\times 0.5}{4\times(0.015)^2\,G\times t};\ \theta_{circle}=\frac{T}{2\pi r^3Gt};$$

From the given conditions,

i. $\tau_{rect} = \tau_{circle}$; $\Rightarrow 2\pi r^2 = 0.015$, or $r = 0.05\,\text{m}$

ii. $\theta_{rect} = \theta_{circle}$; $\Rightarrow \pi r^3 = 9\times10^{-4}$, or $r = 0.066\,\text{m}$.

On increasing r, the shear stress τ will decrease, but following the relation for angle of twist θ, if r is reduced, then θ will increase, which is not desired.

Hence, $r = 0.066$ m or diameter of tube is 132 mm. ...(Ans.)

Example 8.8

A two-compartment thin-walled box section has dimensions as shown in the figure. Torsional shear stress must not exceed 70 MPa. Assuming there is no stress concentration at the corners. Find the maximum allowable torque and the angle of twist under that torque, $G = 80\,\text{GPa}$.

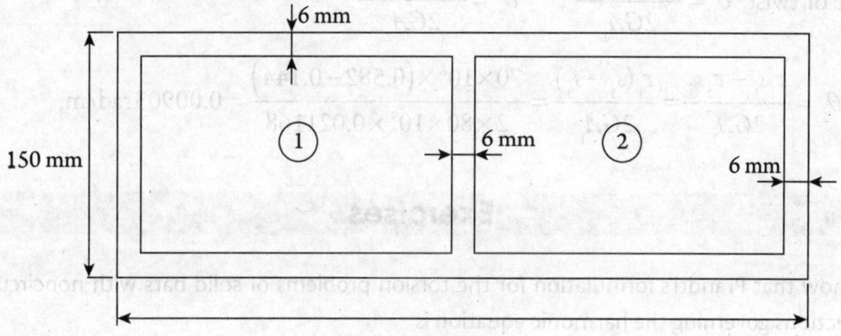

Solution:

Following the membrane analogy, $\tau_1 = \dfrac{h_1}{t_1}$, $\tau_2 = \dfrac{h_2}{t_2}$, $\tau_3 = \dfrac{h_1 - h_2}{t_3}$.

Torque, $T = 2\left(A_1h_1 + A_2h_2\right) = 2\left(A_1t_1\tau_1 + A_2t_2\tau_2\right)$ and, $\tau_1 s_1 - \tau_2 s_3 = 2G\theta A_1$;

$\tau_2 s_2 - \tau_1 s_3 = 2G\theta A_2$.

Here A_1 and A_2 are the enclosed areas by compartments 1 and 2, t_1, t_2, t_3 are the wall thicknesses, and τ_1, τ_2, τ_3 are the shear stresses in the walls.

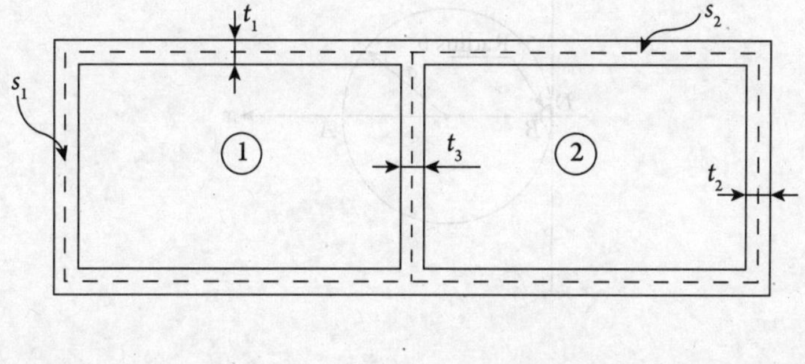

For the symmetrical section, $s_1 = s_2, t_1 = t_2, A_1 = A_2$, and $\tau_3 = 0$, i.e., no stress in the central section.

$s_1 = 2(147+144) = 582\,\text{mm} = 0.582\,\text{m}, \quad s_3 = 144\,\text{mm} = 0.144\,\text{m}$

$t_1 = t_2 = 6\,\text{mm} = 0.006\,\text{m}$

$A_1 = A_2 = 147 \times 144 \times 10^{-6} = 0.021168\,\text{m}^2.$

Due to the symmetrical section, $\tau_1 = \tau_2 = \tau_{allowable} = 70\,\text{MPa}$

Net torque, $T = 2(A_1 h_1 + A_2 h_2) = 2(A_1 t_1 \tau_1 + A_2 t_2 \tau_2) = 4A_1 t_1 \tau_1$

$T = 4 \times 0.021168 \times 0.006 \times 70 \times 10^6 = 35562.2\,\text{Nm} = 35.56\,\text{kNm}$ ... (Ans.)

Angle of twist, $\theta_1 = \dfrac{\tau_1 s_1 - \tau_2 s_3}{2GA_1}, \quad \theta_2 = \dfrac{\tau_2 s_2 - \tau_1 s_3}{2GA_2}$

$$\theta_1 = \theta_2 = \frac{\tau_1 s_1 - \tau_2 s_3}{2GA_1} = \frac{\tau_1 (s_1 - s_3)}{2GA_1} = \frac{70 \times 10^6 \times (0.582 - 0.144)}{2 \times 80 \times 10^9 \times 0.021168} = 0.00905\,\text{rad/m}. \quad \text{... (Ans.)}$$

Exercises

1. Show that Prandtl's formulation for the torsion problems of solid bars with noncircular cross-sections governing the harmonic equation is

   $\nabla^2 \phi = -2G\theta$

   and the boundary condition is $\phi = 0$ along the cross-section boundary, where ϕ is the stress function, G is the shear modulus, and θ is the angle of twist per unit length.

2. Show that the following expression is a permissible stress function for the torsion of a circular shaft with a semicircular keyway, as shown in the figure below.

   $$\phi = C\left[(x^2 + y^2) - 2ax + \frac{2ab^2 x}{x^2 + y^2} - b^2\right].$$

   Find the value of constant C in terms of θ, the angle of twist per unit length.

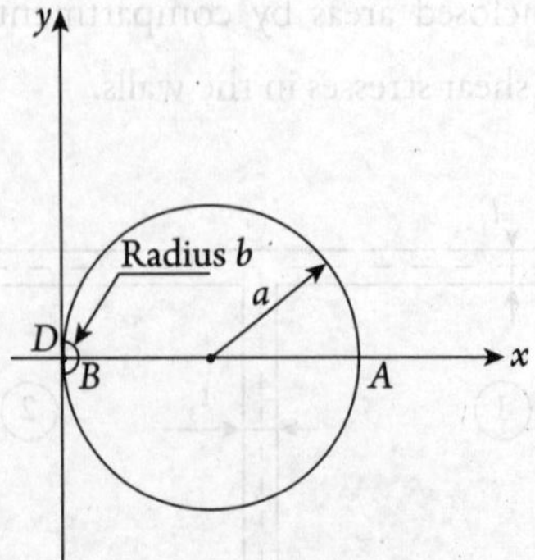

Find also the two nonzero shear stresses τ_{xz} and τ_{yz} at points A, B, and D.

$$[C = -2G\theta; \text{ at point A: } \tau_{xz} = 0, \tau_{yz} = 2G\theta\frac{(4a^2 - b^2)}{2a};$$

$$\text{at point B: } \tau_{xz} = 0,\ \tau_{yz} = -4G\theta(2a - b);\ \text{at point D: } \tau_{xz} = -4G\theta b,\ \tau_{yz} = 0].$$

3. Show that the torque transmitted by a prismatic bar of any cross-section may be given by

$T = \int\int \phi dxdy,$

where ϕ is a suitable stress function.

Find the torque transmitted by a bar with an elliptical cross-section with the major and minor semiaxes a and b and thereby show that when $a = b = r$, the torque equation reduces to a simple formula for torsion for a circular cross-section

$$\frac{T}{J} = \frac{\tau}{r} = \frac{G\theta}{l}.$$

With the assumption of no warping.

Here the polar moment of inertia $J = \pi r^4/2$ and θ is the angle of twist.

4. The section of a thin wall tube is of uniform thickness of 5 mm. The dimensions of the section are shown in the figure below. If the tube is subjected to a torque of 5 KNm, find the shear stresses in the walls and the angle of twist per unit length. Take shear modulus as G = 80 GPa.

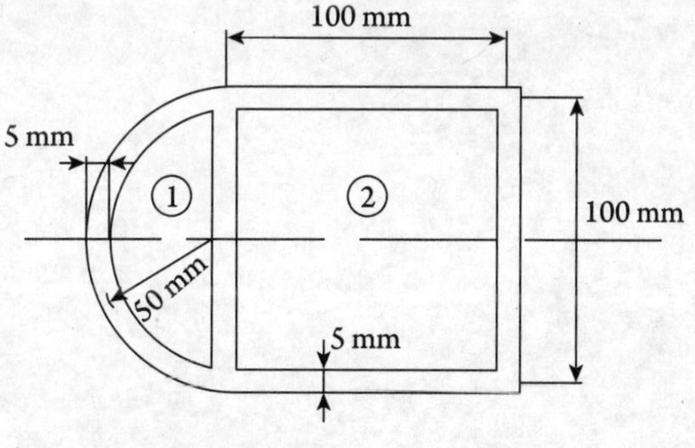

$$[\tau_1 = 34.55\,\text{MPa}, \tau_2 = 37.22\,\text{MPa},\ \ \tau_3 = 2.67\,\text{MPa}, \theta = 0.0075\,\text{radian/m}]$$

5. Analytically show that the shear stress in a thin wall tube of uniform thickness t and the angle of twist per unit length θ are given by

$$\tau = \frac{T}{2\bar{A}t}$$

$$\theta = \frac{TS}{4\bar{A}^2 Gt},$$

where $\bar{A}$ is the mean of the areas enclosed by the inner and outer boundaries, S is the length of the perimeter, and T is the torque transmitted. Show that the same results may be obtained by the membrane analogy method.

6. Solve Example 8.8 if the compartments of the thin-walled cross-section have the dimensions, as shown below.

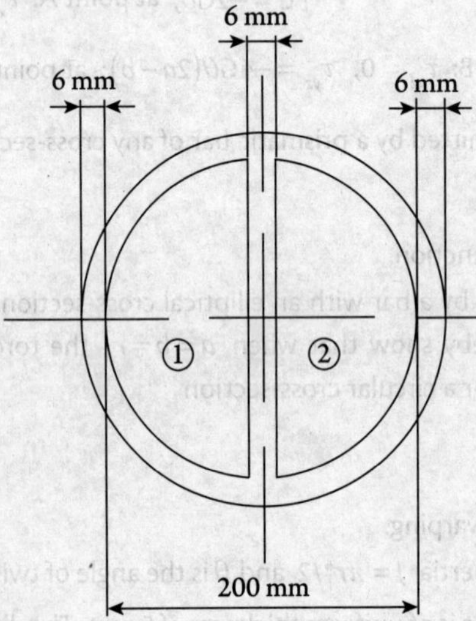

[27.985 KNm, 0.0139 rad/m]

9

Stress Concentration

Learning Objectives

After careful study of this chapter, students should be able to do the following:

LO1: Identify stress concentration in machine members.

LO2: Explain stress concentration from the theory of elasticity approach.

LO3: Calculate stress concentration due to a circular hole in a plate.

LO4: Analyze stress concentration due to an elliptical hole in a plate.

LO5: Evaluate notch sensitivity.

LO6: Create designs for reducing stress concentration.

9.1 INTRODUCTION [LO1]

Stresses given by relatively simple equations in the strength of materials for structures or machine members are based on the assumed continuity of the elastic medium. However, the presence of discontinuity destroys the assumed regularity of stress distribution in a member and a sudden increase in stresses occurs in the neighborhood of the discontinuity. In developing machines, it is impossible to avoid abrupt changes in cross-sections, holes, notches, shoulders, etc. Abrupt changes in cross-section also occur at the roots of gear teeth and threads of bolts. Some examples are shown in Figure 9.1.

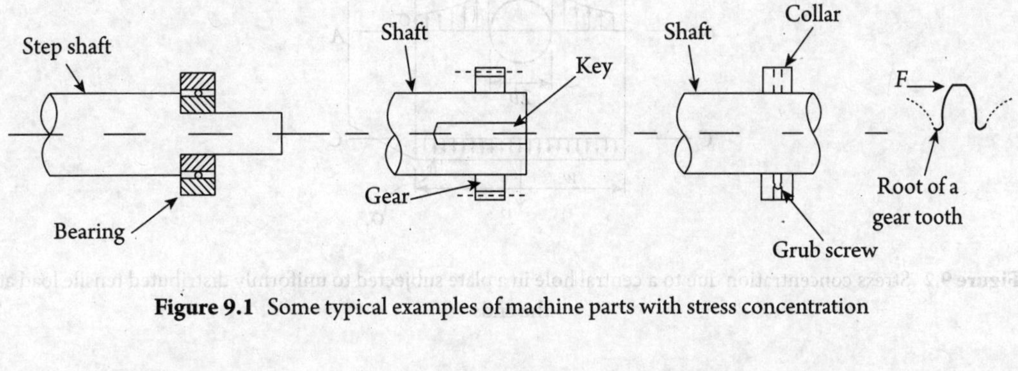

Figure 9.1 Some typical examples of machine parts with stress concentration

Any such discontinuity acts as a stress raiser. Ideally, discontinuity in materials such as non-metallic inclusions in metals, casting defects, residual stresses from welding may also act as stress raisers. In this chapter, however, we shall consider only the geometric discontinuity that arises from design considerations of structures or machine parts.

Many theoretical methods and experimental techniques have been developed to determine stress concentrations in different structural and mechanical systems. In order to understand the concept, we shall begin with a plate with a centrally located hole. The plate is subjected to uniformly distributed tensile loading at the ends, as shown in Figure 9.2.

Stress distributions at sections *BB* and *CC* remote from the hole are uniform and are given by $\sigma_1 = P/wt$ following St. Venant's Principle. At section *AA*, average stress $\sigma_2 = P/(w-2b)t$. However, at section *AA*, stress σ_3 in the vicinity of the hole rises sharply, and this can be predicted using the *Theory of Elasticity* approach. We define here the theoretical stress concentration factor k_t as

$$k_t = \frac{\sigma_3}{\sigma_2}.$$

Here σ_3 is the maximum stress that occurs due to the presence of a hole and σ_2 is the average stress at the section through the hole. Although this is the widely accepted definition of stress concentration, some authors, such as Seely and Smith (1952), have defined it as

$$k_t = \frac{\sigma_3}{\sigma_1},$$

where σ_1 is the nominal stress that would occur at the same point if there were no holes in the plate.

It is possible to predict the stress concentration factors for certain geometric shapes using the *Theory of Elasticity* approach.

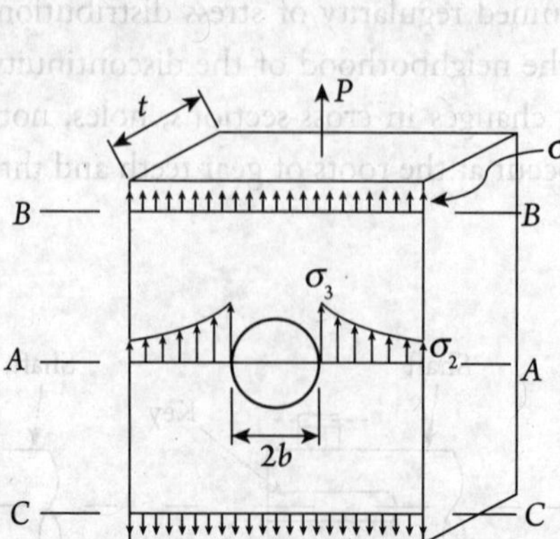

Figure 9.2 Stress concentration due to a central hole in a plate subjected to uniformly distributed tensile load at the ends

9.2 UNDERSTANDING STRESS CONCENTRATION USING THE THEORY OF ELASTICITY APPROACH [LO2]

We shall now consider the basis of stress concentration in a plate with a centrally located hole, which is subjected to uniform tensile stress σ at the ends, as shown in Figure 9.3.

In analyzing this situation, we first consider St. Venant's principle which states that the stress distribution at points remote from the neighborhood of the hole that produces disturbance in the stress distribution, remains unchanged. To illustrate this, we consider an imaginary circle of radius "b" such that $b \gg a$, where "a" is the hole radius. At $r \geq b$, the effect of the hole on the stress distribution is negligible, and stresses at $r \geq b$ is the same as in the plate without any hole.

Now for an element A at $r > b$ from the elementary stress transformation analysis (Figure 9.4), we may express two-dimensional (2D) polar components of stresses σ_r, σ_θ, and $\tau_{r\theta}$ in terms of the Cartesian components of stresses σ_x, σ_y, and τ_{xy} as follows:

$$\begin{aligned}
\sigma_r &= \sigma_x \cos^2\theta + \sigma_y \sin^2\theta + 2\tau_{xy}\cos\theta\sin\theta \\
\sigma_\theta &= \sigma_x \sin^2\theta + \sigma_y \cos^2\theta - 2\tau_{xy}\cos\theta\sin\theta \\
\tau_{r\theta} &= \left(\sigma_y - \sigma_x\right)\cos\theta\sin\theta + \tau_{xy}\left(\cos^2\theta - \sin^2\theta\right).
\end{aligned} \tag{9.2.1}$$

Referring to Figure 9.3, we have $\sigma_x = \sigma$, $\sigma_y = 0$, and $\tau_{xy} = 0$. This gives

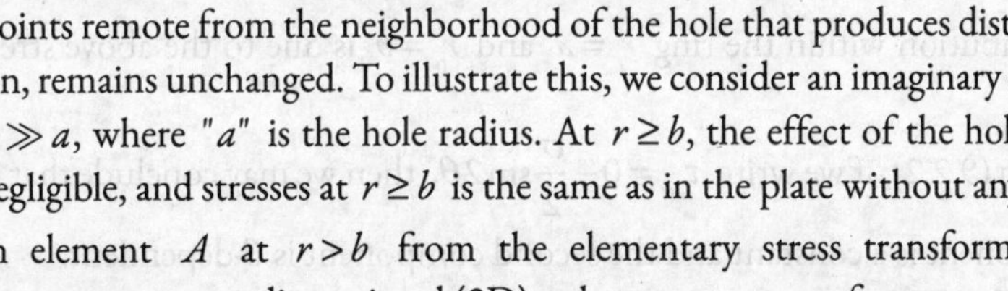

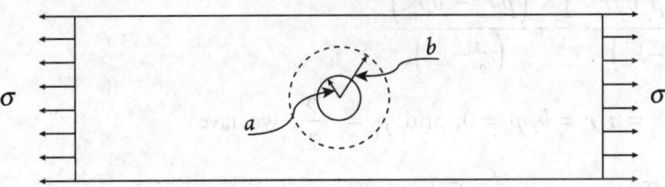

Figure 9.3 Stress concentration due to a hole in a plate subjected to uniform tensile stress σ at the ends

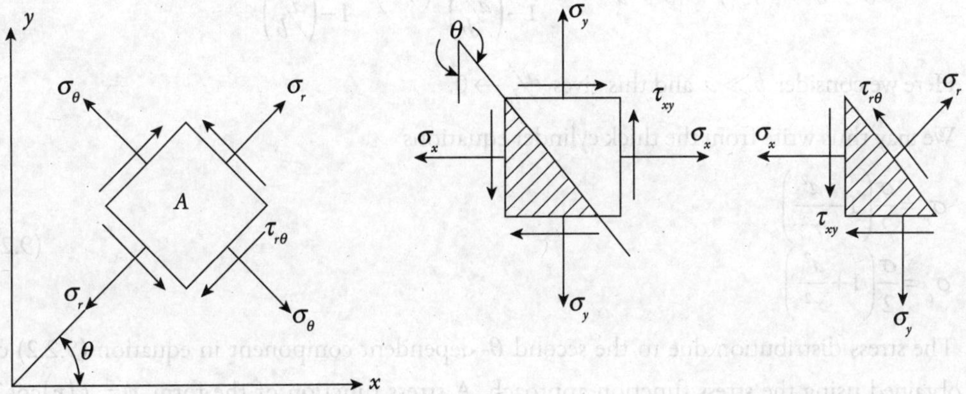

Figure 9.4 Stress on an element at $r \geq b$ in $r-\theta$ and $x-y$ coordinates

$$\sigma_r=\sigma_x\cos^2\theta=\frac{\sigma}{2}+\frac{\sigma}{2}\cos2\theta$$

$$\sigma_\theta=\sigma_x\sin^2\theta=\frac{\sigma}{2}-\frac{\sigma}{2}\cos2\theta \tag{9.2.2}$$

$$\tau_{r\theta}=-\frac{\sigma}{2}\sin2\theta.$$

Stress distribution within the ring $r_i=a$ and $r_0=b$ is due to the above stress acting around the ring.

In equation (9.2.2), if we write $\tau_{r\theta}=0-\frac{\sigma}{2}\sin2\theta$, then we may conclude that in all the stresses, the first component is a constant and the second component is θ-dependent.

The stress distribution within the ring $a<r<b$ due to the first component can be analyzed using the theory of thick cylinder subjected to a uniform external pressure $-\frac{\sigma}{2}$ only.

It was shown in the earlier chapter that for a thick cylinder of inside and outside radii r_i and r_0 respectively, subjected to internal and external pressures p_i and p_0 respectively, the radial and circumferential stresses σ_r and σ_θ are given as

$$\sigma_r=\frac{\left(p_0-p_i\right)r_i^2r_0^2}{\left(r_0^2-r_i^2\right)}\frac{1}{r^2}+\frac{\left(p_ir_i^2-p_0r_0^2\right)}{\left(r_0^2-r_i^2\right)}$$

$$\sigma_\theta=-\frac{\left(p_0-p_i\right)r_i^2r_0^2}{\left(r_0^2-r_i^2\right)}\frac{1}{r^2}+\frac{\left(p_ir_i^2-p_0r_0^2\right)}{\left(r_0^2-r_i^2\right)}.$$

Substituting $r_i=a, r_0=b, p_i=0$, and $p_0=-\frac{\sigma}{2}$, we have

$$\sigma_r=\frac{a^2b^2\left(-\frac{\sigma}{2}\right)}{b^2-a^2}\left(\frac{1}{r^2}\right)+\frac{-b^2\left(-\frac{\sigma}{2}\right)}{b^2-a^2}=-\frac{a^2\frac{\sigma}{2}}{1-\left(a/b\right)^2}\left(\frac{1}{r^2}\right)+\frac{\frac{\sigma}{2}}{1-\left(a/b\right)^2}.$$

Here we consider $b\gg a$ and this gives $a/b\rightarrow0$.

We may thus write from the thick cylinder equations

$$\sigma_r=\frac{\sigma}{2}\left(1-\frac{a^2}{r^2}\right)$$

$$\sigma_\theta=\frac{\sigma}{2}\left(1+\frac{a^2}{r^2}\right). \tag{9.2.3}$$

The stress distribution due to the second θ- dependent component in equation (9.2.2) can be obtained using the stress function approach. A stress function of the form $\phi=f\left(r\right)\cos2\theta$ is suitable in this instance. A stress function of this form that satisfies the governing equation $\nabla^4\phi=0$ would be

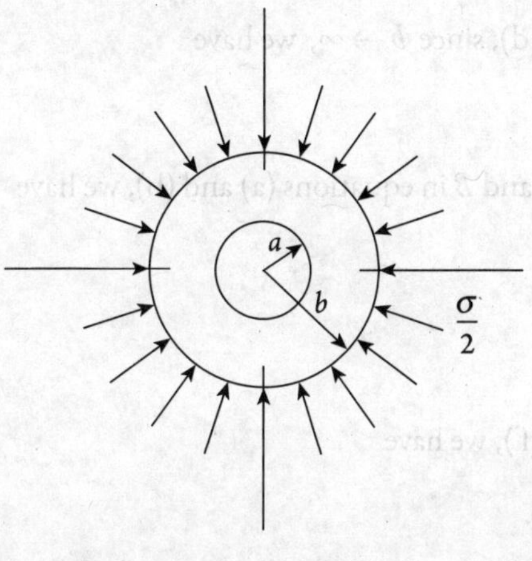

Figure 9.5 Stress within $a<r<b$ in the plate with a hole due to external pressure $\sigma/2$

$$\phi=\left(Ar^2+Br^4+\frac{C}{r^2}+D\right)\cos 2\theta, \tag{9.2.4}$$

where A, B, C, and D are constants.

This gives

$$\sigma_r=\frac{1}{r}\frac{\partial\phi}{\partial r}+\frac{1}{r^2}\frac{\partial^2\phi}{\partial\theta^2}=-\left(2A+\frac{6C}{r^4}+\frac{4D}{r^2}\right)\cos 2\theta$$

$$\sigma_\theta=\frac{\partial^2\phi}{\partial r^2}=\left(2A+12Br^2+\frac{6C}{r^4}\right)\cos 2\theta \tag{9.2.5}$$

$$\tau_{r\theta}=-\frac{\partial}{\partial r}\left(\frac{1}{r}\frac{\partial\phi}{\partial\theta}\right)=\left(2A+6Br^2-\frac{6C}{r^4}-\frac{2D}{r^2}\right)\sin 2\theta.$$

From the above discussion, we may write the boundary conditions as:

at $r=a$, $\sigma_r=\tau_{r\theta}=0$ at all values of θ, since the hole edge is free from external forces (Figure 9.5).

At $r=b$, $\sigma_r=\frac{\sigma}{2}\cos 2\theta$, $\tau_{r\theta}=-\frac{\sigma}{2}\sin 2\theta$, from equation (9.2.2).

Substituting these conditions in equation (9.2.5), we have

$$2A+\frac{6C}{a^4}+\frac{4D}{a^2}=0 \tag{a}$$

$$2A+6Ba^2-\frac{6C}{a^4}-\frac{2D}{a^2}=0 \tag{b}$$

$$2A+\frac{6C}{b^4}+\frac{4D}{b^2}=-\frac{\sigma}{2} \tag{c}$$

and $2A+6Bb^2-\frac{6C}{b^4}-\frac{2D}{b^2}=-\frac{\sigma}{2}.$ (d)

From equations (c) and (d), since $b \to \infty$, we have

$$A = -\frac{\sigma}{4},\ B = 0.$$

Substituting values of A and B in equations (a) and (b), we have

$$-\frac{\sigma}{2} + \frac{6C}{a^4} + \frac{4D}{a^2} = 0 \tag{e}$$

$$-\frac{\sigma}{2} - \frac{6C}{a^4} - \frac{2D}{a^2} = 0. \tag{f}$$

From equations (e) and (f), we have

$$C = -\sigma a^4/4,\ D = \sigma a^2/2.$$

Substituting values of A, B, C, and D in equations (9.2.5), we have

$$\begin{aligned}
\sigma_r &= \frac{\sigma}{2}\left(1 + 3\frac{a^4}{r^4} - 4\frac{a^2}{r^2}\right)\cos 2\theta \\
\sigma_\theta &= -\frac{\sigma}{2}\left(1 + 3\frac{a^4}{r^4}\right)\cos 2\theta \\
\tau_{r\theta} &= -\frac{\sigma}{2}\left(1 - 3\frac{a^4}{r^4} + 2\frac{a^2}{r^2}\right)\sin 2\theta.
\end{aligned} \tag{9.2.6}$$

In order to find the complete stress state at a point within $a < r < b$, we combine equations (9.2.3) and (9.2.6) to give

$$\begin{aligned}
\sigma_r &= \frac{\sigma}{2}\left[\left(1 - \frac{a^2}{r^2}\right) + \left(1 + 3\frac{a^4}{r^4} - 4\frac{a^2}{r^2}\right)\cos 2\theta\right] \\
\sigma_\theta &= \frac{\sigma}{2}\left[\left(1 + \frac{a^2}{r^2}\right) - \left(1 + 3\frac{a^4}{r^4}\right)\cos 2\theta\right] \\
\tau_{r\theta} &= -\frac{\sigma}{2}\left(1 - 3\frac{a^4}{r^4} + 2\frac{a^2}{r^2}\right)\sin 2\theta.
\end{aligned} \tag{9.2.7}$$

If we inspect equations (9.2.7), we find that the stresses at the internal radius $r = a$ are

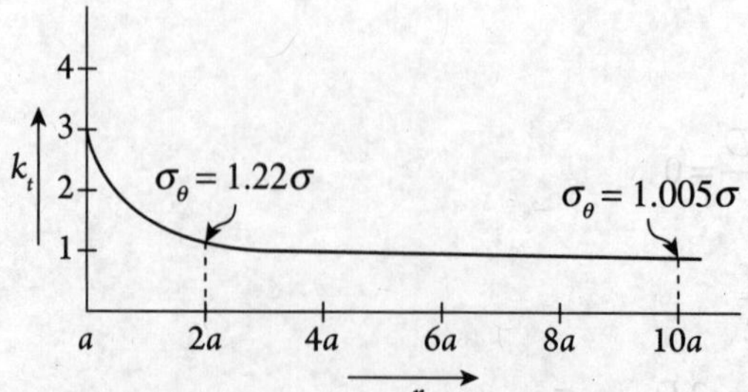

Figure 9.6 Rapid decrease of circumferential stress beyond the edge of the hole in a plate subjected to uniform tensile stress σ, as shown in Figure 9.3

$$\left.\begin{aligned}\sigma_r &= 0\\ \sigma_\theta &= \sigma(1-2\cos 2\theta)\\ \tau_{r\theta} &= 0\end{aligned}\right\}. \tag{9.2.8}$$

If we further consider the stresses at $r=a$, $\theta=\pm\frac{\pi}{2}$, σ_θ is the largest at $\theta=\frac{\pi}{2}$ or $\theta=\frac{3\pi}{2}$ and is equal to 3σ. σ_θ is equal to $-\sigma$ at $\theta=0$ and 2π. This means that the maximum theoretical stress concentration factor $k_t=3$. However, σ_θ decreases rapidly with r up to $r=2a$ and then decreases slowly, as shown in Figure 9.6.

9.3 STRESS CONCENTRATION DUE TO A CIRCULAR HOLE IN A PLATE SUBJECTED TO EQUAL STRESSES IN TWO PERPENDICULAR DIRECTIONS [LO3]

We consider a plate with a hole of radius "a", same as shown in Figure 9.3, except that the plate is subjected to a uniform stress σ in two perpendicular directions, as shown in Figure 9.7. Stresses within a ring $a<r<b$ are influenced by the presence of the hole of radius "a". Following the analysis in Section 9.2, the maximum tangential tensile stress at points A and B due to the vertically applied stress σ at the edge of the plate is 3σ, and the tangential compressive stress at C and D is $-\sigma$.

Similarly, due to applied horizontal stress σ at the edge of the plate, the tangential tensile stress at C and D is 3σ, and compressive stress at A and B is $-\sigma$. Therefore, the maximum stress at both A,B and C,D is

$$\sigma_{max} = 3\sigma - \sigma = 2\sigma. \tag{9.3.1}$$

This is shown in Figure 9.7 (b).

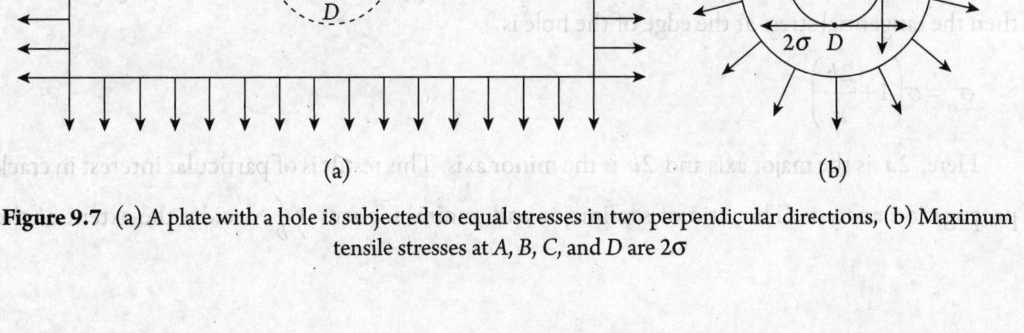

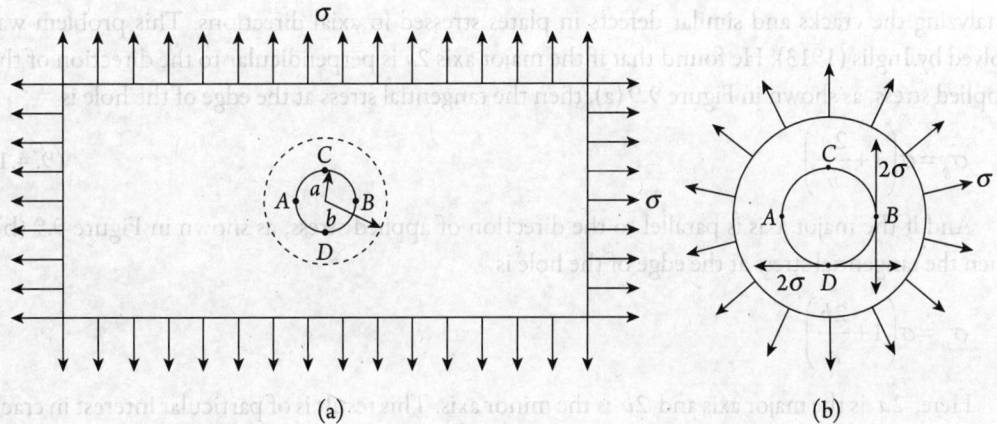

Figure 9.7 (a) A plate with a hole is subjected to equal stresses in two perpendicular directions, (b) Maximum tensile stresses at A, B, C, and D are 2σ

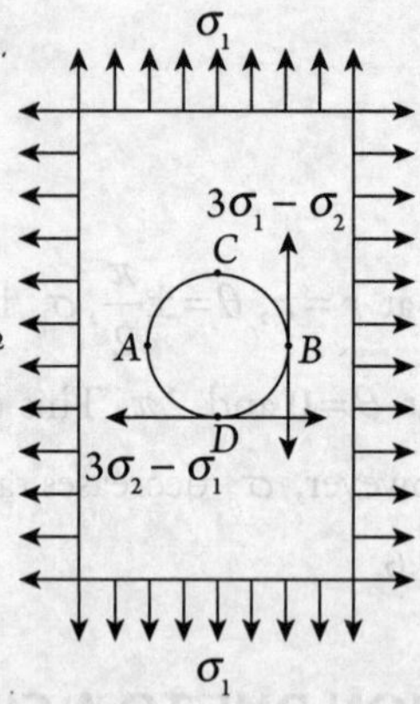

Figure 9.8 Stresses developed at the edge of a hole in a plate subjected to two unequal stresses in two perpendicular directions

The argument may be applied to find the tangential stress at the edge of a hole in a plate subjected to two unequal stresses in perpendicular directions, as shown in Figure 9.8, and we may write at points A and B

$$\sigma_{max} = 3\sigma_1 - \sigma_2 \tag{9.3.2}$$

and at point C and D

$$\sigma_{max} = 3\sigma_2 - \sigma_1. \tag{9.3.3}$$

9.4 STRESS CONCENTRATION DUE TO AN ELLIPTICAL HOLE IN A PLATE SUBJECTED TO A TENSILE STRESS [LO4 AND LO5]

This is an important issue in the theory of elasticity, because these results are of much use in analyzing the cracks and similar defects in plates stressed in axial directions. This problem was solved by Inglis (1913). He found that if the major axis $2a$ is perpendicular to the direction of the applied stress, as shown in Figure 9.9 (a), then the tangential stress at the edge of the hole is

$$\sigma_\theta = \sigma\left(1 + \frac{2a}{b}\right). \tag{9.4.1}$$

And if the major axis is parallel to the direction of applied stress, as shown in Figure 9.9 (b), then the tangential stress at the edge of the hole is

$$\sigma_\theta = \sigma\left(1 + \frac{2b}{a}\right).$$

Here, $2a$ is the major axis and $2b$ is the minor axis. This result is of particular interest in crack propagation analysis. The ratio of semimajor and semiminor axes a/b decides the extent of the stress at the hole edges when the major axis is perpendicular to the applied stress, as seen in Figure 9.9(a).

The increase in tangential stress at the hole edge is without limit and as the hole becomes more and more slender the increase in stress is limited only by yielding. At $a/b = 1$, the hole reduces to a circular one and as we have seen earlier, the maximum tangential stress is 3σ. This is shown in Figure 9.10. Sometimes, it is suggested to drill a hole at the edge of a slit or crack to reduce this high stress.

Theoretical stress concentration factors for different notches and grooves have been evaluated by different authors. There are also many experimental methods available. Some of these methods are:

1. strain gauge method
2. photoelasticity method
3. brittle coating technique
4. grid method

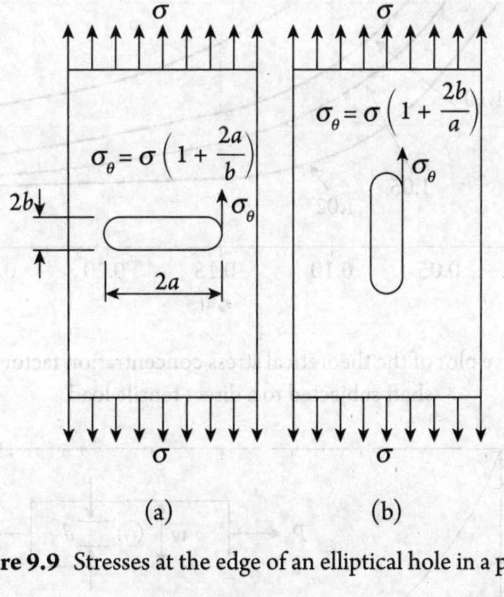

Figure 9.9 Stresses at the edge of an elliptical hole in a plate

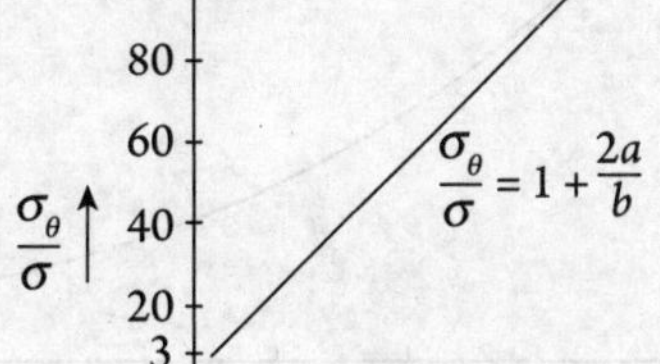

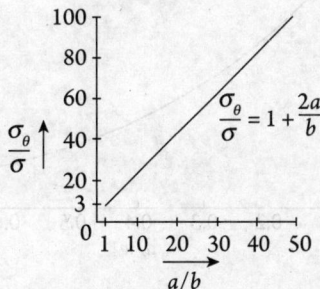

Figure 9.10 Variation of tangential stress as the elliptical hole becomes more and more slender

Out of all these methods, the values of stress concentration factors obtained by the photoelasticity method agree well with the values obtained theoretically. This is why the photoelasticity method is sometimes used to validate the theoretical stress concentration results. Theoretical stress concentration factors for different configurations and loading conditions are available in graphical forms in handbooks. A representative plot of theoretical stress concentration factor k_t vs r/d ratio is shown in Figure 9.11, where r is the fillet radius and d is the smaller diameter in a step shaft subjected to axial tensile load.

Representative stress concentration charts for different geometries and loadings are shown in Figures 9.12–9.17.

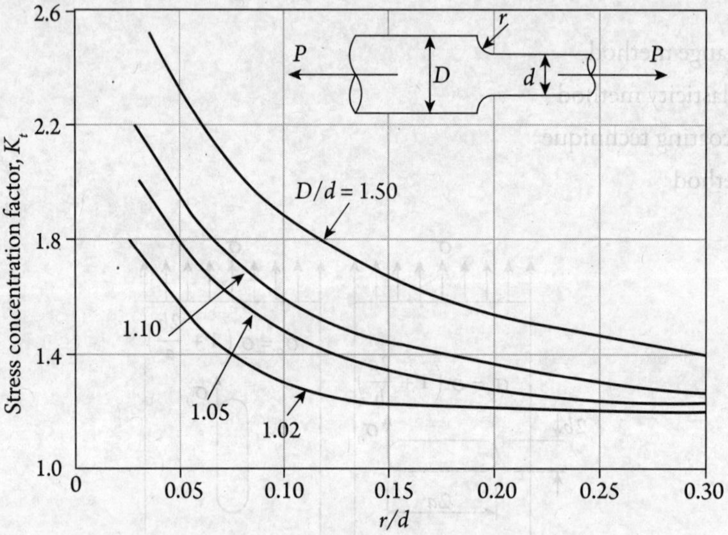

Figure 9.11 A representative plot of the theoretical stress concentration factor k_t versus r/d of a stepped shaft subjected to a direct tensile load

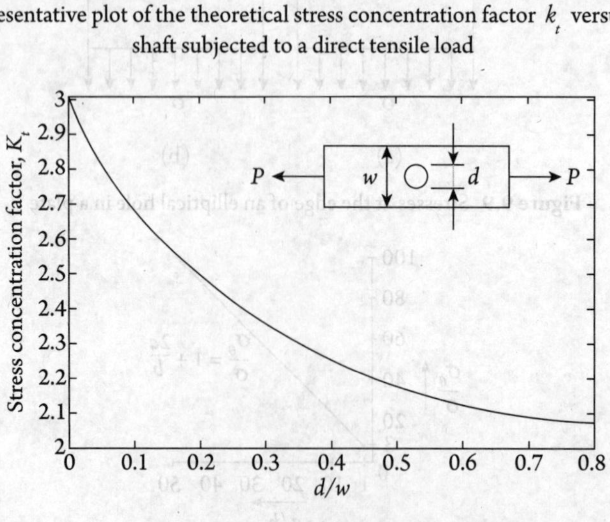

Figure 9.12 Variation of theoretical stress concentration factor with the hole diameter to plate width ratio (d/w)

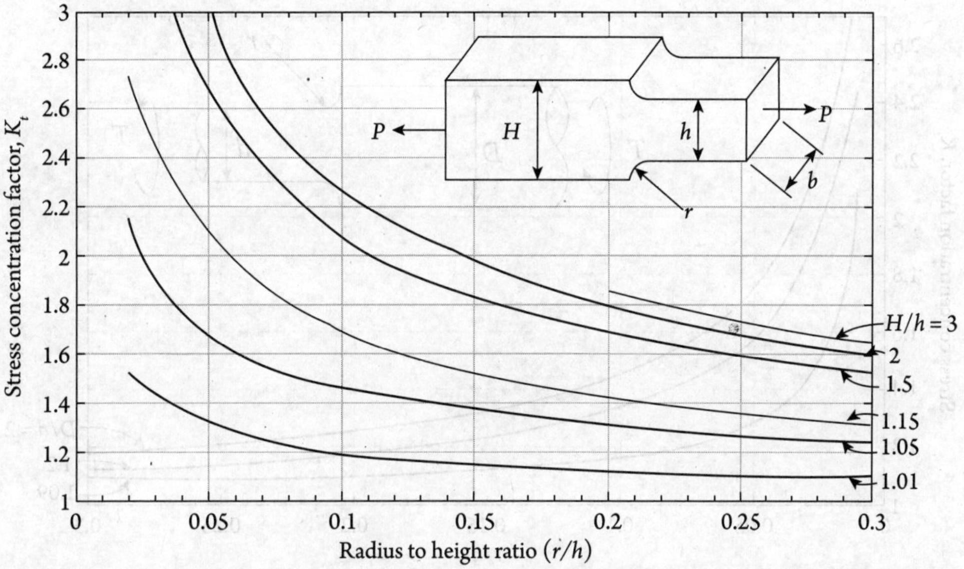

Figure 9.13 Variation of theoretical stress concentration factor with the ratio of fillet radius to smaller plate width (r/h) for stepped member subjected to axial loading

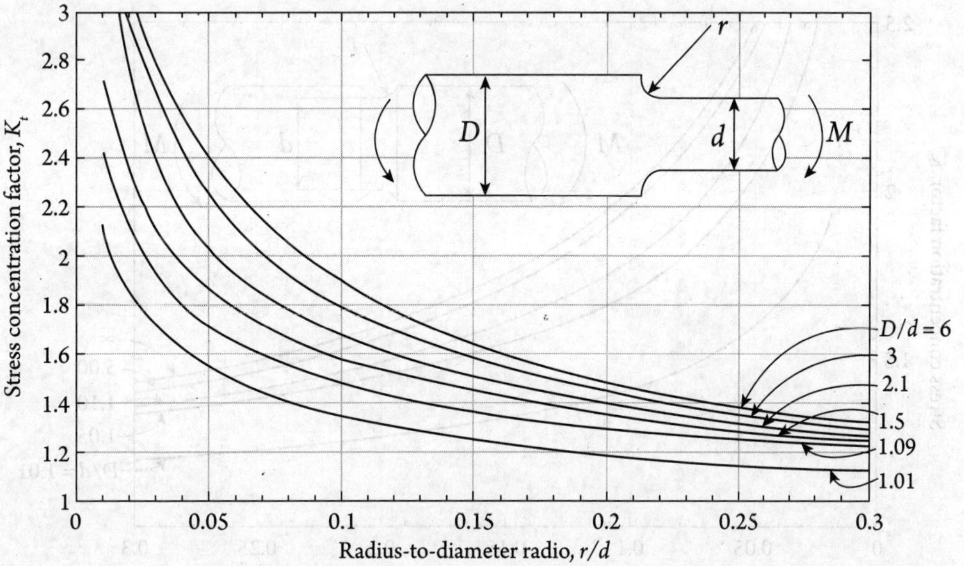

Figure 9.14 Variation of theoretical stress concentration factor with the ratio of fillet radius to smaller diameter (r/d) for stepped shaft subjected to a bending moment

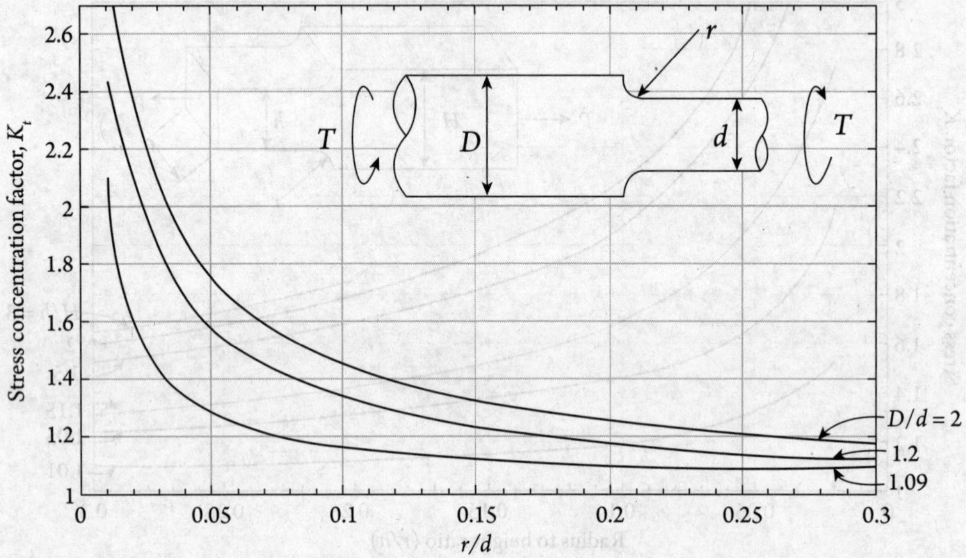

Figure 9.15 Variation of theoretical stress concentration factor with the ratio of fillet radius to smaller diameter (r/d) of stepped shafts subjected to twisting moment

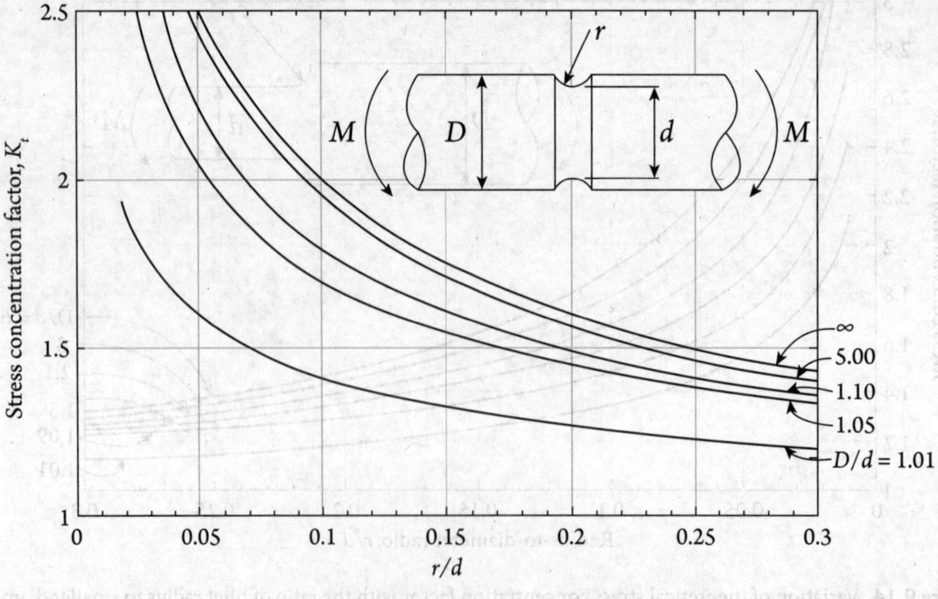

Figure 9.16 Variation of theoretical stress concentration factor with the ratio of fillet radius to smaller diameter in a grooved shaft subjected to a bending moment

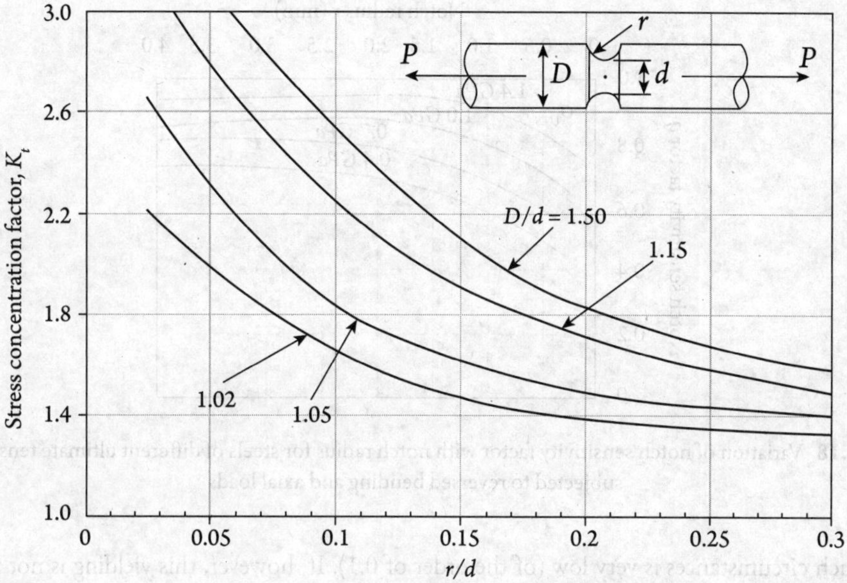

Figure 9.17 Variation of theoretical stress concentration factor with the ratio of fillet radius to smaller diameter in a grooved shaft subjected to axial load

In design under fatigue loading, the stress concentration factor is an important parameter, and the factor is used to modify the values of endurance limit, whereas, for design under static loading, the stress concentration factor is used simply as a stress raiser. This means actual stress = k_t × calculated stress. The stress concentration factor effect is important in the design of components with both ductile and brittle materials, but this is taken to be a very serious issue in the design of components with brittle materials even under static loading.

It has been found that some materials are not sensitive to the existence of notches or discontinuities. In such cases, it is not necessary to use the full value of the stress concentration factor k_t; instead, a reduced value is used. This is given by a factor known as fatigue strength reduction factor k_f, which is defined as

$$k_f = \frac{\textit{endurance limit of notch free specimens}}{\textit{endurance limit of notched specimens}}.$$

Another term called "notch sensitivity factor" q is often used in design, and this is defined as

$$q = \frac{k_f - 1}{k_t - 1}. \tag{9.4.2}$$

The values of q usually lie between 0 and 1. If $q = 0$, $k_f = 1$, and this means there is full notch sensitivity. Design charts for q can be found in design handbooks and knowing k_t, k_f may be found. A typical set of notch sensitivity curves for steel is shown in Figure 9.18.

There are many and varied conditions that affect k_f and q. Under static loading for ductile materials, the localized increase in stress is largely relieved by localized yielding and the value of q

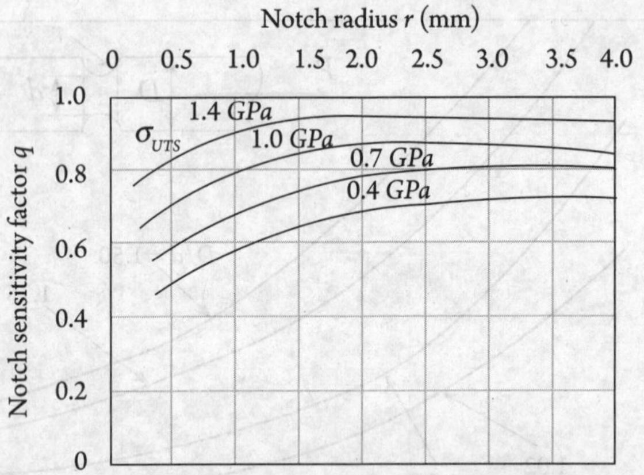

Figure 9.18 Variation of notch sensitivity factor with notch radius for steels of different ultimate tensile stresses subjected to reversed bending and axial loads

under such circumstances is very low (of the order of 0.1). If, however, this yielding is not permitted for certain applications then the value of q would rise. For metal members under static loading at elevated temperature, creep may occur, and in that case, the value of q would rise significantly.

With an abrupt change in the cross-section of members made of brittle materials under static loads, q generally lies between 0.5 and 1. If, however, there are internal stress risers, such as graphite flakes in grey cast iron, or external stress risers such as an abrupt change in the cross-section, there would be only a small influence on the strength, and q value would be relatively small.

Under repeated or cyclic loading, members with an abrupt change in the cross-section usually fail by progressive localized fracture or crack propagation, and the ability of the materials, even for ductile type, to make accommodation by localized yielding is small. This sort of fatigue fracture gives very little warning by yielding before fracture and the value of q is usually large (near unity).

The presence of residual stress in members with an abrupt change in the cross-section would influence the value of q if the static stress is in the same direction as the residual stress. Under such circumstances, q would be relatively large. Under reversed loading, only micro residual stress plays an important role. Macro residual stresses under such circumstances with completely reversed loading are reduced.

If components with localized stress due to an abrupt change in the cross-section are subjected to impact loading, then the zone of localized stress would absorb greater energy before the main body of the component could absorb the energy. Consequently, it is likely that rupture would take place at the area of localized stress.

9.5 METHODS OF REDUCING STRESS CONCENTRATION [LO6]

In engineering practice, it is more important to find methods of reducing stress concentration than to calculate stress concentration factor. Clearly, reduction in stress concentration must constitute methods of reducing abrupt changes in cross-section. This is often implemented by adding or removing small amounts of material, for example, by providing fillets. A fillet radius allows the cross-section to change gradually. Sometimes, elliptical fillets are also used. If a notch is unavoidable in a design, it is better to provide a number of small notches rather than a long notch. Similarly, if a projection is unavoidable from design considerations, then it is preferable to provide a narrow projection rather than a wide projection. Stress relieving grooves are sometimes provided. These are illustrated in Figure 9.19.

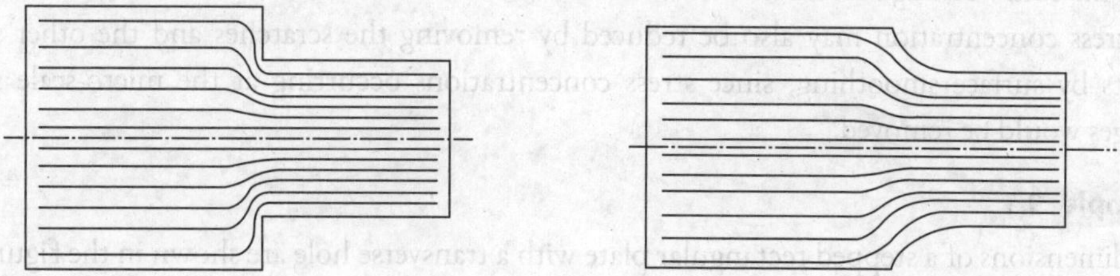

(i) (ii)

(a) (i) Force flow around a sharp corner (flow lines are crowded near corner). (ii) Force flow around a corner with a fillet (less crowding of flow lines – low stress concentration)

(i) (ii)

(b) (i) Force flow around a large notch. (ii) Force flow around a number of small notches. Low stress concentration

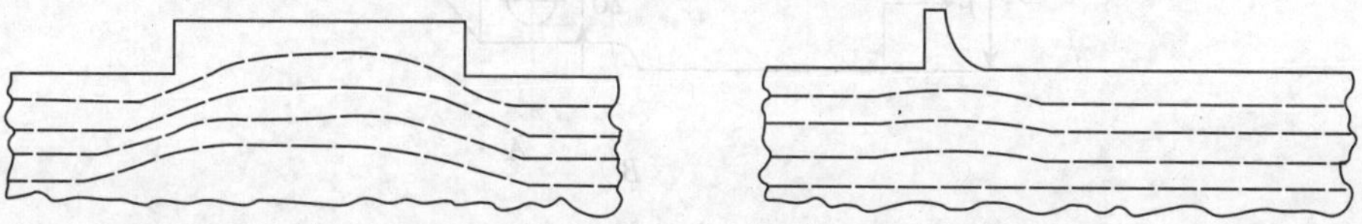

(i) (ii)

(c) (i) Force flow around a wide projection. (ii) Force flow around a narrow projection. Low stress concentration

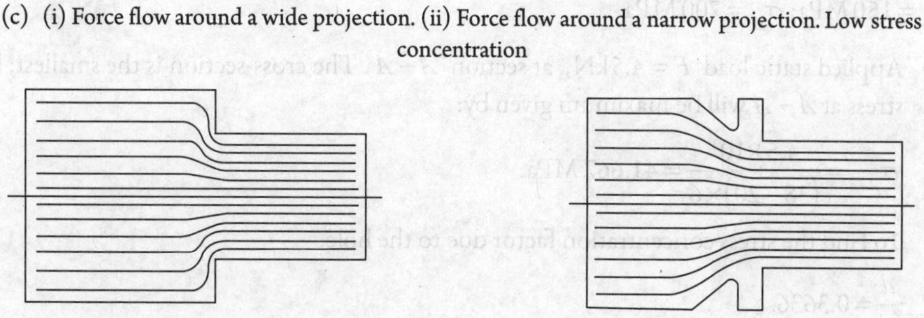

(i) (ii)

(d) (i) Force flow around a sudden change in diameter. (ii) Force flow around stress relieving groove

Figure 9.19 Different methods to reduce stress concentration

Stress concentration may also be reduced by substituting a member made of a material with lower modulus of elasticity. For example, replacing the steel nut in a pair of steel nut and bolt with a bronze nut may reduce the stress concentration in bolt threads.

Sometimes, increased fatigue strength of material near the zone of high stress concentration may be obtained by cold working. This is primarily due to residual compressive stress developed in the cold-worked material by the surrounding elastic material as it comes back to its original shape when the cold working tool is removed.

Stress concentration may also be reduced by removing the scratches and the other surface defects by surface smoothing, since stress concentrations occurring at the micro-scale surface changes would be removed.

Example 9.1

The dimensions of a stepped rectangular plate with a transverse hole are shown in the figure. The allowable tensile stress and the ultimate tensile stress of the plate material are 150 MPa and 0.7 GPa respectively. Find the factor of safety if the plate is subjected to:

(a) a static axial load of $F = 4.5\,\text{kN}$, and

(b) a completely reversed axial load of $F_{variable} = 4.5\,\text{kN}$ and endurance limit of the material in axial loading is 250 MPa.

Solution:

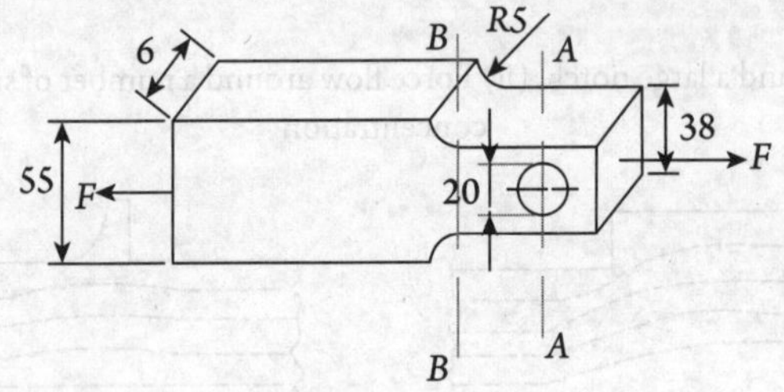

$\sigma_{allow} = 150\,\text{MPa};\ \sigma_{ut} = 700\,\text{MPa}$

(a) Applied static load $F = 4.5\,\text{kN}$, at section $A-A$. The cross-section is the smallest; hence, stress at $A-A$ will be maximum given by:

$$\sigma_{A-A} = \frac{4.5\times10^3}{(38-20)\times6} = 41.667\,\text{MPa}.$$

To find the stress concentration factor due to the hole,

$$\frac{d}{w} = 0.3636.$$

From Figure 9.12, $k_t \approx 2.25$.

Therefore, $\sigma_{max} = k_t \times \sigma_{A-A} = 93.75\,\text{MPa}$.

Factor of safety, $f.s = \frac{allowable\ stress}{max.stress} = \frac{150}{93.75} \approx 1.6.$

At the step, section- $B-B$,

Axial stress without the hole = $\sigma_{B-B} = \frac{4.5 \times 10^3}{38 \times 6} = 19.7368\,\text{MPa}.$

From Figure 9.13, fillet $r/d = 0.13158$ and $D/d = 1.447$

$k_t = 1.5$; hence, maximum stress at fillet = $\sigma_{B-B} \times k_t = 29.61\,\text{MPa}.$

Factor of safety, $(fs)_{fillet} = \frac{150}{29.61} = 5.07.$

(b) Cyclic loading,

$F_{var} = 4.5\,\text{kN};\ \sigma_e = 250\,\text{MPa}\,(endurance\ limit).$

For completely reversed cycle, the variation of F_{var} is

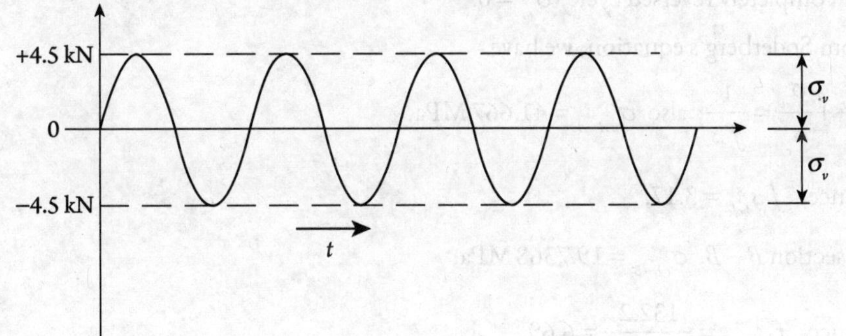

$F_{max} = +4.5\,kN;\ F_{min} = -4.5\,kN;\ F_{avg} = F_m = 0.$

Following Soderberg's equation

$$\frac{\sigma_m}{\sigma_y} + \frac{\sigma_v}{\sigma_e'} = \frac{1}{f.s}$$

Here, modified endurance limit, $\sigma_e' = \sigma_e.C_1.C_2.C_3.C_4.C_5/k_f$

C_1 – Size factor

C_2 – Load factor

C_3 – Surface factor

C_4 – Temperature factor

C_5 – Reliability factor

k_f – Fatigue strength reduction factor

From Figure 9.18; $\sigma_{ut} = 0.7\,\text{GPa}$; $\textit{fillet radius} = \textit{notch radius} = 5\,\text{mm}$;

$q = 0.81(\textit{notch sensitivity})$;

From Figure 9.12,

$\frac{d}{w} = \frac{20}{38} = 0.526$; $k_t = 2.1$.

From equation (9.4.2), $q = \frac{k_f - 1}{k_t - 1}$.

Hence, $k_f = 1.891$.

Considering C_1 to C_5 to be insignificant, the modified endurance limit is

$$\sigma_e' = \sigma_e / k_f = 132.2\,\text{MPa}.$$

For completely reversed cycle, $\sigma_m = 0$.

From Soderberg's equation, we have

$\frac{\sigma_m}{\sigma_y} + \frac{\sigma_v}{\sigma_e'} = \frac{1}{f.s}$; also $\sigma_{vA-A} = 41.667\,\text{MPa}$.

Hence, $f.s_{hole} = 3.17$.

At section $B - B$, $\sigma_{vB-B} = 19.7368\,\text{MPa}$,

Hence, $f.s_{fillet} = \frac{132.2}{19.7368} = 6.69$.

Example 9.2

A thin-walled cylindrical tank of diameter d and wall thickness t is subjected to an internal pressure p. If there is a small circular hole on the wall, find the theoretical maximum longitudinal and circumferential stress around the hole in terms of p, d, and E.

Solution:

Thin-walled pressure vessel ($t << d$)

Pressure – p

Hoop stress, $\sigma_t = \frac{pd}{2t}$.

Longitudinal stress, $\sigma_l = \frac{pd}{4t}$.

As the hole is very small, the problem can be restructured as,

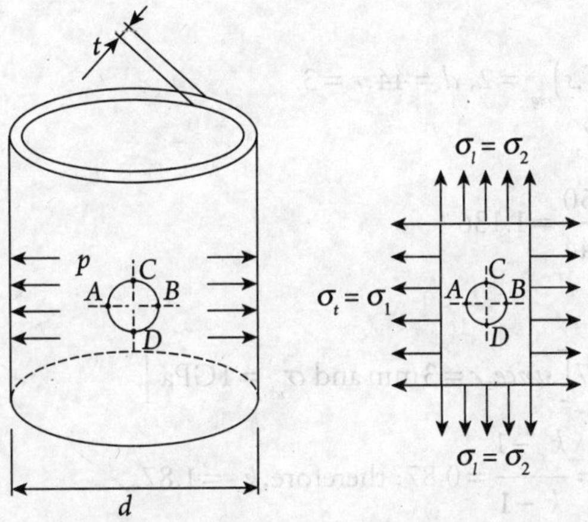

For horizontal section $A-B$,

$$\sigma_{max} = 3\sigma_1 - \sigma_2 = 3\left(\frac{pd}{2t}\right) - \frac{pd}{4t} = \frac{5}{4}\frac{pd}{t}.$$

For vertical section $C-D$,

$$\sigma_{max} = 3\sigma_2 - \sigma_1 = 3\left(\frac{pd}{4t}\right) - \frac{pd}{2t} = \frac{1}{4}\frac{pd}{t}.$$

Therefore,

$$\sigma_{max\,A-B} = \frac{5}{4}\frac{pd}{t};\ \sigma_{max\,C-D} = \frac{pd}{4t}.$$

Example 9.3

A 50 mm diameter grooved shaft, as shown in the figure, is subjected to completely reversed bending load due to a fluctuating force *F*. The shaft material has an ultimate tensile stress $\sigma_{ut} = 1.0\,\text{GPa}$ and an endurance limit in bending $= 0.9\,\text{GPa}$. How great can *F* be for a factor of safety of 2? For modifying endurance limit, use only fatigue strength reduction factor K_f.

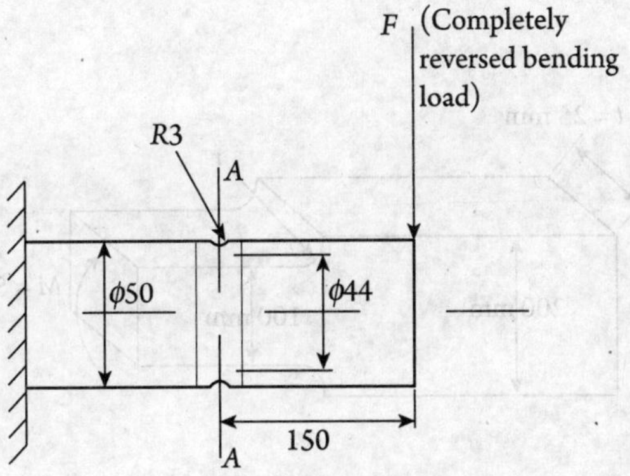

Solution:

$\sigma_{ut} = 1\,\text{GPa}, \sigma_e = 0.9\,\text{GPa}, (f.s)_{req} = 2, d = 44, r = 3.$

At the thin section $A - A'$,

$$r/d = \frac{3}{44} = 0.068;\ D/d = \frac{50}{44} = 1.136$$

From Figure 9.16, $k_t = 2$.

From Figure 9.18, $q = 0.87\left[since, r = 3\,\text{mm and } \sigma_{ut} = 1\,\text{GPa}\right]$.

From equation (9.4.2), $q = \dfrac{k_f - 1}{k_t - 1} = 0.87$; therefore, $k_f = 1.87$.

Modified endurance limit, $\sigma_e' = \sigma_e / k_f = \dfrac{0.9}{1.87} = 0.481\,\text{GPa}$.

Bending stress at the groove

$$\sigma_b = \frac{F \times 0.15}{\left(\frac{\pi}{64} d^4\right) \times \left(2/d\right)} = 17945F\ Pa = \sigma_a\ \left(stress\ amplitude\right)\ \left[using \frac{M}{I} = \frac{\sigma_b}{y}\right].$$

From Soderberg's equation for completely reversed cycle

$$\frac{\sigma_a}{\sigma_e'} = \frac{1}{f.s},$$

i.e., $\dfrac{17945\text{F}}{0.481 \times 10^9} = \dfrac{1}{2}$

or, $F = 13.4\,\text{kN}$.

Example 9.4

The plate shown in the figure below is made of steel with yield stress 350 MPa. The plate is subjected to a bending moment of 5 kNm. Find the fillet radius *r* suitable for applied moment if the plate is 25 mm thick.

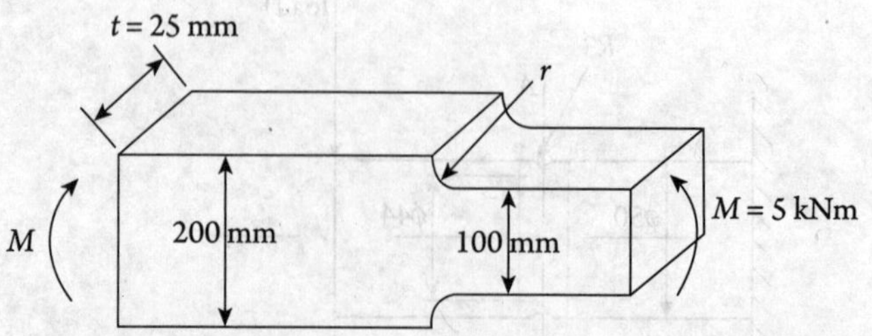

Solution:

$\sigma_{yt} = 350\,\text{MPa}; t = 25\,\text{mm}; M = 5\,\text{kNm}; d = 100\,\text{mm}; D = 200\,\text{mm}.$

We need to find the fillet radius r.

At the fillet section, $\frac{r}{d} = \frac{r}{100}; \frac{D}{d} = \frac{200}{100} = 2$. Here, we need to find k_t.

To find the bending stress

$$\frac{M}{I} = \frac{\sigma_b}{y}$$

$$\Rightarrow \frac{5 \times 10^3}{\frac{1}{12} t d^3} = \frac{\sigma_b}{d/2}$$

$$\Rightarrow \sigma_b = 120\,\text{MPa}.$$

Therefore,

$$k_t = \frac{\sigma_{yt}}{\sigma_b} = \frac{350}{120} = 2.9167.$$

From Figure 9.13 for $k_t = 2.9167$ and $\frac{D}{d} = 2$, $\frac{r}{d} = 0.05$.

Hence, $r = 5\,\text{mm}$. ... (Ans.)

Example 9.5

The stepped shaft is supported at two ends and is loaded at the center as shown. Find the length of the shaft if the bending stress at the fillet equals the bending stress at the shaft center.

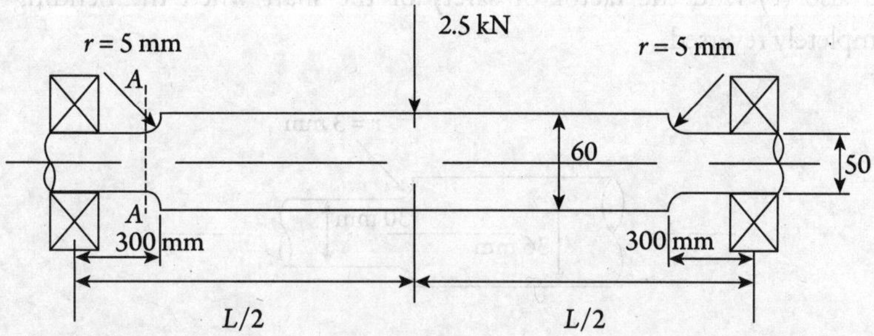

Solution:

$D = 60\,\text{mm}, d = 50\,\text{mm}, r = 5\,\text{mm}.$

Given:

Bending stress at section $A-A$ = bending stress at shaft center.

The moment at section $A-A = \left(\frac{2.5\times10^3}{2}\right)\times 0.3\,\text{Nm} = 375\ \text{Nm}$

$$\sigma_{A-A} = \frac{32M_{A-A}}{\pi d^3} = \frac{32\times375}{\pi\times(0.05)^3}\text{N/m}^2 = 30.56\ \text{MPa}$$

$$\frac{D}{d} = \frac{60}{50} = 1.2; \frac{r}{d} = \frac{5}{50} = 0.1.$$

From Figure 9.14, $k_t = 1.55$

$$\sigma_{max} = k_t \times \sigma_{A-A} = 3.56\times1.55 = 47.37\ \text{MPa}.$$

Bending stress at the shaft center

$$\sigma_b = \frac{32M_b}{\pi d^3} = \frac{32\times2.5\times10^3\left(L/4\right)}{\pi\times(0.06)^3} = 29.49\text{L MPa}.$$

By the given condition,

$$\sigma_b = \sigma_{max}.$$

Solving this, we get

$L = 1.606\,\text{m}$. ... (Ans.)

Example 9.6

Dimensions of a steel stepped shaft of ultimate tensile stress of 1 GPa and endurance limit of 500 MPa are shown in the figure below. Find the maximum stress developed in the shaft when (a) it is subjected to a bending moment of 200 Nm, (b) it is subjected to a twisting moment of 100 Nm, and also (c) find the factor of safety of the shaft when the bending moment of 200 Nm is completely reversed.

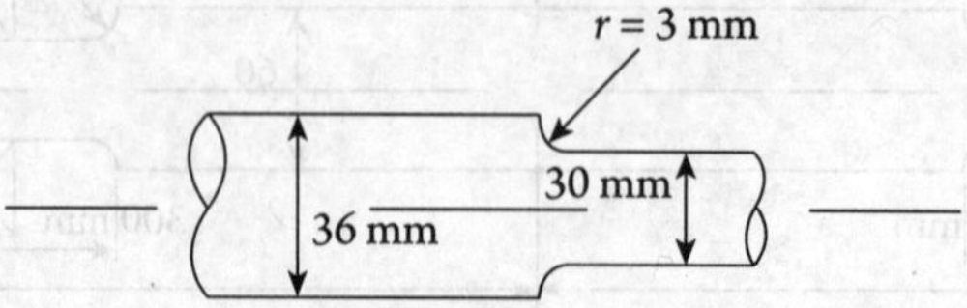

Solution:

$\sigma_{ut} = 1\text{GPa}; \sigma_e = 500\,\text{MPa}; d = 30\,\text{mm}; D = 36\,\text{mm}, r = 3\,\text{mm}; D/d = 1.2; r/d = 3/30 = 0.1.$

(a) When the bending moment is 200 Nm

$$\sigma_b = \frac{32M_b}{\pi d^3} = 75.48\,\text{MPa}.$$

From Figure 9.14, $k_t = 1.55$

$\sigma_{max} = k_t \times \sigma_b = 1.55 \times 75.48 = 117\,\text{MPa}.$

(b) Twisting moment of 100 Nm

$$\sigma_t = \frac{16M_t}{\pi d^3} = 18.87\,\text{MPa}.$$

From Figure 9.15, $k_t = 1.4$

$\sigma_{max} = k_t \times \sigma_t = 26.42\,\text{MPa}.$

(c) Completely reversed bending,

Mean stress : $\sigma_m = 0$; *Stress amplitude* : $\sigma_a = \sigma_b = 75.48\,\text{MPa}$; $k_t = 1.55$.

From Figure 9.18: $q = 0.87$;

$$q = \frac{k_f - 1}{k_t - 1} \Rightarrow 0.87 = \frac{k_f - 1}{1.55 - 1} \Rightarrow k_f = 1.478.$$

Modified endurance limit, $\sigma_e' = \sigma_e / k_f = \dfrac{500}{1.478} = 338.3\,\text{MPa}.$

From Soderberg's equation $\dfrac{\sigma_a}{\sigma_e'} = \dfrac{1}{f.s}$.

We get, $f.s = 4.48.$ ... (Ans.)

Example 9.7

The flat plate shown in the figure is 10-mm thick and is subjected to an axial force of 20 kN. Determine the maximum stress developed in the plate.

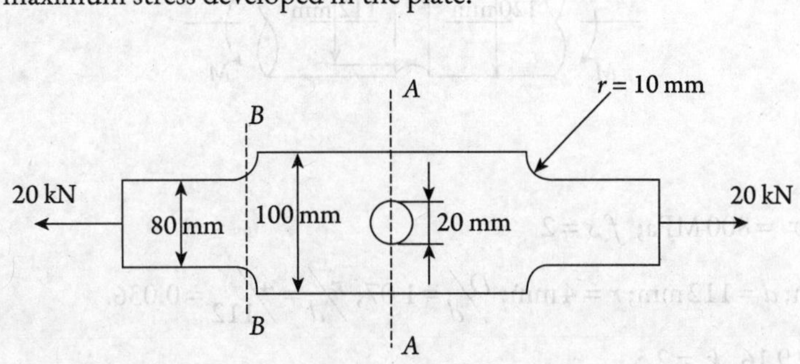

Solution:

$w = 100\,\text{mm}; d = 20\,\text{mm}; r = 10\,\text{mm}; t = 10\,\text{mm}.$

(a) At section $A-A$, $\sigma_{A-A} = \dfrac{20 \times 10^3}{(100-20) \times 10} = 25\,\text{MPa}.$

Stress concentration factor due to hole, $d/w = 20/100 = 0.2$.

From Figure 9.12, $k_t = 2.5$

$$\sigma_{max\,hole} = 25 \times 2.5 = 62.5\,\text{MPa} \quad \text{... (i)}$$

(b) At section $B-B$, $\sigma_{B-B} = \dfrac{20\times10^3}{80\times10} = 25$ MPa.

Stress concentration factor due to fillet radius,

$r/h = 10/80 = 0.125$

$H/h = 100/80 = 1.25$.

From Figure 9.13,

$k_{t\,fillet} = 1.72;$

$$\sigma_{max\,fillet} = 25 \times 1.72 = 43\,\text{MPa} \quad \text{... (ii)}$$

Therefore, the maximum stress induced, $\sigma_{max} = \sigma_{max_{hole}} = 62.5\,\text{MPa}$. ...(Ans.)

Example 9.8

A grooved shaft, as shown in the figure, is subjected to a completely reversed bending moment M. The shaft material has an ultimate tensile stress of 1.4 GPa and endurance limit of 800 MPa. To maintain a factor of safety of 2, what is the maximum value of the bending moment?

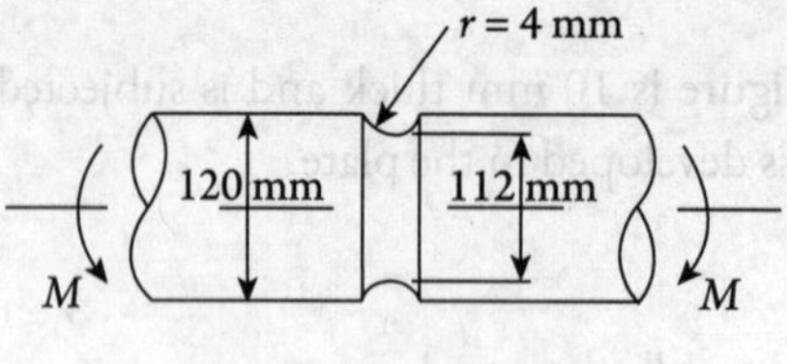

Solution:

$\sigma_{ut} = 1.4\,\text{GPa}$; $\sigma_e = 800\,\text{MPa}$; $f.s = 2$

$D = 120\,\text{mm}$; $d = 112\,\text{mm}$; $r = 4\,\text{mm}$; $D/d = 1.07$; $r/d = 4/112 = 0.036$.

From Figure 9.16, $k_t = 2.5$.

From Figure 9.17, $q = 0.95$;

$$q = \frac{k_f - 1}{k_t - 1} = 0.95.$$

Therefore, $k_f = 2.425$.

Modified endurance limit,

$$\sigma_e' = \sigma_e / k_f = \frac{800}{2.425} = 329.89\,\text{MPa}.$$

From Soderberg's equation for completely reversed bending moment,

Mean stress: $\sigma_m = 0$;

Stress amplitude: $\sigma_a = \sigma_b$;

$$\frac{\sigma_v}{\sigma_e'} = \frac{1}{f.s}$$

$$\Rightarrow \sigma_a = \sigma_b = 164.945\,\text{MPa}.$$

Also, from the equation of pure bending, $\sigma_b = \dfrac{32M_b}{\pi d^3}$

$$\Rightarrow M_b = 22740\,\text{Nm} = 22.74\,\text{kNm}.$$ ... (Ans.)

Exercises

1. A plate with a centrally located hole of radius "*a*" is subjected to a uniform tensile stress "σ" at the two longitudinal ends. Show that the maximum tangential stress at the hole radius is 3σ and at radius 2a is 1.22σ.
2. A stepped shaft has the dimensions as shown in the figure below. Find the maximum stresses:
   (a) If only a direct tensile load *P* = 300 kN acts.
   (b) If a bending moment *M* = 15 kNm acts at the step in addition to the direct tensile load *P* = 300 kN.
   (c) If a torque *T* = –20 kNm acts in addition to the moment *M* = 15 kNm at the step and a direct tensile load *P* = 300 kN.

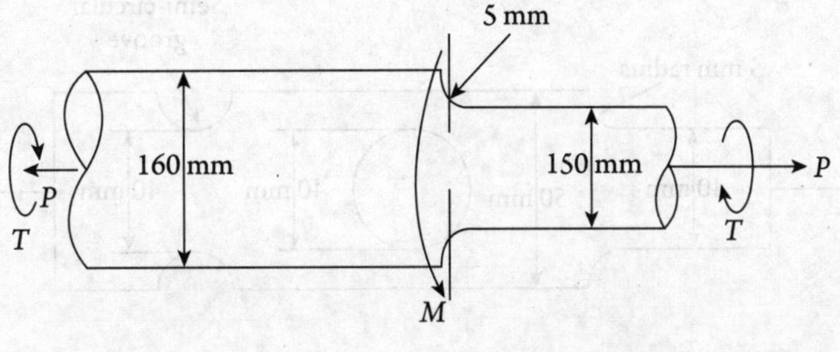

[34.65 MPa, 116.18 MPa, 155.43 MPa]

3. A cylindrical shaft with a circumferential semicircular groove of 5-mm radius, as shown in the figure below, is subjected to a stable bending moment *M*. If the allowable working stress of the material is 175 MPa, find the allowable bending moment *M*.

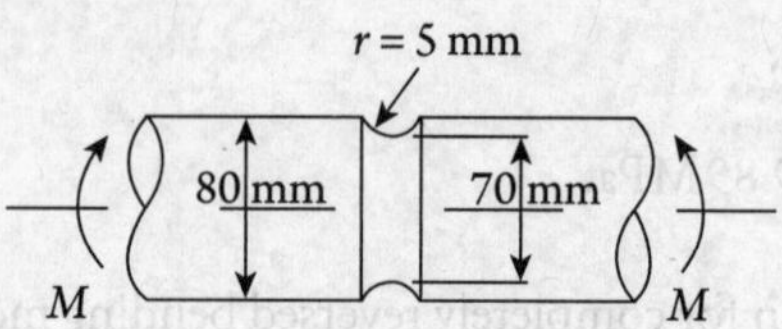

[2.8 kNm]

4. Discuss the limitations of the use of stress concentration factors, determined theoretically or by using photoelastic methods in cases:
   (a) Where machine parts are gradually stressed beyond elastic limit.
   (b) Where machine parts are subjected to fluctuating loads, and they fail by fatigue.
5. A cantilever beam with a circumferential semi-circular groove (as shown in the figure) is subjected to an axial load P, bending moment M at the grooved section, and torque T at the free end. Determine the maximum principal stress at the root of the notch, point A.

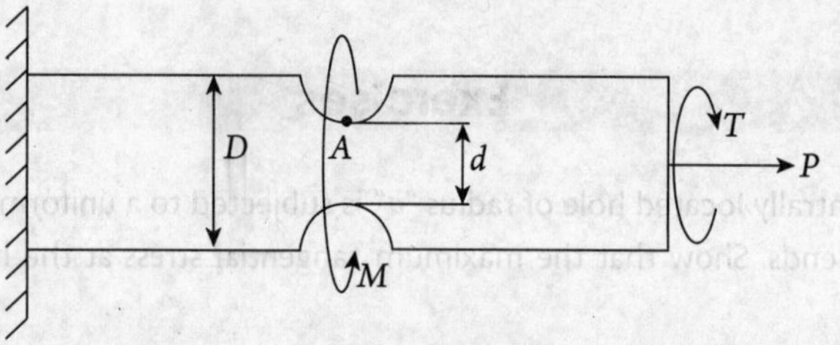

$\left[\frac{2}{\pi d^3}[(k_1 pd + 8k_2 M) + (\sqrt{(k_1 pd + 8k_2 M)^2 + 64k_3^2 T^2}\,],\right.$ where k_1, k_2, and k_3 are stress concentration factors due to axial bending and torsional loadings]

6. A rectangular plate of mild steel with a yield strength of 250 MPa and thickness of 10 mm is axially loaded as shown below:

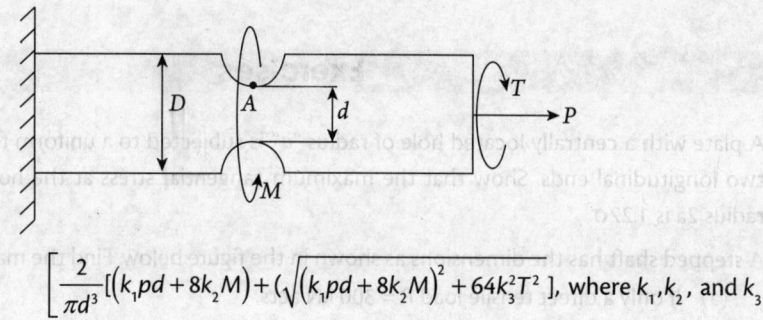

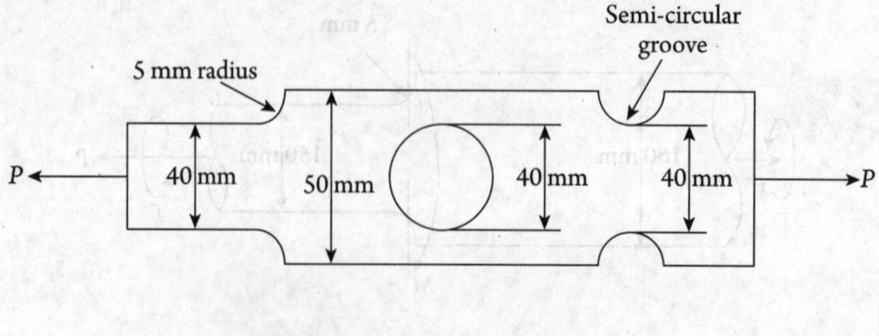

Determine the value of P that would cause yielding at some point in the bar.

[Yielding would first be initiated at the hole section with P = 12 kN]

10

Thermoelasticity

Learning Objectives

After careful study of this chapter, students should be able to do the following:

LO1: Define thermal stress and thermal strain.

LO2: Describe equilibrium equation in the presence of thermal stresses.

LO3: Analyze plane strain and plane stress compatibility in thermoelasticity problems.

LO4: Evaluate stress function formulation in thermoelasticity problems.

LO5: Plan polar coordinate formulation for thermoelasticity problems.

10.1 INTRODUCTION [LO1]

There are many applications where structures or machine parts are subjected to significant changes in temperature, for example, turbine blades, high-speed rotating machinery, and boilers in thermal power plants. Large thermal stresses may be developed in such applications, and sometimes such stresses may exceed the yield limit. It is therefore necessary to make provisions in the design of components to avoid failure due to thermal stresses. If the ends of a rod or any other machine parts are rigidly fixed such that the expansion or compression is prevented and the temperature is changed, tensile or compressive stress would be set up, and in simple terms, these stresses are called thermal stresses. In a long steam pipe, expansion joints are sometimes inserted, and in bridges, one end may be rigidly fastened to the main structure while the other end rests on rollers to avoid thermal stresses. In simple terms, this may be demonstrated considering a rod of length l and cross-sectional area A fixed at both ends (Figure 10.1) and temperature is raised by ΔT. This would produce a thermal strain $\in_t$ in the rod such that

$$\in_t = \alpha \Delta T, \tag{10.1.1}$$

where α is the coefficient of thermal expansion. Since the rod is not free to expand, a compressive stress σ_c would be developed in the rod, and this is given by

$$\sigma_c = E\alpha\Delta T. \tag{10.1.2}$$

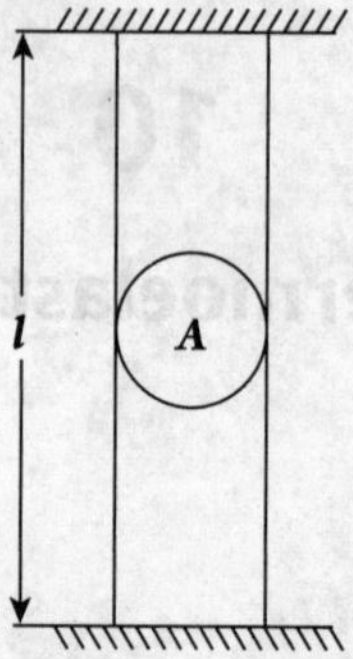

Figure 10.1 Rod fixed at both ends

If the rise in temperature is significant, the rod may buckle, which is of serious consideration in the design of machine parts or structures. To avoid this, we need to find the critical force P_{cr} for buckling. Using the basic buckling criterion, this can be given for a column pin ended at both ends, by

$$P_{cr} = \frac{\pi^2 EI}{l^2} = AE\alpha\Delta T, \tag{10.1.3}$$

where I is the least moment of inertia of the constant cross-section column (rod) and A is the cross-sectional area of the column.

Thermal stresses may also be induced due to nonuniform heating. One common example of this type of stress is the fracture of a thick glass when hot water is poured into it. This is because of the nonuniform expansion of the glass across the wall thickness. Nonuniform volume expansion can also cause thermal stress. If a tightly capped bottle filled with water is heated, then there may be a fracture in the glass. This is because the thermal expansion coefficient of water is greater than that of glass. There may be many different cases of thermal stresses in engineering applications, but we will discuss only general cases of thermal effects on stress–strain relations of materials.

10.2 GENERAL FORMULATIONS [LO2]

For general three-dimensional (3D) thermoelastic problems, many of our previous equations, such as strain–displacement, strain–compatibility equations, and equilibrium equations, are still valid. These are to be used with Hooke's Law modified for thermal effect. Here the strain field consists of an extra term shown in equation (10.2.1) as follows:

$$\epsilon_x = \frac{1}{E}\left[\sigma_x - \upsilon\sigma_y - \upsilon\sigma_z\right] + \alpha T$$

$$\epsilon_y = \frac{1}{E}\left[\sigma_y - \upsilon\sigma_x - \upsilon\sigma_z\right] + \alpha T$$

$$\epsilon_z = \frac{1}{E}\left[\sigma_z - \upsilon\sigma_x - \upsilon\sigma_y\right] + \alpha T \tag{10.2.1}$$

$$\gamma_{xy} = \frac{\tau_{xy}}{G}\ \gamma_{yz} = \frac{\tau_{yz}}{G}\ \gamma_{zx} = \frac{\tau_{zx}}{G}. \tag{10.2.2}$$

Here the change in temperature is given by T for simplicity. In equation (10.2.2), shear strains are not affected by temperature change, since free thermal expansion does not cause any distortion.

Adding up the strain components in equation (10.2.1), we have

$$\epsilon_x + \epsilon_y + \epsilon_z = \frac{1-2\upsilon}{E}\left(\sigma_x + \sigma_y + \sigma_z\right) + 3\alpha T. \tag{10.2.3}$$

Manipulating equations (10.2.1) and (10.2.3), we have

$$\sigma_x = \left(\epsilon_x + \epsilon_y + \epsilon_z\right)\frac{E\upsilon}{(1-2\upsilon)(1+\upsilon)} + 2\frac{E}{2(1+\upsilon)}\epsilon_x - \frac{E\alpha T}{(1-2\upsilon)}.$$

As before, if we define

$$\lambda = \frac{E\upsilon}{(1-2\upsilon)(1+\upsilon)} \text{ and } G = \frac{E}{2(1+\upsilon)},$$

we have

$$\sigma_x = \left(\epsilon_x + \epsilon_y + \epsilon_z\right)\lambda + 2G\epsilon_x - \frac{E\alpha T}{(1-2\upsilon)}.$$

In similar arguments, we may write two more such equations and write all the stress components in terms of strains as follows:

$$\sigma_x = \left(\epsilon_x + \epsilon_y + \epsilon_z\right)\lambda + 2G\epsilon_x - \frac{E\alpha T}{(1-2\upsilon)}. \tag{10.2.4}$$

$$\sigma_y = \left(\epsilon_x + \epsilon_y + \epsilon_z\right)\lambda + 2G\epsilon_y - \frac{E\alpha T}{(1-2\upsilon)}$$

$$\sigma_z = \left(\epsilon_x + \epsilon_y + \epsilon_z\right)\lambda + 2G\epsilon_z - \frac{E\alpha T}{(1-2\upsilon)}$$

$$\tau_{xy} = G\gamma_{xy}\quad \tau_{yz} = G\gamma_{yz}\quad \tau_{zx} = G\gamma_{zx}. \tag{10.2.5}$$

Equations (10.2.4) and (10.2.5) are similar to those derived in boundary value problems shown in equation (4.3.1) except that one extra term due to thermal effect appears.

Now equations of equilibrium in 3D in the absence of body force, as shown in equation (3.6.2), are reproduced here for convenience

$$\frac{\partial \sigma_x}{\partial x} + \frac{\partial \tau_{xy}}{\partial y} + \frac{\partial \tau_{xz}}{\partial z} = 0$$

$$\frac{\partial \tau_{xy}}{\partial x} + \frac{\partial \sigma_y}{\partial y} + \frac{\partial \tau_{yz}}{\partial z} = 0$$

$$\frac{\partial \tau_{xz}}{\partial x}+\frac{\partial \tau_{yz}}{\partial y}+\frac{\partial \sigma_z}{\partial z}=0. \tag{10.2.6}$$

Substituting σ_x, τ_{xy}, and τ_{xz} from equations (10.2.4) and (10.2.5) in the first equilibrium equation, shown above, we have

$$\left[\lambda\frac{\partial}{\partial x}\left(\epsilon_x+\epsilon_y+\epsilon_z\right)+2G\frac{\partial \epsilon_x}{\partial x}-\frac{E\alpha}{1-2\upsilon}\frac{\partial T}{\partial x}\right]+G\frac{\partial \gamma_{xy}}{\partial y}+G\frac{\partial \gamma_{xz}}{\partial z}=0.$$

Writing the strain components in terms of displacements, we have

$$\left[\lambda\frac{\partial}{\partial x}\left(\frac{\partial u}{\partial x}+\frac{\partial v}{\partial y}+\frac{\partial w}{\partial z}\right)+2G\frac{\partial^2 u}{\partial x^2}-\frac{E\alpha}{1-2\upsilon}\frac{\partial T}{\partial x}\right]+G\left[\frac{\partial^2 u}{\partial y^2}+\frac{\partial^2 v}{\partial x\partial y}\right]+G\left[\frac{\partial^2 u}{\partial z^2}+\frac{\partial^2 w}{\partial x\partial z}\right]=0.$$

On rearrangement, this gives

$$\left[(\lambda+G)\frac{\partial}{\partial x}\left(\frac{\partial u}{\partial x}+\frac{\partial v}{\partial y}+\frac{\partial w}{\partial z}\right)+G\nabla^2 u-\frac{E\alpha}{1-2\upsilon}\frac{\partial T}{\partial x}\right]=0, \tag{10.2.7}$$

where the Laplacian operation $\nabla^2 u$ is given here as

$$\nabla^2 u=\frac{\partial^2 u}{\partial x^2}+\frac{\partial^2 u}{\partial y^2}+\frac{\partial^2 u}{\partial z^2}.$$

Considering all the three equilibrium equations shown above and following the same procedure, we have the following three equations in the absence of any body forces but in the presence of thermal stresses:

$$\left[(\lambda+G)\frac{\partial}{\partial x}\left(\frac{\partial u}{\partial x}+\frac{\partial v}{\partial y}+\frac{\partial w}{\partial z}\right)+G\nabla^2 u-\frac{E\alpha}{1-2\upsilon}\frac{\partial T}{\partial x}\right]=0 \tag{10.2.8}$$

$$\left[(\lambda+G)\frac{\partial}{\partial y}\left(\frac{\partial u}{\partial x}+\frac{\partial v}{\partial y}+\frac{\partial w}{\partial z}\right)+G\nabla^2 v-\frac{E\alpha}{1-2\upsilon}\frac{\partial T}{\partial y}\right]=0$$

$$\left[(\lambda+G)\frac{\partial}{\partial z}\left(\frac{\partial u}{\partial x}+\frac{\partial v}{\partial y}+\frac{\partial w}{\partial z}\right)+G\nabla^2 w-\frac{E\alpha}{1-2\upsilon}\frac{\partial T}{\partial z}\right]=0.$$

These are similar to Navier–Lame equations (4.3.2) derived for problems on the displacement boundary condition except that an extra term appears due to thermal stresses. It can be shown that the last terms in equation (10.2.7) may be considered to be thermal body forces in Navier–Lame equations (4.3.2). We may therefore consider that the temperature change contributes to the generation of stress, strain, and displacement fields.

Thermoelastic stress formulation in line with the Beltrami–Michell equation based on equation (10.2.1) may be developed. Equation (10.2.1) introduces an extra thermal strain term, and thus, the thermoelastic behavior is taken care of. It is more useful to carry out the exercises for two-dimensional (2D) problems, and this is taken up in the next section.

10.3 TWO-DIMENSIONAL THERMOELASTIC PROBLEMS [LO3]

As we have seen earlier, two main 2D assumptions in elasticity problems are plane strain and plane stress. We consider the plane strain problems first.

10.3.1 Plane Strain

Plane strain assumptions apply to long prismatic bodies where temperature varies across any cross-section but remains constant along the longitudinal axis. Therefore, temperature *T* is independent of *z* in Figure 10.2.

Basic assumption of plain strain gives the following displacement field

$$u = u(x, y),\ v = v(x, y),\ w = 0$$

and this gives the following strain field:

$$\epsilon_x = \frac{\partial u}{\partial x}, \epsilon_y = \frac{\partial v}{\partial y}, \gamma_{xy} = \frac{\partial u}{\partial y} + \frac{\partial v}{\partial x}, \epsilon_z = \gamma_{xz} = \gamma_{yz} = 0. \tag{10.3.1}$$

Substituting these in equations (10.2.4) and (10.2.5), we get the 2D stress equations in terms of displacements as

$$\sigma_x = \lambda\left(\frac{\partial u}{\partial x} + \frac{\partial v}{\partial y}\right) + 2G\frac{\partial u}{\partial x} - \frac{E\alpha T}{1-2\upsilon} \tag{10.3.2}$$

$$\sigma_y = \lambda\left(\frac{\partial u}{\partial x} + \frac{\partial v}{\partial y}\right) + 2G\frac{\partial v}{\partial y} - \frac{E\alpha T}{1-2\upsilon}$$

$$\sigma_z = \upsilon\left(\sigma_x + \sigma_y\right) - E\alpha T$$

$$\tau_{xy} = G\left(\frac{\partial u}{\partial y} + \frac{\partial v}{\partial x}\right)$$

$$\tau_{xz} = \tau_{yz} = 0.$$

Equations of equilibrium in 2D reduce to the following simple forms in the absence of body forces

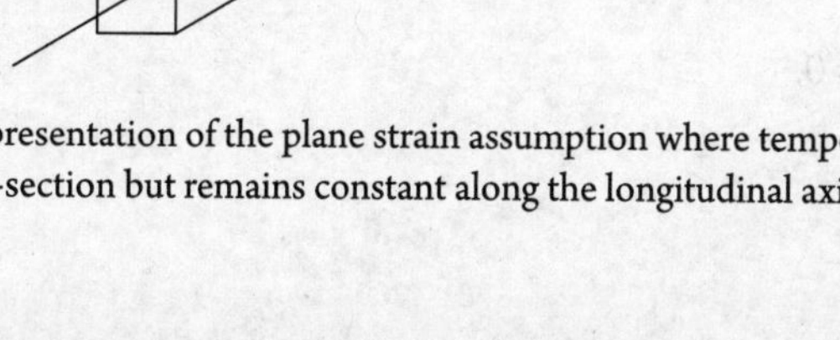

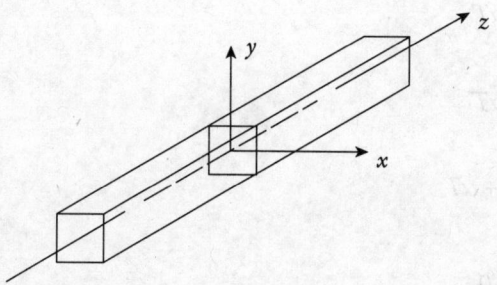

Figure 10.2 Typical representation of the plane strain assumption where temperature varies across any cross-section but remains constant along the longitudinal axis "*z*"

$$\frac{\partial \sigma_x}{\partial x} + \frac{\partial \tau_{xy}}{\partial y} = 0 \tag{10.3.3}$$

$$\frac{\partial \tau_{xy}}{\partial x} + \frac{\partial \sigma_y}{\partial y} = 0.$$

Combining equations (10.3.2) and (10.3.3), we get equations of equilibrium in terms of displacements similar to 3D equations in (10.2.8)

$$(\lambda + G)\frac{\partial}{\partial x}\left(\frac{\partial u}{\partial x} + \frac{\partial v}{\partial y}\right) + G\nabla^2 u - \frac{E\alpha}{1-2\upsilon}\frac{\partial T}{\partial x} = 0 \tag{10.3.4}$$

$$(\lambda + G)\frac{\partial}{\partial y}\left(\frac{\partial u}{\partial x} + \frac{\partial v}{\partial y}\right) + G\nabla^2 v - \frac{E\alpha}{1-2\upsilon}\frac{\partial T}{\partial y} = 0.$$

Here,

$$\nabla^2 = \frac{\partial^2}{\partial x^2} + \frac{\partial^2}{\partial y^2}.$$

Here again, comparing with 3D Navier–Lame equations (4.3.2), extra thermoelastic terms are noted.

Using Hooke's Law in the only nonzero compatibility equation, we get

$$\nabla^2\left(\sigma_x + \sigma_y\right) + \frac{E\alpha}{1-v}\nabla^2 T = 0. \tag{10.3.5}$$

Here again, an additional thermal term is noted.

10.3.2 Plane Stress

As discussed before, in the plane stress state

$$\sigma_z = \tau_{xz} = \tau_{yz} = 0$$

and this applies to bodies that are thin in the z-direction compared to other directions. The thermoelastic strain field may then be given as

$$\epsilon_x = \frac{1}{E}\left[\sigma_x - v\sigma_y\right] + \alpha T$$

$$\epsilon_y = \frac{1}{E}\left[\sigma_y - v\sigma_x\right] + \alpha T$$

$$\epsilon_z = -\frac{1}{E}\left[v\sigma_x + v\sigma_y\right] + \alpha T$$

$$\gamma_{xy} = \frac{\tau_{xy}}{G}; \gamma_{yz} = \gamma_{xz} = 0. \tag{10.3.6}$$

The stress field may be derived from equations (10.3.6) as

$$\sigma_x = \frac{E}{1-v^2}\left[\in_x + v\in_y\right] - \frac{E\alpha T}{1-v} \tag{10.3.7}$$

$$\sigma_y = \frac{E}{1-v^2}\left[\in_y + v\in_x\right] - \frac{E\alpha T}{1-v}$$

$$\tau_{xy} = G\gamma_{xy}.$$

The equilibrium equations and the only nonzero compatibility equation remain the same as in the case of plane strain. Combining these with equations (10.3.7) and (10.3.6), we write the equilibrium equations in terms of displacements as

$$\frac{E}{2(1-v)}\frac{\partial}{\partial x}\left(\frac{\partial u}{\partial x}+\frac{\partial v}{\partial y}\right)+G\nabla^2 u-\frac{E\alpha}{1-v}\frac{\partial T}{\partial x}=0 \tag{10.3.8}$$

$$\frac{E}{2(1-v)}\frac{\partial}{\partial y}\left(\frac{\partial u}{\partial x}+\frac{\partial v}{\partial y}\right)+G\nabla^2 v-\frac{E\alpha}{1-v}\frac{\partial T}{\partial y}=0.$$

And the plane stress compatibility relations reduce to

$$\nabla^2\left(\sigma_x+\sigma_y\right)+E\alpha\nabla^2 T=0. \tag{10.3.9}$$

Boundary conditions for both plain strain and stress thermoelasticity problems would be as discussed in Chapter 4 on boundary value problems, either in terms of traction or displacements at the edges of the domain, and the temperature field must also depend only on in-plane coordinates, such as $T = T(x, y)$.

10.4 STRESS FUNCTION FORMULATION FOR TWO-DIMENSIONAL THERMOELASTICITY PROBLEMS [LO4]

It has been discussed earlier that the stress function ϕ automatically satisfies the equilibrium equations, and in 2D situations, they are given as

$$\sigma_x = \frac{\partial^2\phi}{\partial y^2}, \sigma_y = \frac{\partial^2\phi}{\partial x^2}, \tau_{xy} = -\frac{\partial^2\phi}{\partial x\partial y}.$$

Thermoelastic strain field is given in equation (10.3.6). These equations may be written in terms of the stress function ϕ as

$$\in_x = \frac{1}{E}\left[\frac{\partial^2\phi}{\partial y^2} - v\frac{\partial^2\phi}{\partial x^2}\right] + \alpha T \tag{10.4.1}$$

$$\in_y = \frac{1}{E}\left[\frac{\partial^2\phi}{\partial x^2} - v\frac{\partial^2\phi}{\partial y^2}\right] + \alpha T$$

$$\in_z = -\frac{v}{E}\left[\frac{\partial^2 \phi}{\partial y^2} + \frac{\partial^2 \phi}{\partial x^2}\right] + \alpha T$$

$$\gamma_{xy} = -\frac{2(1+v)}{E}\frac{\partial^2 \phi}{\partial x \partial y}; \gamma_{yz} = \gamma_{xz} = 0.$$

For the plane stress state, the following compatibility equation needs to be satisfied

$$\frac{\partial^2 \in_x}{\partial y^2} + \frac{\partial^2 \in_y}{\partial x^2} = \frac{\partial^2 \gamma_{xy}}{\partial x \partial y}.$$

Substituting the strain components from equation (10.4.1) in the above compatibility equation, we have

$$\frac{1}{E}\left[\frac{\partial^4 \phi}{\partial y^4} - v\frac{\partial^4 \phi}{\partial x^2 \partial y^2}\right] + \alpha\frac{\partial^2 T}{\partial y^2} + \frac{1}{E}\left[\frac{\partial^4 \phi}{\partial x^4} - v\frac{\partial^4 \phi}{\partial x^2 \partial y^2}\right] + \alpha\frac{\partial^2 T}{\partial x^2} = -\frac{2(1+v)}{E}\frac{\partial^4 \phi}{\partial x^2 \partial y^2}.$$

This may be written in a simplified form as

$$\nabla^4 \phi + E\alpha \nabla^2 T = 0. \quad (10.4.2)$$

The equilibrium and compatibility equations are identical for plane strain and stress states, but because of the difference in the form of Hooke's Law, plane stress and plane strain theories give slightly different displacement equilibrium and stress compatibility relations. However, by simple interchange of elastic moduli, as shown in Table 10.1, one theory can be transformed into other. Following the transformation relations in Table 10.1, the compatibility relation for plane strain may be obtained.

In the absence of both body forces and thermal stresses, equations (10.4.2) and that for plane strain reduce to

$$\nabla^4 \phi = 0,$$

which is the previously discussed biharmonic equation that needs to be satisfied for any elasticity problem. In a more general form, the governing equation is given by

Table 10.1 Elastic moduli and thermal coefficient transformation relations for plane stress and plane strain thermoelastic problems

Plane Stress	*Plane Strain*
$E^* = E$	$E^* = \frac{E}{(1-v^2)}$
$v^* = v$	$v^* = \frac{v}{(1-v)}$
$\alpha^* = \alpha$	$\alpha^* = (1+v)\alpha$

$$\nabla^4\phi+\left(1-v^*\right)\nabla^2\Omega+E^*\alpha^*\nabla^2 T=0. \tag{10.4.3}$$

Here body forces are given in terms of potential function Ω as

$$X=-\frac{\partial\Omega}{\partial x},\ Y=-\frac{\partial\Omega}{\partial y},\ Z=-\frac{\partial\Omega}{\partial z}$$

and for elastic moduli and thermal coefficient, Table 10.1 applies.

10.5 POLAR COORDINATE FORMULATIONS OF TWO-DIMENSIONAL THERMOELASTICITY PROBLEMS USING STRESS FUNCTION APPROACH [LO5]

As discussed in Chapter 4, Airy's stress function ϕ in polar coordinates (r, θ) is defined as

$$\sigma_r=\frac{1}{r}\frac{\partial\phi}{\partial r}+\frac{1}{r^2}\frac{\partial^2\phi}{\partial\theta^2}$$
$$\sigma_\theta=\frac{\partial^2\phi}{\partial r^2} \tag{10.5.1}$$
$$\tau_{r\theta}=-\frac{\partial}{\partial r}\left(\frac{1}{r}\frac{\partial\phi}{\partial\theta}\right).$$

This definition satisfies the following equations of equilibrium in 2D cases in the absence of body forces:

$$\frac{\partial\sigma_r}{\partial r}+\frac{1}{r}\frac{\partial\tau_{r\theta}}{\partial\theta}+\frac{\sigma_r-\sigma_\theta}{r}=0$$
$$\frac{\partial\tau_{r\theta}}{\partial r}+\frac{1}{r}\frac{\partial\sigma_\theta}{\partial\theta}+\frac{2\tau_{r\theta}}{r}=0.$$

Governing stress function equations for plane stress (10.4.2) still hold, where

$$\nabla^2=\frac{\partial^2}{\partial r^2}+\frac{1}{r}\frac{\partial}{\partial r}+\frac{1}{r^2}\frac{\partial^2}{\partial\theta^2} \tag{10. 5.2}$$

$$\nabla^4=\nabla^2\nabla^2=\left(\frac{\partial^2}{\partial r^2}+\frac{1}{r}\frac{\partial}{\partial r}+\frac{1}{r^2}\frac{\partial^2}{\partial\theta^2}\right)\left(\frac{\partial^2}{\partial r^2}+\frac{1}{r}\frac{\partial}{\partial r}+\frac{1}{r^2}\frac{\partial^2}{\partial\theta^2}\right).$$

For radially symmetric problems in plane stress states, all the stress components and temperature depend only on radius, and therefore equation (10.5.1) reduces to

$$\sigma_r=\frac{1}{r}\frac{d\phi}{dr}$$
$$\sigma_\theta=\frac{d^2\phi}{dr^2}=\frac{d}{dr}\left(r\sigma_r\right) \tag{10.5.3}$$
$$\tau_{r\theta}=0$$

and using equation (10.5.2) for the plane stress state, equation (10.4.2) reduces to

$$\frac{1}{r}\frac{d}{dr}\left[r\frac{d}{dr}\left\{\frac{1}{r}\frac{d}{dr}\left(r\frac{d\phi}{dr}\right)\right\}\right]+E\alpha\frac{1}{r}\frac{d}{dr}\left(r\frac{dT}{dr}\right)=0. \tag{10.5.4}$$

Combining with equation (10.5.3), we may write equation (10.5.4) in terms of radial stress σ_r as

$$\frac{1}{r}\frac{d}{dr}\left[r\frac{d}{dr}\left\{\frac{1}{r}\frac{d}{dr}\left(r^2\sigma_r\right)\right\}\right]+E\alpha\frac{1}{r}\frac{d}{dr}\left(r\frac{dT}{dr}\right)=0. \tag{10.5.5}$$

On repeated integration, this gives

$$\sigma_r=\frac{A}{r^2}+B+\frac{C}{4}(2\log r-1)-\frac{E\alpha}{r^2}\int Trdr, \tag{10.5.6}$$

where A, B, and C are constants, which may be determined from the boundary conditions. From equation (10.5.1), considering the radially symmetrical cases, we have

$$\sigma_\theta=\frac{d^2\phi}{dr^2}=\frac{d}{dr}\left(r\sigma_r\right)=-\frac{A}{r^2}+B+\frac{C}{4}(1+2\log r)+\frac{E\alpha}{r^2}\left(\int Trdr-Tr^2\right). \tag{10.5.7}$$

For an example, let us consider a thin circular disk of radius a. The edges are traction free but raised to a temperature $T = T_o r^2 \sin^2(\theta)$, where T_o is a constant. Considering this as a plane stress axisymmetric problem, the governing equation is equation (10.4.2)

$$\nabla^4\phi+E\alpha\nabla^2 T=0.$$

Substituting the temperature distribution in equation (10.4.2) we have $\nabla^4\phi=-2E\alpha T_0$ and a particular solution

$$\phi^{(p)}=-\frac{E\alpha T_o r^4}{32}. \tag{10.5.8}$$

Substituting this in equation (10.5.3), we get the stresses due to $\phi^{(p)}$ and this is given as

$$\sigma_r=-\frac{E\alpha T_o r^2}{8}$$

$$\sigma_\theta=-\frac{3E\alpha T_o r^2}{8} \tag{10.5.9}$$

$$\tau_{r\theta}=0.$$

Now, in order to keep the edges stress free, we need to superimpose a uniform tension of $\frac{E\alpha T_o a^2}{8}$, as shown in Figure 10.3. This gives the resultant stress field as

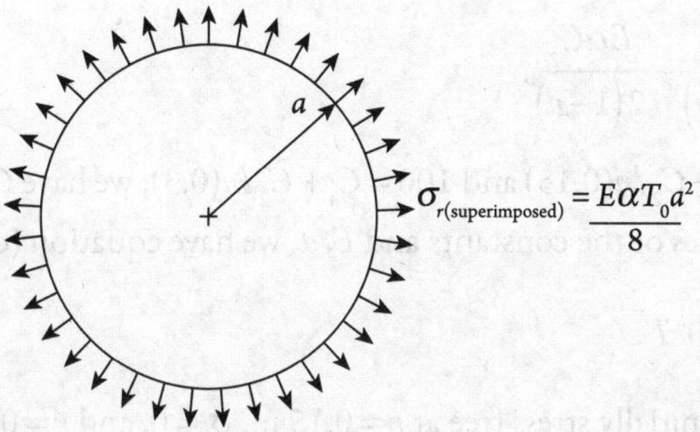

Figure 10.3 A thin disk of radius a is initially traction-free at the edges and its temperature is raised to $T = T_0 r^2 \sin^2\theta$

$$\sigma_r = \frac{E\alpha T_o\left(a^2 - r^2\right)}{8}$$

$$\sigma_\theta = \frac{E\alpha T_o\left(a^2 - 3r^2\right)}{8} \qquad (10.5.10)$$

$$\tau_{r\theta} = 0.$$

Example 10.1

A steel cylinder with an internal radius of 150 mm and external radius 300 mm is subjected to a temperature distribution of the form $T = C_1 + C_2 logr$, where r is the radius. The cylinder is initially unstressed and if the temperatures at the inside and outside surfaces are maintained at 150°C and 100°C respectively, find the radial and circumferential stresses setup in the cylinder. Take E = 200 GPa, $\alpha = 11\times10^{-6}$/°C, and v = 0.3.

Solution:

Here we may use equations (10.5.6) and (10.5.7). However, since the stresses developed from the displacement solutions for isothermal cases do not contain the logarithmic term, we may neglect the terms here. This ensures finite stresses at the origin. The equations therefore reduce to

$$\sigma_r = B + \frac{A}{r^2} - \frac{E\alpha}{r^2}\int Trdr \qquad \text{...(a)}$$

$$\sigma_\theta = B - \frac{A}{r^2} + \frac{E\alpha}{r^2}\left(\int Trdr - Tr^2\right). \qquad \text{...(b)}$$

With a steady heat flow in cases of thick cylinders and with the assumptions of plane strain, we have $E\alpha$ as $E\alpha/(1 - v)$. This gives $\frac{rdT}{dr} = C_2$, a constant leading to $T = C_1 + C_2 ln\ r$, and the equations (a) and (b) become

$$\sigma_r = B + \frac{A}{r^2} - \frac{E\alpha T}{2(1-v)} \qquad \text{...(c)}$$

$$\sigma_\theta = B - \frac{A}{r^2} - \frac{E\alpha T}{2(1-v)} - \frac{E\alpha C_2}{2(1-v)}. \quad \text{...(d)}$$

Now since $150 = C_1 + C_2\, ln(0.15)$ and $100 = C_1 + C_2\, ln(0.3)$, we have $C_1 = 285.7$ and $C_2 = 71.43$. Substituting the values of the constants and E, α, we have equation (c) as follows:

$$\sigma_r = B + \frac{A}{r^2} - 1.6\times10^6 T.$$

Since the cylinder is initially stress free at $r = 0.15$ m, $\sigma_r = 0$, and $r = 0.3$ m, $\sigma_r = 0$.

This gives $A = 2.4\times10^6$ and $B = 133.3\times10^6$ and we have

$$\sigma_r = 133.3\times10^6 + \frac{2.4\times10^6}{r^2} - 1.6\times10^6 T$$

$$\sigma_\theta = 133.3\times10^6 - \frac{2.4\times10^6}{r^2} - 1.6\times10^6 T - 628.5\times10^6.$$

Example 10.2

Show that in the thermoelastic plane strain state, the only nonzero compatibility equation reduces to

$$\nabla^2\left(\sigma_x + \sigma_y\right) + \frac{E\alpha}{1-v}\nabla^2 T = 0.$$

Solution:

For the plane strain problem, the only nonzero strain compatibility condition is

$$\frac{\partial^2 \in_x}{\partial y^2} + \frac{\partial^2 \in_y}{\partial x^2} = \frac{\partial^2 \gamma_{xy}}{\partial x \partial y}.$$

Considering the thermal effect, the constitutive relations for the plane strain case are given by

$$\in_x = \frac{1-\nu^2}{E}\sigma_x - \frac{\nu(1+\nu)}{E}\sigma_y + (1+\nu)\alpha T$$

$$\in_y = \frac{1-\nu^2}{E}\sigma_y - \frac{\nu(1+\nu)}{E}\sigma_x + (1+\nu)\alpha T$$

$$\gamma_{xy} = \frac{2(1+\nu)}{E}\tau_{xy}.$$

Putting the strain expressions in the compatibility equation and on simplification, we get

$$\frac{1-\nu}{E}\left(\frac{\partial^2\sigma_x}{\partial y^2} + \frac{\partial^2\sigma_y}{\partial x^2}\right) - \frac{\nu}{E}\left(\frac{\partial^2\sigma_x}{\partial x^2} + \frac{\partial^2\sigma_y}{\partial y^2}\right) + \alpha\left(\frac{\partial^2 T}{\partial x^2} + \frac{\partial^2 T}{\partial y^2}\right) = \frac{2}{E}\frac{\partial^2\tau_{xy}}{\partial x \partial y}. \quad \text{...(i)}$$

Equilibrium equations for the plane strain case in the absence of body force as given by equation (10.3.3) are

$$\frac{\partial \sigma_x}{\partial x}+\frac{\partial \tau_{xy}}{\partial y}=0$$

$$\frac{\partial \tau_{xy}}{\partial x}+\frac{\partial \sigma_y}{\partial y}=0.$$

Differentiating the first equilibrium equation with respect to x and the second equilibrium equation by y and adding them, we get

$$2\frac{\partial^2 \tau_{xy}}{\partial x \partial y}=-\left(\frac{\partial^2 \sigma_x}{\partial x^2}+\frac{\partial^2 \sigma_y}{\partial y^2}\right).$$

Using the above relation in equation (i), we obtain

$$\nabla^2\left(\sigma_x+\sigma_y\right)+\frac{E\alpha}{1-\nu}\nabla^2 T=0.$$

Example 10.3

Develop equations for radial and circumferential stresses in a solid disk of radius a subjected to a linear thermal gradient. Also, find the position for maximum stresses.

Solution:

We begin with equations (10.5.6) and (10.5.7)

$$\sigma_r=B+\frac{A}{r^2}+\frac{C}{4}(2\log r-1)-\frac{E\alpha}{r^2}\int Trdr$$

$$\sigma_\theta=B-\frac{A}{r^2}+\frac{C}{4}(1+2\log r)+\frac{E\alpha}{r^2}\left(\int Trdr-Tr^2\right).$$

To avoid infinite stress at the origin, we neglect the logarithmic term and also the term $\frac{A}{r^2}$, since here r appears in the denominator. This reduces the equations to

$$\sigma_r=B-\frac{E\alpha}{r^2}\int Trdr$$

$$\sigma_\theta=B+\frac{E\alpha}{r^2}\left(\int Trdr-Tr^2\right).$$

Integrating $\int Trdr$ with $T=Kr$, K being an arbitrary constant and clubbing the constant of integration with B, we have

$$\sigma_r=B-\frac{EK\alpha r}{3};\ \sigma_\theta=B-\frac{2EK\alpha r}{3}.$$

In both cases, maximum stress occurs at $r=0$. Constant B may be evaluated using the boundary condition $\sigma_r=0$ at $r=a$. This gives $B=\frac{EK\alpha a}{3}$.

Therefore, both the maximum stresses σ_r and σ_θ equal $\frac{EK\alpha a}{3}$.

Example 10.4

Develop equations for radial and circumferential stresses in a circular ring with inside and outside radii a and b respectively, where the inner and outer boundaries being stress free and the thermal boundary conditions are at $r = a$, $T = T_0$, and $r = b$, $T = 0$.

Solution:

We begin with the equations for radially symmetric problems as given in equations (10.5.6) and (10.5.7), ignoring the logarithmic terms:

$$\sigma_r = B + \frac{A}{r^2} - \frac{E\alpha}{r^2}\int Trdr \qquad \text{...(a)}$$

$$\sigma_\theta = B - \frac{A}{r^2} + \frac{E\alpha}{r^2}\left(\int Trdr - Tr^2\right). \qquad \text{...(b)}$$

Before we apply the stress boundary conditions, we need to consider the thermal boundary conditions.

For radially symmetric cases, the conduction equation reduces to

$$\frac{1}{r}\frac{d}{dr}\left(r\frac{dT}{dr}\right) = 0.$$

This, on integration, gives $T = C_1 + C_2 ln\, r$.

Since at $r = a$, $T = T_0$ and $r = b$, $T = 0$, we have

$$T = \frac{T_0}{\ln\left(\frac{a}{b}\right)}\ln\left(\frac{r}{b}\right)$$

For convenience, we may also write this as: $T = \frac{T_0}{\ln\left(\frac{b}{a}\right)}\ln\left(\frac{b}{r}\right)$.

This gives $\sigma_r = B + \frac{A}{r^2} - \frac{E\alpha}{2}\frac{T_0}{\ln\left(\frac{b}{a}\right)}\ln\left(\frac{r}{b}\right)$.

Now we apply the stress boundary conditions: $\sigma_r = 0$, at $r = a, b$.

This gives: $B = \frac{E\alpha}{2}T_0\left(\frac{a^2}{b^2 - a^2}\right)$ and $A = \frac{E\alpha}{2}T_0\left(\frac{a^2 b^2}{b^2 - a^2}\right)$.

Substituting these in equations (a) and (b), we have

$$\sigma_r = \frac{E\alpha}{2}\frac{T_0}{\ln\left(\frac{b}{a}\right)}\left[-\left(\frac{a^2}{b^2 - a^2}\right)\left(1 - \frac{b^2}{a^2}\right)\ln\left(\frac{b}{a}\right) - \ln\left(\frac{b}{r}\right)\right]$$

$$\sigma_\theta = \frac{E\alpha}{2}\frac{T_0}{\ln\left(\frac{b}{a}\right)}\left[-\left(\frac{a^2}{b^2 - a^2}\right)\left(1 + \frac{b^2}{a^2}\right)\ln\left(\frac{b}{a}\right) + \left(1 - \ln\left(\frac{b}{r}\right)\right)\right].$$

Example 10.5

A thin circular disk is raised to a temperature T such that the stresses and displacements due to heating do not vary over the thickness. Considering axisymmetry, the stresses σ_r and σ_θ satisfy the equilibrium equation

$$\frac{d\sigma_r}{dr}+\frac{\sigma_r-\sigma_\theta}{r}=0.$$

Write the stress components σ_r and σ_θ in terms of strains $\in_r$ and $\in_\theta$, considering the thermal effect on the strains. Write also the equation of equilibrium.

Solution:

Considering the thermal effect, the strain components $\in_r$ and $\in_\theta$ in the polar coordinate system are given by

$$\in_r=\frac{1}{E}\left(\sigma_r-\nu\sigma_\theta\right)+\alpha T$$

$$\in_\theta=\frac{1}{E}\left(\sigma_\theta-\nu\sigma_r\right)+\alpha T.$$

Solving the above set of equations for the stress components σ_r and σ_θ, we get

$$\sigma_r=\frac{E}{1-\nu^2}\left[\in_r+\nu\in_\theta-(1+\nu)\alpha T\right]$$

$$\sigma_\theta=\frac{E}{1-\nu^2}\left[\in_\theta+\nu\in_r-(1+\nu)\alpha T\right].$$

And the equation of equilibrium becomes

$$r\frac{d}{dr}\left(\in_r+r\in_\theta\right)+(1-\nu)\left(\in_r-\in_\theta\right)=(1+\nu)\alpha r\frac{dT}{dr}.$$

Example 10.6

A thin rectangular bar of cross-section 6 mm × 3 mm is pinned at both ends and 200 mm long. The assembly is now heated. At what temperature rise is the bar expected to start buckling? Assume E and α for the bar material to be 200 GPa and 15×10^{-6} per °C respectively.

Solution:

For the rectangular bar under consideration, the geometric and material properties are given by

Length $l=200\text{ mm}=0.2\text{ m}$

Cross-sectional area $A=6\times3\times10^{-6}=18\times10^{-6}\text{ m}^2$

Least moment of inertia of the cross-section $I=\frac{1}{12}\times6\times3^3\times10^{-12}=13.5\times10^{-12}\text{ m}^4$;

$E=200\times10^9$ Pa; $\alpha=15\times10^{-6}/°\text{C}$.

For the bar pinned at both ends, we have the following relation as given by equation (10.1.3)

$$AE\alpha\Delta T = \frac{\pi^2 EI}{l^2}.$$

Solving the above equation for the temperature rise ΔT

$$\Delta T = \frac{\pi^2 I}{A\alpha l^2}.$$

Putting the numerical values in the above equation, we get

$$\Delta T = 12.3 \text{ °C}.$$

Exercises

1. The governing thermoelastic equation in terms of stress function ϕ in the plane stress state in the absence of body forces is given by: $\nabla^4\phi + E\alpha\nabla^2 T = 0$. Show that the polar coordinate formulation for radially symmetric problems is

$$\frac{1}{r}\frac{d}{dr}\left[r\frac{d}{dr}\left(r\frac{d\phi}{dr}\right)\right] + E\alpha\frac{1}{r}\frac{d}{dr}\left(r\frac{dT}{dr}\right) = 0.$$

2. For the 3D thermoelastic problems with no body forces, develop the compatibility equations in terms of stresses, commonly known as Beltrami–Mitchell compatibility equations.

3. A body traction free at the edges is subjected to temperature T (x, y, z). If the coefficient of linear expansion α remains unchanged both in position and direction, show that the body remains stress free only if T is constant or a linear function of x, y, and z. [Hints: Use equations (10.2.1) and (2.7.2) in the absence of traction force and body forces.]

4. Develop equations for radial and tangential stresses in a solid disk of radius a when it is subjected to a thermal gradient of the form $T = C_1 + C_2 r$, where C_1 and C_2 are constants.

$$\left[\sigma_r = B + \frac{A}{r^2} - \frac{E\alpha}{r^2}\left(\frac{C_1 r^2}{2} + \frac{C_2 r^3}{3}\right);\ \sigma_\theta = B - \frac{A}{r^2} + \frac{E\alpha}{r^2}\left(\frac{C_1 r^2}{2} + \frac{C_2 r^3}{3}\right) - E\alpha T\right]$$

5. Find the radial and tangential stresses in the solid disk of Problem 4 if temperature at the center of the disk is 20 °C and the temperature rise across the disk is 100°C. Find also the location of maximum stress in the disk taking a = 200 mm, E = 200 GPa, $\alpha = 11 \times 10^{-6}$/°C.

[58.66 MPa at the center of the disk]

6. A thin rectangular plate, as shown in the figure, is subjected to a parabolic temperature distribution in the y-direction as follows: $T = T_0\left(1 - \frac{4y^2}{h^2}\right)$. The plate is free to expand laterally, but the thermal expansion in the x-axis is completely suppressed. Show that the thermal stress distribution in the x-direction is given by

$$\sigma_x = \frac{2}{3}E\alpha T_0 - E\alpha T_0\left(1 - \frac{4y^2}{h^2}\right),$$

where T_0 is a constant, α is the coefficient of thermal expansion, and E is the modulus of elasticity.

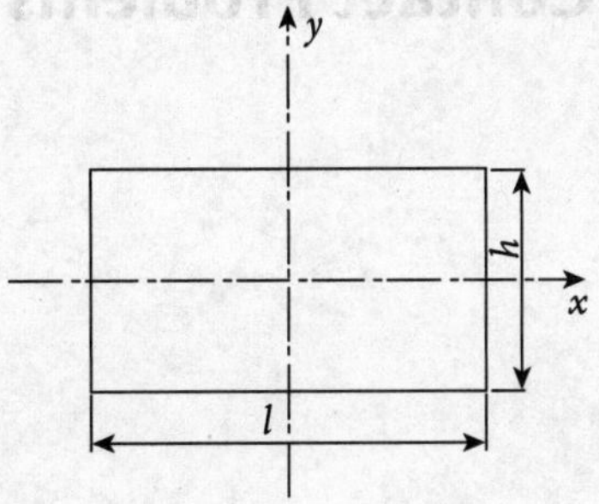

11
Contact Problems

Learning Objectives

After careful study of this chapter, students should be able to do the following:

LO1: Describe the importance of contact stress analysis.

LO2: Describe different types of contact surfaces.

LO3: Solve plane contact problems.

LO4: Explain pressure distribution between curved bodies in contact.

LO5: Evaluate contact area and pressure in spherical contacts.

11.1 INTRODUCTION [LO1]

Stresses developed at the contact between two loaded elastic bodies are generally localized and most machine parts or structures are designed based on the stresses in the main body. However, there are many important machine members where the localized stresses developed at the contact between curved surfaces with initially limited contact area play an important role in their design. Ball or roller bearings, gears, cams, and valve tappets of internal combustion engines are some of the examples of machine parts where contact stresses must be taken into account in order to predict their failure probability.

The localized contact stresses that develop between two curved bodies as they are loaded with small deformations are often referred to as Hertzian stresses, following the work of H. Hertz (1881), who first solved these contact problems elegantly more than a century ago. Since then the topic has received a good deal of attention by the researchers due to its importance in engineering practice and science. Much work has been done on the stress distribution at the Hertzian contact surfaces and sub-surfaces. Ball bearings and gear teeth often fail by pitting. Hertzian stress analysis can precisely locate the depth at which maximum shear stress occurs where cracks may initiate and propagate leading to failure. Thus, a remedy to such failures may be prescribed in terms of limiting stresses. In many rolling contact problems, failure occurs with the initiation of a tiny crack that eventually grows due to repeated contacts. Analysis of crack initiation and growth is often based on Hertzian stress analysis. In this chapter, we shall consider the basics and application of contact stress analysis, beginning with some basic elasticity theory necessary for such analyses.

11.2 TYPES AND GEOMETRY OF CONTACT SURFACES [LO2]

11.2.1 Types of Contacts

Contacts between curved surfaces under load initially start with a point or line contact and then as the deformation grows, contact area between regular curved surfaces, such as spheres and cylinders would develop into circular, elliptical, or rectangular shape. Hertzian contacts have special significance in tribology, which deals with the friction and wear between rubbing solids. Many tribological testing machines have been developed based on Hertzian contact theory. Some typical contacts in friction and wear measurement are shown in Figure 11.1. In many engineering applications and also in the tribological measurements, the key parameters necessary are often the contact area, maximum contact pressure, and contact stresses. Certain basic elasticity theorems are used to arrive at these parameters. We shall discuss these theorems in subsequent sections, but first we shall consider the geometry of the surfaces in contact.

11.2.2 Geometry of Some Common Contacting Surfaces

Contact between spherical bodies is a widely used model in the localized stress analysis of many machine elements. We first consider the method of approach for finding the pressure distribution in this type of geometry of contacting surfaces.

Let us now consider a typical contact between two spherical bodies under load P (Figure 11.2). Contact starts at a point O with no load. With the increasing load P, corresponding points M and N on the two surfaces, lying perpendicular to the common tangent at point O, approach each other. For a given load and material properties of the surfaces, let the approach be α. As the surfaces approach, there is also local deformation due to load P. Let displacement of points M and N away from the common tangent be w_1 and w_2 respectively.

Considering that the points M and N are at a small distance r away from the initial point of contact O, we may write the distance z_1 and z_2 of points M and N respectively from the common tangent as

$$z_1 = \frac{r^2}{2R_1};\; z_2 = \frac{r^2}{2R_2},$$

where R_1 and R_2 are the radii of the bodies 1 and 2 respectively, as shown in Figure 11.2. These are obtained from geometrical considerations, neglecting the higher powers of binomial expansion.

This gives $z_1 + z_2 = \dfrac{r^2\left(R_1 + R_2\right)}{2R_1R_2} = \beta r^2,$

where $\beta = \dfrac{R_1 + R_2}{2R_1R_2}.$

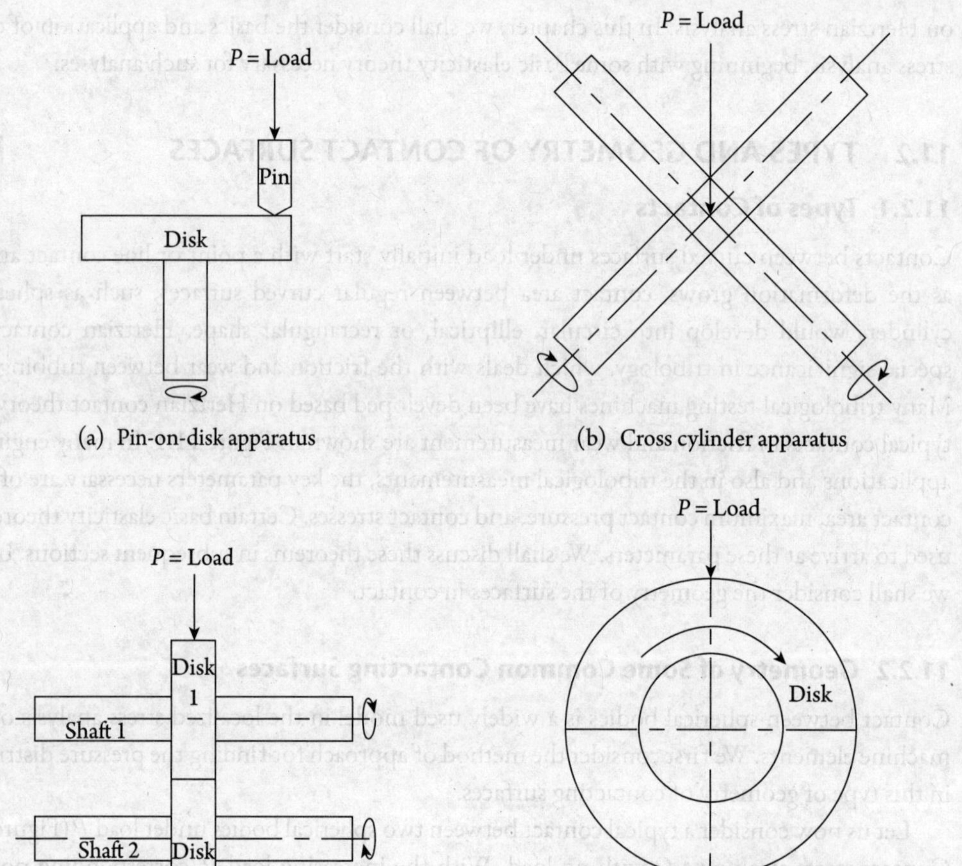

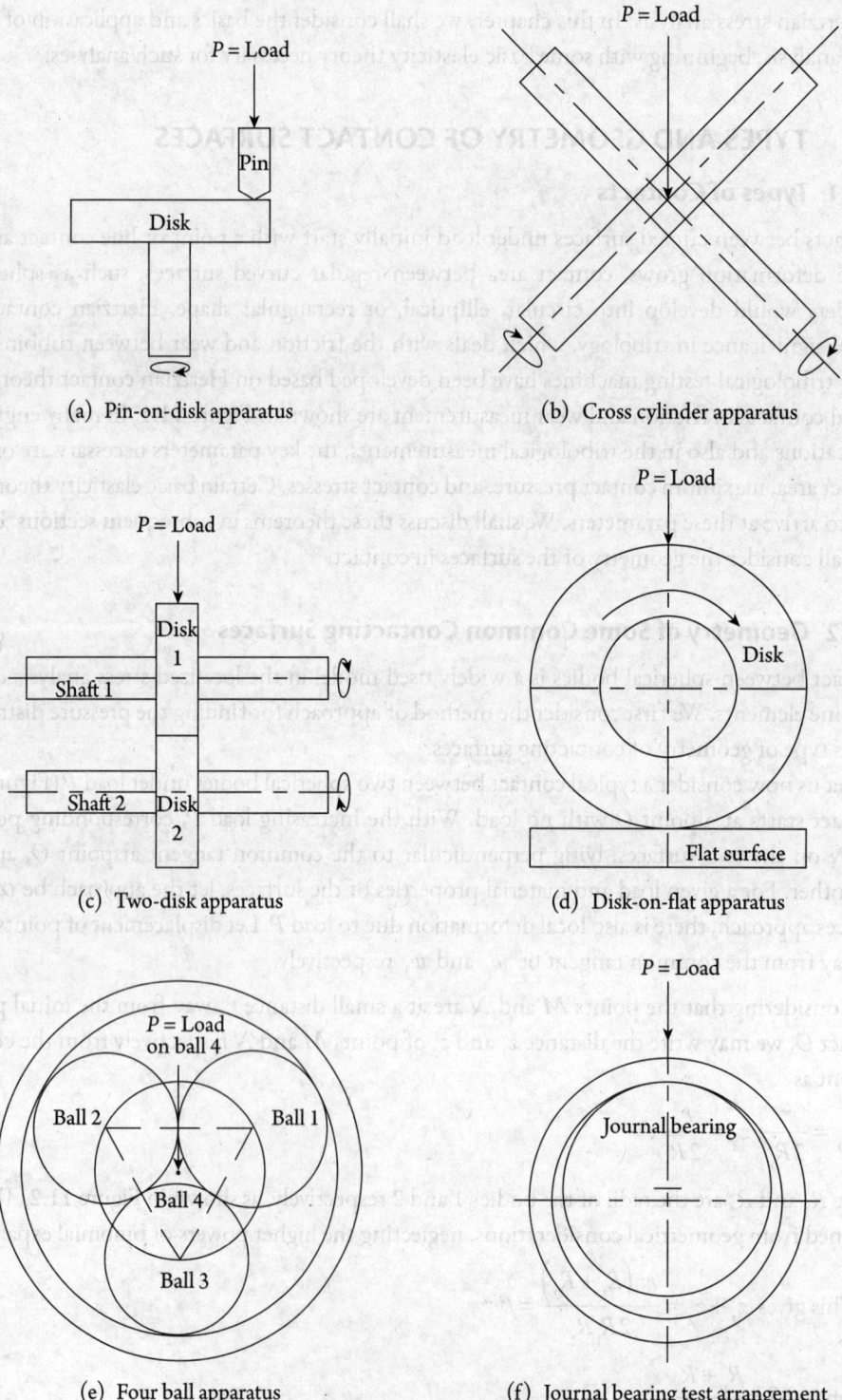

Figure 11.1 Different types of Hertzian contacts: Tribological testing apparatus

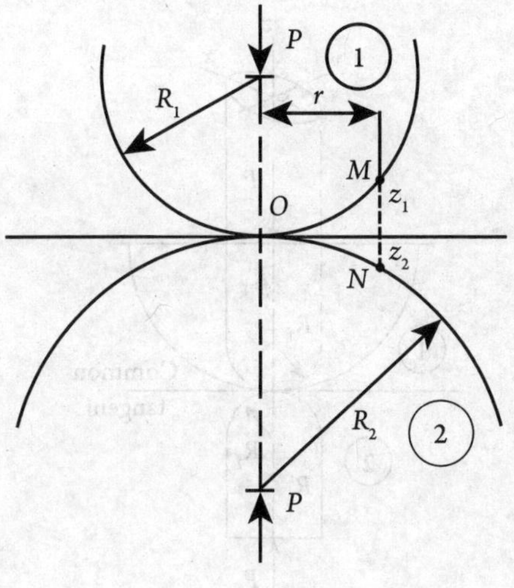

Figure 11.2 Contact between two spherical bodies under load P

We may therefore write

$$\alpha-(w_1+w_2)=z_1+z_2=\beta r^2.$$

Simplifying this, we have

$$w_1+w_2=\alpha-\beta r^2. \tag{11.2.2.1}$$

Equation (11.2.2.1) is a useful equation in understanding the contact mechanics and is a first step in finding contact approach, contact area, and also surface and sub-surface stresses. This will be followed up in subsequent sections.

In the general case of two curved bodies in contact, we may proceed in the same manner, as discussed above, except that the total distance between corresponding points on any two surfaces may be approximately given by

$$z_1+z_2=Ax^2+By^2, \tag{11.2.2.2}$$

where x and y are the coordinates with reference to the coordinate axes x and y with origin at the point of contact, and A and B are constants depending on the principal curvatures of the two bodies at the point of contact, and on the angle between the planes of principal curvatures of the two surfaces. For example, we now consider the contact between two semicircular elastic disks under a force P with an initial point contact, as shown in Figure 11.3. Following Timoshenko and Goodier (1970), the constants A and B are given by the following relations

$$B+A=\frac{1}{2}\left(\frac{1}{R_1}+\frac{1}{R_1'}+\frac{1}{R_2}+\frac{1}{R_2'}\right) \tag{11.2.2.3}$$

$$B-A=\frac{1}{2}\left[\left(\frac{1}{R_1}-\frac{1}{R_1'}\right)^2+\left(\frac{1}{R_2}-\frac{1}{R_2'}\right)^2+2\left(\frac{1}{R_1}-\frac{1}{R_1'}\right)\left(\frac{1}{R_2}-\frac{1}{R_2'}\right)\cos 2\psi\right]^{\frac{1}{2}},$$

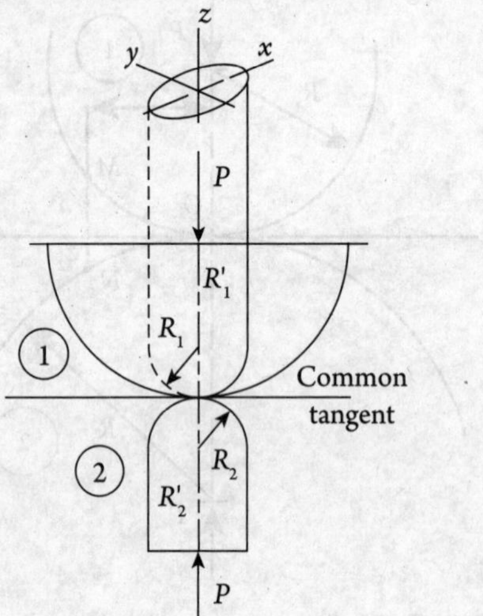

Figure 11.3 Contact between two elastic semicircular disks under a force P

where R_1 and R_1' are the principal radii of curvatures at the point of contact of body 1, and R_2, R_2' are those of body 2, and ψ is the angle between the normal planes containing the curvatures. Since $z_1 + z_2$ is positive, constants A and B are both positive. We may therefore conclude that all points with the same mutual distance $z_1 + z_2$ would lie on an ellipse.

We can now relate the approach and displacement in the same manner as shown for two spherical bodies and write

$$w_1 + w_2 + z_1 + z_2 = \alpha.$$

This may be written as

$$w_1 + w_2 = \alpha - Ax^2 - By^2. \tag{11.2.2.4}$$

This is the more general form of equation (11.2.2.1). We shall discuss the method of finding the approach and contact area in the following sections.

11.3 SOME BASIC THEOREMS OF ELASTICITY FOR CONTACT PROBLEMS [LO3]

11.3.1 Plane Contact Problems

There are many engineering components where the state of stress can be considered as two-dimensional (2D), such as plane stress and plane strain state. These have been discussed in earlier chapters. In plane contact problems, it is necessary to apply these methodologies. In the absence of body forces, the equations to be satisfied for 2D problems reduce to

$$\frac{\partial \sigma_x}{\partial x}+\frac{\partial \tau_{xy}}{\partial y}=0$$

$$\frac{\partial \tau_{yx}}{\partial x}+\frac{\partial \sigma_y}{\partial y}=0 \tag{11.3.1.1}$$

$$\nabla^2\left(\sigma_x+\sigma_y\right)=0.$$

It has also been discussed earlier that the stress function approach is suitable in these problems, and they are defined as

$$\sigma_x=\frac{\partial^2 \phi}{\partial y^2}, \sigma_y=\frac{\partial^2 \phi}{\partial x^2}, \tau_{xy}=-\frac{\partial^2 \phi}{\partial x \partial y} \tag{11.3.1.2}$$

and the governing equation here is the biharmonic equation

$$\nabla^4 \phi=0. \tag{11.3.1.3}$$

Since many contact problems require the polar coordinate approach, it is important to revisit equations similar to equations (11.3.1.1)–(11.3.1.3) in polar coordinates, and they are given as follows:

$$\frac{\partial \sigma_r}{\partial r}+\frac{1}{r}\frac{\partial \tau_{r\theta}}{\partial \theta}+\frac{\sigma_r-\sigma_\theta}{r}=0$$

$$\frac{1}{r}\frac{\partial \sigma_\theta}{\partial \theta}+\frac{\partial \tau_{r\theta}}{\partial r}+\frac{2\tau_{r\theta}}{r}=0 \tag{11.3.1.4}$$

$$\sigma_r=\frac{1}{r}\frac{\partial \phi}{\partial r}+\frac{1}{r^2}\frac{\partial^2 \phi}{\partial \theta^2}$$

$$\sigma_\theta=\frac{\partial^2 \phi}{\partial r^2} \tag{11.3.1.5}$$

$$\tau_{r\theta}=-\frac{\partial}{\partial r}\left(\frac{1}{r}\frac{\partial \phi}{\partial \theta}\right)$$

$$\left(\frac{\partial^2}{\partial r^2}+\frac{1}{r}\frac{\partial}{\partial r}+\frac{1}{r^2}\frac{\partial^2}{\partial \theta^2}\right)\left(\frac{\partial^2 \phi}{\partial r^2}+\frac{1}{r}\frac{\partial \phi}{\partial r}+\frac{1}{r^2}\frac{\partial^2 \phi}{\partial \theta^2}\right)=0. \tag{11.3.1.6}$$

Equations (11.3.1.1) to (11.3.1.6) are re-numbered for ease of referencing in this chapter.

11.3.2 Normal Line Load on the Surface of a Semi-infinite Solid

Now we proceed to consider the most basic 2D contact problem, which is that of a normal line load acting on the surface of a semiinfinite solid, as shown in Figure (11.4).

Here, $y=r\cos\theta;\ x=r\sin\theta.$ The solution of this problem is obtained by using the Boussinesq stress function, leading to

$$\phi=-\frac{Pr\theta}{\pi}\sin\theta.$$

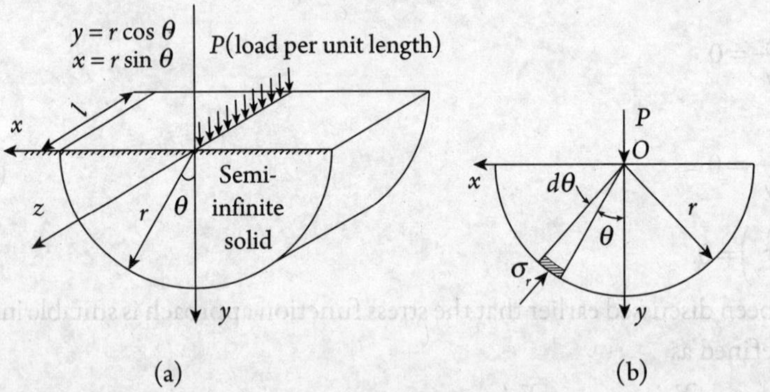

Figure 11.4 (a) Line load on the surface of a semi-infinite surface, (b) Equilibrium of an element at radius of r subtending an angle $d\theta$ from the point of application of load P and at an angle θ from the vertical

The problem is sometimes approached as a concentrated load per unit length on a semi-infinite plane (Figure 11.4 (a)), where we consider no stresses on the planes $\theta = \pm\pi/2$ for $r > 0$ and stresses become smaller as r increases, and they are inversely proportional to r.

To achieve stresses that are proportional to $1/r$, we need stress function of the order of r. There may be many functions that satisfy this condition, but a simple function would be of the following form:

$$\phi = c\, r\, \theta \sin\theta,$$

where c is a constant.

Corresponding stresses from equation (11.3.1.5) are

$$\sigma_r = \frac{2c\cos\theta}{r};\ \sigma_\theta = 0;\ \tau_{r\theta} = 0.$$

This seems to be the unique solution to the problem.

Now, we consider the equilibrium of an element, shaded in Figure (11.4 (b)), we have

$$P = -\int_{-\frac{\pi}{2}}^{\frac{\pi}{2}} \sigma_r\, r\, d\theta \cos\theta = -\pi c.$$

This gives $c = -\dfrac{P}{\pi}$ and $\phi = -\dfrac{P}{\pi} r\,\theta \sin\theta$ as assumed before.

This gives the stress field using equation (11.3.1.5)

$$\sigma_r = -\frac{2P}{\pi}\frac{\cos\theta}{r};\ \sigma_\theta = 0;\ \tau_{r\theta} = 0. \qquad (11.3.2.1)$$

This represents a field of radial compressive stress directed toward the line of application of the load. Using the methods of superposition, this solution forms the basis of the solutions to a number of important engineering contact problems, such as stress distribution under distributed load, in circular disks or rollers transversely loaded. As stated before, for these results to be valid, there are two conditions to be satisfied:

(a) No stresses on the plane $\theta = \pm\pi/2$ at $r > 0$.

(b) Stresses become smaller as r increases.

At $\theta = \pm\pi/2$, equation (11.3.2.1) shows $\sigma_r = 0$, $\sigma_\theta = 0$, $\tau_{r\theta} = 0$, and therefore condition (a) is satisfied.

Also, at $r = 0$, σ_r is infinite since a finite load acts on an infinitesimally small area.

Condition (b) is also satisfied, since in equation (11.3.2.1), σ_r decreases as r increases because here r appears in the denominator. Considering again, a small element at a radius r from the load line and the radial stress obtained from the Boussinesq solution, the total load necessarily works out to be P, as shown in Figure 11.5.

$$\int_{-\frac{\pi}{2}}^{\frac{\pi}{2}} \sigma_r r d\theta \cos\theta = \int_{-\frac{\pi}{2}}^{\frac{\pi}{2}} -\frac{2P}{\pi}\cos^2\theta d\theta = -P.$$

It can also be shown that $\phi = -\frac{Pr\theta}{\pi}\cos\theta$ satisfies the biharmonic equation (11.3.1.6).

It is clear from equation (11.3.2.1) that a constant radial stress $\sigma_r = \frac{2P}{\pi d}$ acts along the circumference of an imaginary circle of $d = \frac{r}{\cos\theta}$. The circle touches the point O of the application of the load, and the radial stresses point toward the origin O. This is shown in Figure 11.6.

At point O, $\sigma_r \to \infty$ since a finite load acts on an infinitely small area. This physically means that the point of contact would deform either elastically or plastically to produce a finite area.

The above stress distribution would apply equally well for a horizontal line load, as shown in Figure 11.7.

The only difference here is that in finding the total load, angle θ must be measured from the line of action of the load, which is horizontal in this case. We may therefore write

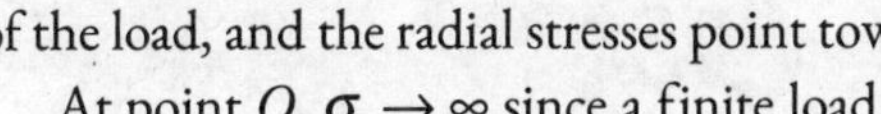

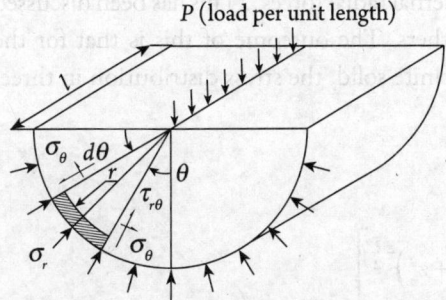

Figure 11.5 Stresses on an element r distance away from the load line and at an angle θ from vertical

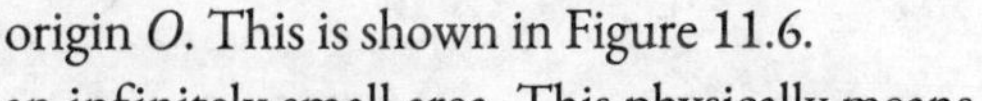

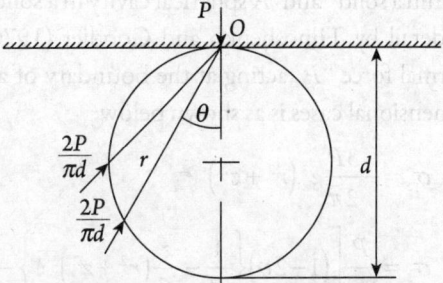

Figure 11.6 Constant stress distribution along a boundary of a circle of diameter d such that $d = \frac{r}{\cos\theta}$, where the circle touches the semi-infinite surface at the point of application of the load

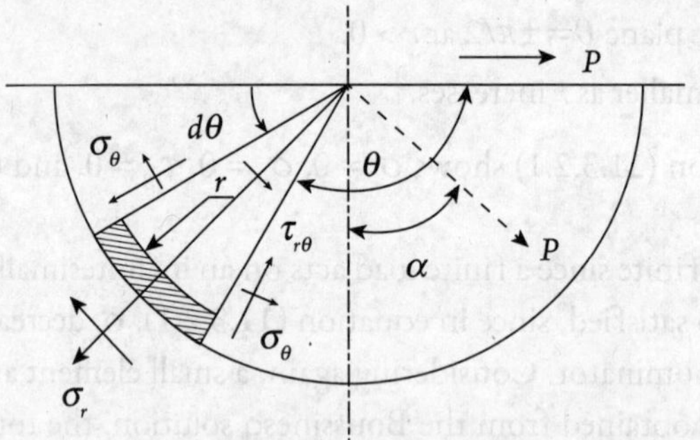

Figure 11.7 Stresses on an element r distance away from the load line at an angle θ from the horizontal load line and at an angle α from the inclined load line

$$\int_0^{\pi} \sigma_r \cos\theta . r d\theta = \int_0^{\pi} -\frac{2P}{\pi}\cos^2\theta d\theta = -P.$$

If we consider a line load inclined at an angle α with the vertical, then the radial stress works out to be

$$\sigma_r = -\frac{2}{\pi r}\left[P\cos\alpha\cos\theta + P\sin\alpha\cos\left(\frac{\pi}{2}+\theta\right)\right] = -\frac{2P}{\pi r}\cos(\alpha+\theta).$$

11.3.3 Force on the Boundary of a Semi-infinite Solid

To proceed further in contact stress analysis, we need to know the sub-surface stress distribution due to a force on the boundary of a semi-infinite solid. In general, to approach a problem of this type, the results of more basic problems are superimposed and the coefficients are adjusted until the required boundary conditions are met.

The present problem was solved by the superposition of the solutions for "A force at a point within a solid" and "A spherical cavity in a solid with internal radial forces". This has been discussed in detail by Timoshenko and Goodier (1970) and others. The outcome of this is that for the normal force "P" acting at the boundary of a semi-infinite solid, the stress distribution in three-dimensional cases is as shown below:

$$\sigma_z = -\frac{3P}{2\pi}z^3\left(r^2+z^2\right)^{-\frac{5}{2}}$$

$$\sigma_r = \frac{P}{2\pi}\left[(1-2\upsilon)\left\{\frac{1}{r^2}-\frac{z}{r^2}\left(r^2+z^2\right)^{-\frac{1}{2}}\right\}-3r^2z\left(r^2+z^2\right)^{-\frac{5}{2}}\right]$$

$$\sigma_\theta = \frac{P}{2\pi}\left[(1-2\upsilon)\left\{-\frac{1}{r^2}+\frac{z}{r^2}\left(r^2+z^2\right)^{-\frac{1}{2}}+z\left(r^2+z^2\right)^{-\frac{3}{2}}\right\}\right]$$

$$\tau_{rz} = -\frac{3P}{2\pi}rz^2\left(r^2+z^2\right)^{-\frac{5}{2}}. \qquad (11.3.3.1)$$

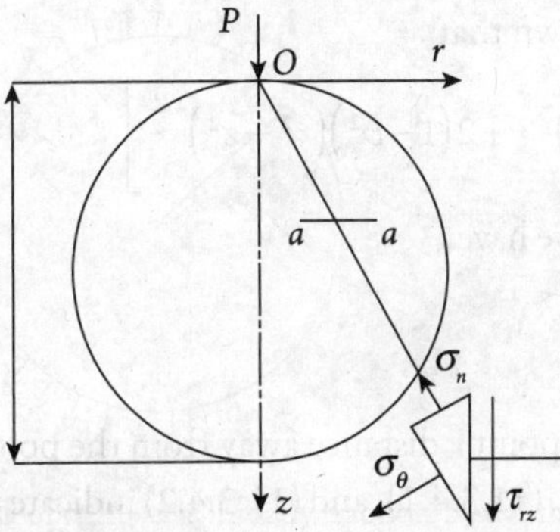

Figure 11.8 Constant stress of $\frac{3P}{2\pi d^2}$ at the boundary of an imaginary sphere of diameter d, tangential to the loading point O, at the surface of a semi-infinite body

These equations are similar to the 2D stress field in equation (11.3.2.1).

For this loading condition, if we consider an elemental area at a–a perpendicular to the z-direction (Figure 11.8), it can be shown from equation (11.3.3.1) that the resultant stress passes through the origin O and its magnitude is inversely proportional to r^2. By a similar argument to that used for the 2D case, it can be shown that by considering a spherical domain of diameter d tangential to the free surface at point O, the resultant stress on horizontal planes is constant and is $\frac{3P}{2\pi d^2}$.

To solve the contact problems, we need the stress field in equation (11.3.3.1) in addition to the displacements within the semi-infinite body subjected to a concentrated force. We shall now discuss this in the next section.

11.3.4 Displacements within a Semi-infinite Body Subjected to a Concentrated Force

The displacements are obtained from the stress–displacement relations.

The radial displacement u_r is given by

$$u_r = \epsilon_\theta r = \frac{r}{E}\left[\sigma_\theta - \upsilon\left(\sigma_r + \sigma_z\right)\right].$$

Substituting σ_θ, σ_r, and σ_z from (11.3.3.1) and at the surface, i.e., at $z = 0$, we have

$$u_{r(z=0)} = -\frac{(1-2\upsilon)(1+\upsilon)}{2\pi E r}P. \tag{11.3.4.1}$$

The vertical displacement ω may be obtained using the following stress–displacement relations:

$$\frac{\partial \omega}{\partial z} = \frac{1}{E}\left[\sigma_z - \upsilon\left(\sigma_r + \sigma_\theta\right)\right] \text{ and } \frac{\partial \omega}{\partial r} = \frac{2(1+\upsilon)}{E}\tau_{rz} - \frac{\partial u_r}{\partial z}.$$

On integration, it can be shown that

$$\omega = \frac{P}{2\pi E}\left[\left(1+\upsilon\right)z^2\left(r^2+z^2\right)^{-\frac{3}{2}}+2\left(1-\upsilon^2\right)\left(r^2+z^2\right)^{-\frac{1}{2}}\right]$$

at the free surface, where $z = 0$, we have

$$\omega_{z=0} = \frac{P\left(1-\upsilon^2\right)}{\pi E r}. \tag{11.3.4.2}$$

This gives the deflection at a point r distance away from the point of load P and at the surface, where $z = 0$. Equations (11.3.3.1), (11.3.4.1), and (11.3.4.2) indicate that at small values of r stresses and displacements are infinite. This physically impossible mathematical condition is avoided by assuming that the point loading is replaced by a hemispherical stress distribution that is statically equivalent to the force P.

11.3.5 Distributed Loading on a Semi-infinite Body

It is necessary to obtain the stresses and displacement on the semi-infinite solid due to the distributed load before solving the contact problems. This is because when two curved bodies are in contact with an applied load, the initial point contact grows into an area and the applied load is uniformly distributed over the contact area.

Let us consider a circular contact area of radius "a" over which a uniformly distributed pressure p acts, as shown in Figure 11.9.

Deflection at a point M on the surface at a distance "r" from the center of the circle in the direction of the load may be obtained by integration of the effect of the distributed load. Consider an element of width ds at a distance of s from the point M subtending an angle $d\psi$. Here the area of the element is $s.ds.d\psi$ and the load on the element is $p.s.ds.d\psi$. Therefore, following equation (11.3.4.2), the vertical deflection at M due to this elementary load may be given as

$$\frac{p.s.ds.d\psi\left(1-\upsilon^2\right)}{\pi E s}.$$

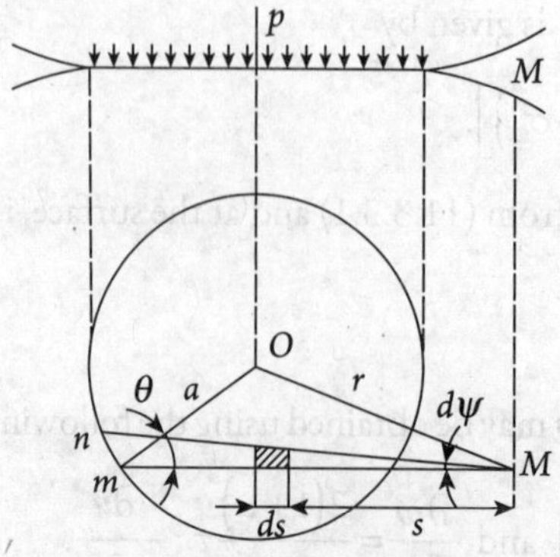

Figure 11.9 Deflection at a point on the surface of a contacting body due to distributed load at the contact

Therefore, the total deflection at M due to all the elements within the loaded circular area is

$$\omega = \frac{1-\upsilon^2}{\pi E}\iint p ds d\psi. \tag{11.3.5.1}$$

This equation holds for any shape of the loaded area, regardless of whether it is circular or not. The equation also holds for deflection at a point within or outside the contact zone.

We shall now consider the deflection at a point within the contact zone. This is more useful for solving contact problems. Figure 11.10 shows the details of such a situation.

All the details are the same as Figure 11.9 except that point M lies within the contact circle.

From the triangle Omq in Figure 11.10, it can be seen that $mq = 2a\cos\theta$, and this gives $r.\sin\psi = a.\sin\theta$.

Now, $\int ds$ over the length $mq = \int_0^{2a\cos\theta} ds = 2a\cos\theta = 2a\sqrt{1-\frac{r^2}{a^2}\sin^2\psi}$.

This gives $\iint ds.d\psi = 2a\int_{-\frac{\pi}{2}}^{\frac{\pi}{2}}\left(1-\frac{r^2}{a^2}\sin^2\psi\right)^{\frac{1}{2}} d\psi.$

Combining this with equation (11.3.5.1), we have

$$\omega = \frac{4(1-\upsilon^2)}{\pi E} pa\int_0^{\frac{\pi}{2}}\left(1-\frac{r^2}{a^2}\sin^2\psi\right)^{\frac{1}{2}} d\psi. \tag{11.3.5.2}$$

The integral may be evaluated numerically or using the tables of elliptical integrals for any particular value of r/a. It is clear from equation (11.3.5.2) that at $r = 0$, i.e., at the center of the circle, the deflection is maximum, and we may write

$$\omega_{max} = \frac{4(1-\upsilon^2)}{\pi E} pa\int_0^{\frac{\pi}{2}} d\psi = \frac{2pa(1-\upsilon^2)}{E}. \tag{11.3.5.3}$$

The deflection along the boundary of the loaded circular area, i.e., at $r = a$, is given by

$$\omega_{r=a} = \frac{4(1-\upsilon^2)}{\pi E} pa\int_0^{\frac{\pi}{2}} \cos\psi\, d\psi = \frac{4(1-\upsilon^2)}{\pi E} pa. \tag{11.3.5.4}$$

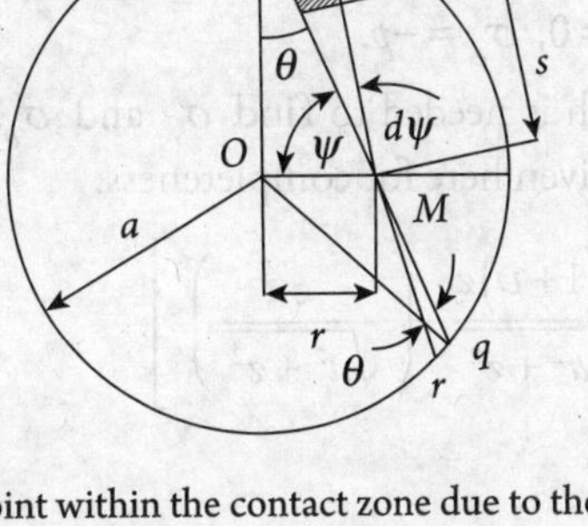

Figure 11.10 Deflection at a point within the contact zone due to the distributed load at the contact

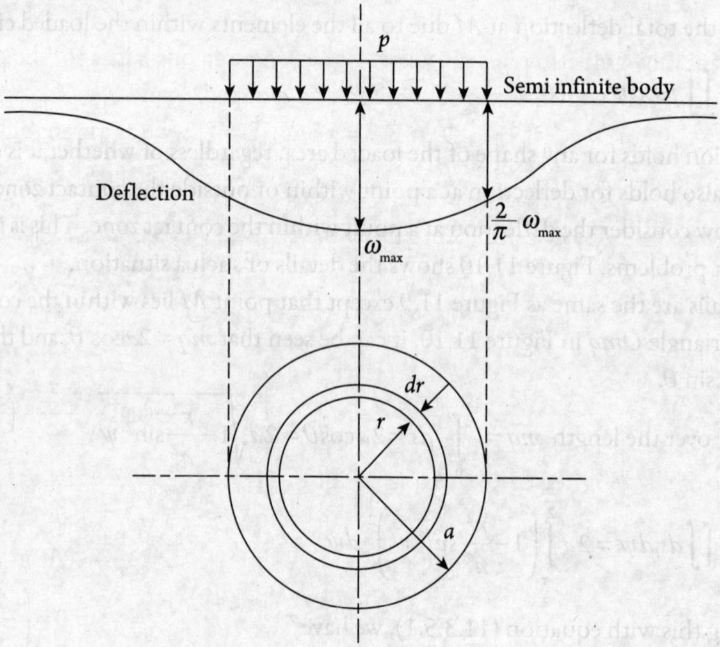

Figure 11.11 Displacement and stresses developed within the semi-infinite body due to a uniformly distributed load

The stresses developed within the semi-infinite solid due to distributed load over it can be obtained by the superposition approach (Figure 11.11).

The figure shows an elemental load ring of thickness dr at a distance of radius r from the center. We now consider the stresses σ_r, σ_θ, and σ_z due to this load ring. From equation (11.3.3.1), we can write σ_z due to elemental load ring as

$$\sigma_z = -3prz^3\left(r^2+z^2\right)^{-\frac{5}{2}}dr$$

taking load P due to the elemental load ring as $2.\pi.r.p.dr$. This gives σ_z due to the entire distributed load as

$$\sigma_z = -3pz^3\int_0^a r\left(r^2+z^2\right)^{-\frac{5}{2}}dr = p\left[z^3\left(a^2+z^2\right)^{-\frac{3}{2}}-1\right]. \tag{11.3.5.5}$$

Therefore, at the surface, $z = 0$, $\sigma_z = -p$.

A slightly different approach is needed to find σ_r and σ_θ, and this is discussed in detail in Section 11.4.2. The results are given here for completeness.

$$\sigma_r = \sigma_\theta = \frac{p}{2}\left[-(1+2\upsilon)+\frac{2(1+\upsilon)z}{\sqrt{a^2+z^2}}-\left(\frac{z}{\sqrt{a^2+z^2}}\right)^3\right]. \tag{11.3.5.6}$$

11.4 STRESSES AND DISPLACEMENTS AT THE CONTACT BETWEEN TWO CURVED BODIES [LO4]

11.4.1 Pressure Distribution Between Two Elastic Bodies in Contact

As discussed in Section 11.2.2, the total displacement and approach at the contact are given by equation (11. 2.2.1), which is rewritten here for convenience.

$$w_1 + w_2 = \alpha - \beta r^2.$$

Here, w_1 and w_2 are the displacements at a point of body 1 and body 2 respectively and $\beta = \dfrac{R_1 + R_2}{2R_1R_2}$, R_1 and R_2 are the radii of the two bodies in contact.

If we consider different materials for the two bodies, E_1, υ_1 for body 1 and E_2, υ_2 for body 2, then following equation (11.3.5.1), we may write

$$w_1 + w_2 = \left(\frac{1-\upsilon_1^2}{\pi E_1} + \frac{1-\upsilon_2^2}{\pi E_2}\right)\iint p\, ds d\psi$$

and if for convenience, we write $\dfrac{1-\upsilon_1^2}{\pi E_1} = K_1$ and $\dfrac{1-\upsilon_2^2}{\pi E_2} = K_2$, then the above equation may be written in a simplified form as

$$w_1 + w_2 = \left(K_1 + K_2\right)\iint p\, ds d\psi = \alpha - \beta r^2. \tag{11.4.1.1}$$

Now when the two bodies are pressed together, we consider both the pressure and the local distribution to be symmetric about point O, as shown in Figure 11.12.

Now, we need to find an expression for the pressure distribution p that would satisfy the equation (11.4.1.1). It is seen that a hemispherical pressure distribution over a circular contact surface area of radius "a" meets this requirement. Let the maximum pressure p_o at the center of the contact surface be given as

$$p_o = a.c, \tag{11.4.1.2}$$

where c is the scale factor to relate the contact radius with contact pressure. Therefore, c has the unit of N/m^3. This is shown in Figure 11.12.

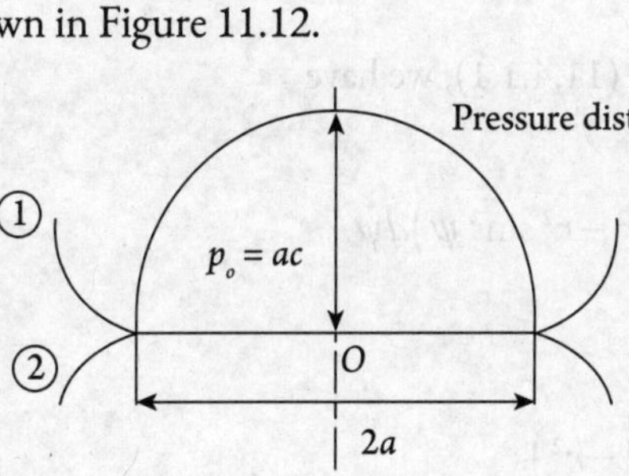

Figure 11.12 Symmetric pressure distribution at the contact between two curved bodies

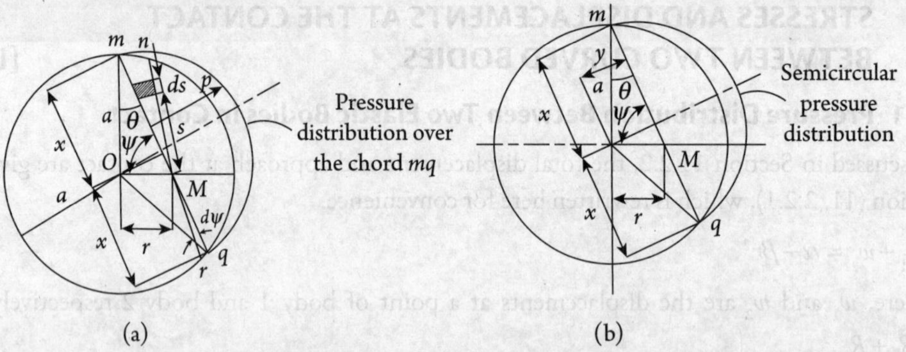

Figure 11.13 (a) Pressure distribution along the chord *mq* over the circular contact area, (b) Pressure distribution along chord *mq* showing the coordinates *x* and *y* to relate *x, c,* and ψ

Now since the pressure distribution over the circular contact area is hemispherical, the pressure along the chord *mq* is semicircular, as shown in Figure 11.13.

Integrating along the chord *mq*, we have

$\int pds =$ *Area of the semicircular shape in appropriate unit*.

Now, referring to Figure 11.13(b), we have

$a^2 = x^2 + y^2$, where $y = r\sin\psi$

Therefore, $x = \sqrt{a^2 - r^2\sin^2\psi}$.

If we take $mq = 2x$ as shown in Figures 11.13(a) and (b), we may write

Area under the semicircle, $A = \frac{1}{2}\pi x^2 = \frac{1}{2}\pi\left(a^2 - r^2\sin^2\psi\right)$.

Therefore, combining with equation (11.4.1.2), we have $\int pds = A\frac{p_o}{a} = A.c$ (area of the semicircle *X* scale factor). This gives

$$\int pds = \frac{p_o}{a}\left[\frac{\pi}{2}\left(a^2 - r^2\sin^2\psi\right)\right].$$

Substituting $\int pds$ in equation (11.4.1.1), we have

$$\alpha - \beta r^2 = 2\left(K_1 + K_2\right)\int_0^{\frac{\pi}{2}} \frac{p_o\pi}{2a}\left(a^2 - r^2\sin^2\psi\right)d\psi.$$

On integration, this gives

$$\alpha - \beta r^2 = \left(K_1 + K_2\right)\frac{p_o\pi^2}{4a}\left(2a^2 - r^2\right). \qquad (11.4.1.3)$$

Equating the coefficients of *r*, we have the expressions for the approach α and contact radius "*a*".

$$\alpha = (K_1 + K_2)\frac{p_o \pi^2 a}{2} \tag{11.4.1.4}$$

$$a = (K_1 + K_2)\frac{p_o \pi^2}{4\beta}$$

$$= (K_1 + K_2)\frac{\pi^2 p_o R_1 R_2}{2(R_1 + R_2)}. \tag{11.4.1.5}$$

Now the total force P is the volume under the hemisphere over the contact area of diameter $2a$ and therefore including the scale factor ($\frac{p_o}{a}$), we may write

$$P = \frac{2}{3}\pi a^3 \left(\frac{p_o}{a}\right).$$

This gives the expression for maximum pressure p_o as

$$p_o = \frac{3P}{2\pi a^2}. \tag{11.4.1.6}$$

This means that the maximum contact pressure is $1\frac{1}{2}$ times the average pressure.

11.4.2 Stress Distribution at the Area of Contact Between Two Solids

This problem can be treated as a problem of distributed load over an area at the boundary of a semi-infinite body (Figure 11.14).

This, in turn, can be obtained from the solution of a concentrated force at the boundary of a semi-infinite body. The stress distribution for this case is given in equation (11.3.3.1). Stresses developed within the contacting bodies due to distributed loading at the contact zone are similar to those for the boundary of semi-infinite solids, and this was discussed in Section 11.3.5 briefly. Here we shall discuss this in further detail.

It has been shown that the stress distribution at any point on the z-axis may be obtained by considering an elemental load ring and substituting $P = 2\pi r dr p$ and integrating $d\sigma_z$ for limits of r between "0" and "a", where "a" is the contact radius. This gives, from equation (11.3.3.1),

$$\sigma_z = p\left[\frac{z^3}{\left(a^2 + z^2\right)^{\frac{3}{2}}} - 1\right]. \tag{11.4.2.1}$$

This was shown earlier in equation (11.3.5.5). This is rewritten here as a separate equation for clarity.

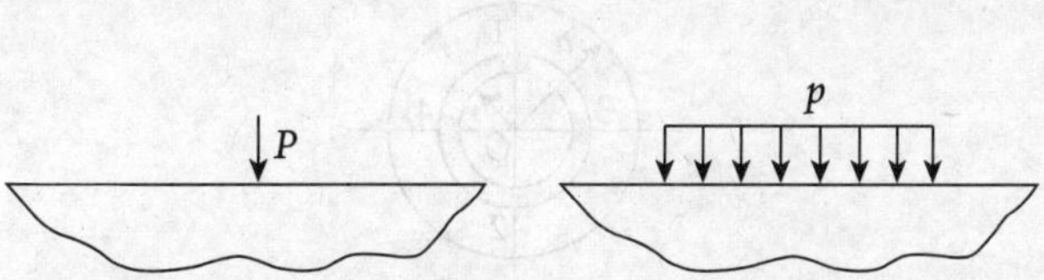

Figure 11.14 Concentrated and distributed load at the boundary of a semi-infinite body

At the surface $z = 0$,

$$\sigma_z = -p. \tag{11.4.2.2}$$

To find σ_r and σ_θ, we need a slightly different approach since σ_r and σ_θ are interchangeable at $\pi/2$. This was discussed in detail by Timoshenko and Goodier (1970).

Consider the loaded elements 1 and 2 as shown in Figure 11.15. Referring to equation (11.3.3.1) and substituting $prd\phi dr$ for P, p being the contact pressure, the stresses at any point on the z-axis due to these two elements are

$$d\sigma_r' = 2p\frac{rd\phi dr}{2\pi}\left[(1-2\upsilon)\left\{\frac{1}{r^2}-\frac{z}{r^2}\left(r^2+z^2\right)^{-\frac{1}{2}}\right\}-3r^2z\left(r^2+z^2\right)^{-\frac{5}{2}}\right]$$

$$d\sigma_\theta' = 2p\frac{rd\phi dr}{2\pi}\left[(1-2\upsilon)\left\{-\frac{1}{r^2}+\frac{z}{r^2}\left(r^2+z^2\right)^{-\frac{1}{2}}+z\left(r^2+z^2\right)^{-\frac{3}{2}}\right\}\right].$$

Similarly, the stresses on the same plane due to the loaded elements 3 and 4

$$d\sigma_r'' = 2p\frac{rd\phi dr}{2\pi}\left[(1-2\upsilon)\left\{-\frac{1}{r^2}+\frac{z}{r^2}\left(r^2+z^2\right)^{-\frac{1}{2}}+z\left(r^2+z^2\right)^{-\frac{3}{2}}\right\}\right]$$

$$d\sigma_\theta'' = 2p\frac{rd\phi dr}{2\pi}\left[(1-2\upsilon)\left\{\frac{1}{r^2}-\frac{z}{r^2}\left(r^2+z^2\right)^{-\frac{1}{2}}\right\}-3r^2z\left(r^2+z^2\right)^{-\frac{5}{2}}\right].$$

Considering these, we get the stresses due to all four elements as

$$d\sigma_r = \frac{prd\phi dr}{\pi}\left[(1-2\upsilon)z\left(r^2+z^2\right)^{-\frac{3}{2}}-3r^2z\left(r^2+z^2\right)^{-\frac{5}{2}}\right] = d\sigma_\theta.$$

The stresses due to the entire loaded zone are given as

$$\sigma_r = \sigma_\theta = \int_0^{\frac{\pi}{2}}\int_0^{a}\frac{p}{\pi}\left[(1-2\upsilon)z\left(r^2+z^2\right)^{-\frac{3}{2}}-3r^2z\left(r^2+z^2\right)^{-\frac{5}{2}}\right]rd\phi dr.$$

Integrating, we have

$$\sigma_r = \sigma_\theta = \frac{p}{2}\left[2(1+\upsilon)\left\{z\left(a^2+z^2\right)^{-\frac{1}{2}}-1\right\}-z^3\left(a^2+z^2\right)^{-\frac{3}{2}}+1\right]. \tag{11.4.2.3}$$

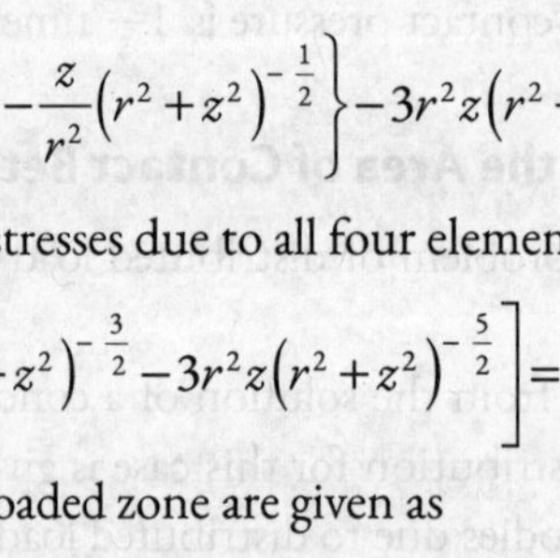

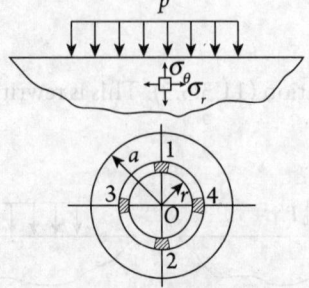

Figure 11.15 Stress distribution at the sub-surface of a semi-infinite solid due to load elements 1, 2, 3, and 4

At the surface $z = 0$ and at point O,

$$\sigma_r = \sigma_\theta = -\frac{p(1+2\upsilon)}{2}. \tag{11.4.2.4}$$

$$\text{Maximum shear stress} = \frac{1}{2}(\sigma_\theta - \sigma_z) = \frac{1}{2}\left[-\frac{p(1+2\upsilon)}{2} + p\right] = p\left(\frac{1-2\upsilon}{4}\right). \tag{11.4.2.5}$$

If $\upsilon = 0.3$ maximum shear stress reduces to $0.1\,p$ at the surface.

In general,

$$\tau_{max} = \frac{1}{2}(\sigma_\theta - \sigma_z) = \frac{p}{4}\left[2(1+\upsilon)\frac{z}{\sqrt{a^2+z^2}} - 3\left(\frac{z}{\sqrt{a^2+z^2}}\right)^3 + (1-2\upsilon)\right].$$

To find the maximum of τ_{max}, we differentiate τ_{max} with respect to $\frac{z}{\sqrt{a^2+z^2}}$ and equate to zero. This gives the maximum shear stress at any point on the z-axis.

$$\frac{d\tau_{max}}{d\left(\frac{z}{\sqrt{a^2+z^2}}\right)} = 0. \text{ This gives } 2(1+\upsilon) - 9\left(\frac{z}{\sqrt{a^2+z^2}}\right)^2 = 0,$$

$$\text{i.e., } \frac{z}{\sqrt{a^2+z^2}} = \frac{1}{3}\sqrt{2(1+\upsilon)}. \tag{11.4.2.6}$$

Substituting this in the expression for τ_{max}, we have

$$\tau_{max} = \frac{p}{2}\left[\frac{1-2\upsilon}{2} + \frac{2}{9}(1+\upsilon)\sqrt{2(1+\upsilon)}\right]. \tag{11.4.2.7}$$

Therefore, from equations (11.4.2.6) and (11.4.2.7), for $\upsilon = 0.3$, $\tau_{max} = 0.33p$ and it occurs at $z = 0.638a$. This result is of interest in the failure analysis of many machine parts.

11.4.3 General Case of Two Curved Bodies in Contact

Hertz showed that the pressure distribution between two curved bodies in contact can be represented by a semi-ellipsoid over the surface of contact, given by the equation

$$\frac{x^2}{a^2} + \frac{y^2}{b^2} + \frac{z^2}{c^2} = 1, \tag{11.4.3.1}$$

where a, b, and c are the semi-axes of the ellipsoid.

The semi-ellipsoidal pressure distribution and a section through the pressure profile in the xz-plane are shown in Figure 11.16 (a) and (b) respectively. Referring to Figure 11.16(b), consider a point (x, z) on the pressure profile. Here the pressure at the point is given by $p = zk$, k being the scale factor and $p_o = ck$, c being the semi-axis of the ellipsoid. Referring to equation (11.4.3.1), we can also write

$$p = zk = c\sqrt{1 - \frac{x^2}{a^2} - \frac{y^2}{b^2}}\,k = p_o\sqrt{1 - \frac{x^2}{a^2} - \frac{y^2}{b^2}}, \tag{11.4.3.2}$$

where a and b are the semi-axes of the contact ellipse.

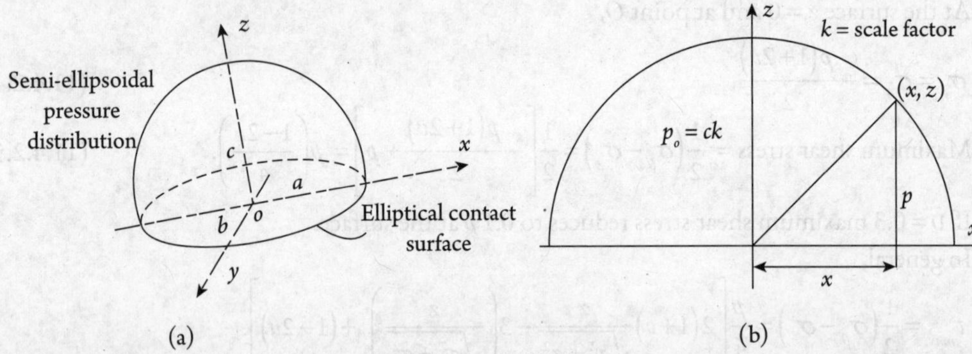

Figure 11.16 (a) Semi-ellipsoid pressure profile over the ellipse of contact between two curved bodies in contact, (b) Section through any plane of the pressure profile passing through the center

The previous case discussed in Section 11.4.1 was a special case, where $a = b$ and this yielded a circular contact area with hemispherical pressure distribution.

Total load P is the volume of the semi-ellipsoid, and this is given by

$$P = \frac{1}{2}.\frac{4}{3}\pi abc.k = \frac{2\pi ab}{3}.p_o, \text{ since } p_o = ck.$$

This gives maximum pressure

$$p_o = \frac{3P}{2\pi ab}. \qquad (11.4.3.3)$$

11.5 CONTACT PROBLEMS IN SOME IMPORTANT MACHINE PARTS [LO5]

As discussed earlier, the analysis of contact stresses, contact area, and approaches carried out in this chapter have great significance in the design and development of many machine parts, such as bearings, gears, and cams. In this section, we shall discuss some details of ball and roller bearings.

In ball bearings, the analysis of contact geometry between two balls is important. Consider the contact between two spherical balls under load P as shown in Figure 11.17. Let the balls 1 and 2 have the radii, elastic moduli and Poisson's ratio R_1, E_1, υ_1 and R_2, E_2, υ_2 respectively. Following the results of Section 11.4.1, we have

$$\text{Contact radius } a = \left(K_1 + K_2\right)\frac{\pi^2 p_o R_1 R_2}{2\left(R_1 + R_2\right)}.$$

$$\text{Approach } \alpha = \left(K_1 + K_2\right)P_o\frac{\pi^2}{2}a.$$

Substituting $p_o = \dfrac{3P}{2\pi a^2}$, we may concisely write

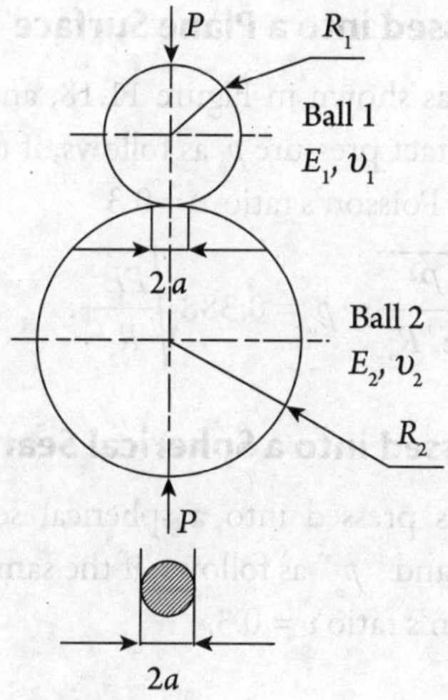

Figure 11.17 Contact between two spherical balls under load P

$$a = \sqrt[3]{\frac{3\pi}{4}\frac{P\left(K_1+K_2\right)R_1R_2}{\left(R_1+R_2\right)}}.$$

$$\alpha = \sqrt[3]{\frac{9\pi^2}{16}\frac{P^2\left(K_1+K_2\right)^2\left(R_1+R_2\right)}{\left(R_1R_2\right)}}. \tag{11.5.1}$$

If we assume the same material for both the balls and take $\upsilon_1 = \upsilon_2 = 0.3$ and $E_1 = E_2 = E$, then we have

$$a = 1.109\sqrt[3]{\frac{PR_1R_2}{E\left(R_1+R_2\right)}}$$

$$\alpha = 1.23\sqrt[3]{\frac{P^2\left(R_1+R_2\right)}{E^2R_1R_2}}. \tag{11.5.2}$$

Here the maximum contact pressure is given by

$$p_o = \frac{3}{2}\frac{P}{\pi a^2} = 0.388\sqrt[3]{\frac{PE^2\left(R_1+R_2\right)^2}{\left(R_1R_2\right)^2}}. \tag{11.5.3}$$

Equations (11.5.1), (11.5.2), and (11.5.3) are useful in the design of ball bearings. However, more useful contact configurations are a ball pressed into a plane surface and a ball pressed into a spherical seat. We shall now consider these two situations one by one.

11.5.1 A Spherical Ball Pressed into a Plane Surface

A flat plane is given as $R_1 \to \infty$, as shown in Figure 11.18, and this gives the contact area "a", approach "α", and maximum contact pressure p_o as follows, if the same material is used for both bodies with elastic modulus E and Poisson's ratio $\upsilon = 0.3$

$$a = 1.109\sqrt[3]{\frac{PR_2}{E}};\quad \alpha = 1.23\sqrt[3]{\frac{P^2}{E^2 R_2}};\quad p_o = 0.388\sqrt[3]{\frac{PE^2}{R_2^{\,2}}}. \tag{11.5.1.1}$$

11.5.2 A Spherical Ball Pressed into a Spherical Seat

For the situation where a ball is pressed into a spherical seat, R_1 is negative, as shown in Figure 11.19. This gives "a", "α", and "p_o" as follows, if the same material is used for both bodies with elastic modulus E and Poisson's ratio $\upsilon = 0.3$.

$$a = 1.1\sqrt[3]{\frac{PR_1R_2}{E\left(R_1 - R_2\right)}}$$

$$\alpha = 1.23\sqrt[3]{\frac{P^2\left(R_1 - R_2\right)}{E^2 R_1 R_2}} \tag{11.5.2.1}$$

$$p_o = 0.388\sqrt[3]{\frac{PE^2\left(R_1 - R_2\right)^2}{\left(R_1 R_2\right)^2}}.$$

In a typical ball bearing, balls are pressed against spherical surfaces and also in the absence of a cage, spherical balls are pressed against each other. In view of this, equations (11.5.1.1) and (11.5.2.1) are relevant.

Stresses at the contact surface and sub-surface have been discussed earlier in Sections 11.3.5 and 11.4.2. However, for this case, it is important to understand the stress distribution and we consider it in some different manner.

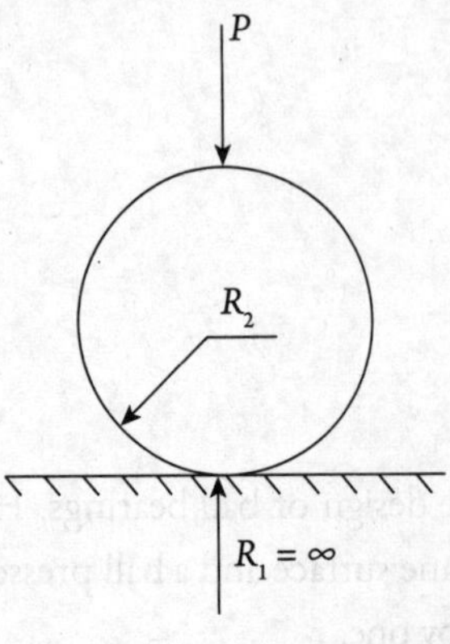

Figure 11.18 A spherical ball pressed into a plane surface

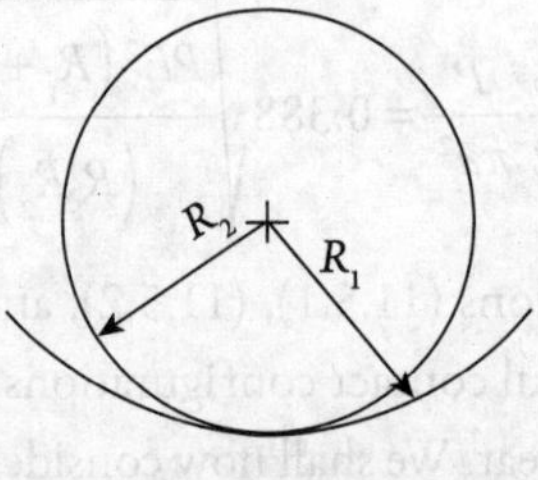

Figure 11.19 A spherical ball pressed into a spherical seat

Referring to Figure 11.16(b) and taking $x=r$ and $z=p$, we may write the pressure distribution as $p=\sqrt{a^2-r^2}\times$scale factor.

Since scale factor $=\frac{p_o}{a}$, where a being the contact radius and p_o denotes the maximum pressure, the semicircular pressure distribution is given as $p=\frac{p_o}{a}\sqrt{a^2-r^2}$, and referring to equation (11.3.3.1), the stresses in the z-direction within the bodies at every point are given by

$$\sigma_z=-\frac{3P}{2\pi}z^3\left(r^2+z^2\right)^{-\frac{5}{2}}.$$

As earlier, considering a circular load element of radius r and of thickness dr (Figure 11.11), the stress due to this load element is given by

$$d\sigma_z=-\frac{3}{2\pi}z^3\left(r^2+z^2\right)^{-\frac{5}{2}}\left[2\pi r\frac{p_o}{a}\sqrt{a^2-r^2}\,dr.\right]$$

Therefore, total stress in the z-direction is given by

$$\sigma_z=-3z^3\frac{p_o}{a}\int_{r=0}^{r=a}\left(r^2+z^2\right)^{-\frac{5}{2}}.r.\left(a^2-z^2\right)^{\frac{1}{2}}dr. \tag{11.5.2.2}$$

Similar expressions can be obtained for σ_r, σ_θ, and τ_{rz}, and these stresses are given below graphically for better understanding:

Figure 11.20 is of great importance in understanding the failure of ball bearings that normally fail by pitting at the inner or outer races. Existing cracks grow due to repetitive loading or crack may be initiated at the point of maximum shear stress. $\tau_{rz(\max)}$, which is here 0.31 p_o and occurs at 0.47a depth. For ductile materials, this position and the value of maximum shear stress are critical from yielding point of view. For brittle materials, the failure is governed by the maximum tensile stress, and therefore, it occurs at the surface and at the circular boundary. This occurs in a radial direction, where $\sigma_r=\frac{(1-2v)}{3}p_o$. At this point σ_θ is also $-\frac{(1-2v)}{3}p_o$ since $\sigma_r=-\sigma_\theta$ and there exists a state of pure shear at this position.

In rolling element bearings, contact geometry analysis of two cylinders under load is important. Equation (11.4.3.2) represents the general case of contact between two curved bodies under load. If we let $a/b\rightarrow\infty$, we have the case of two cylinders with parallel axes for which contact area is a long rectangle, as shown in Figure 11.21.

Here the pressure distribution is semi-elliptical. If $p'=P/l$ is the load per unit length, then the average pressure $=\frac{P}{2bl}$.

Since the area of the semi-ellipse of pressure $=\frac{\pi b p_o}{2}$ in arbitrary unit, the average height of the semi-ellipse is

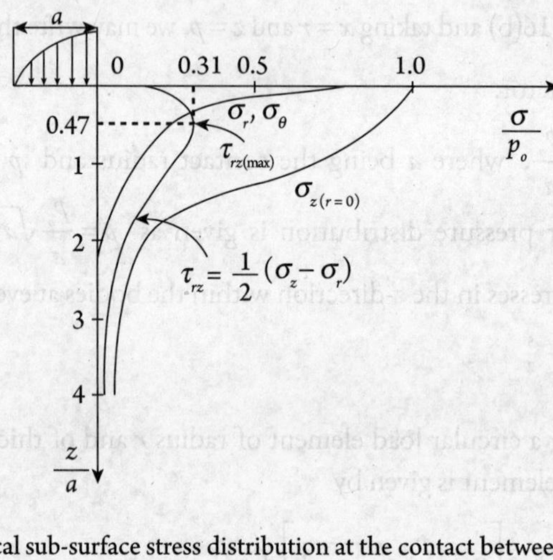

Figure 11.20 Typical sub-surface stress distribution at the contact between two spherical bodies

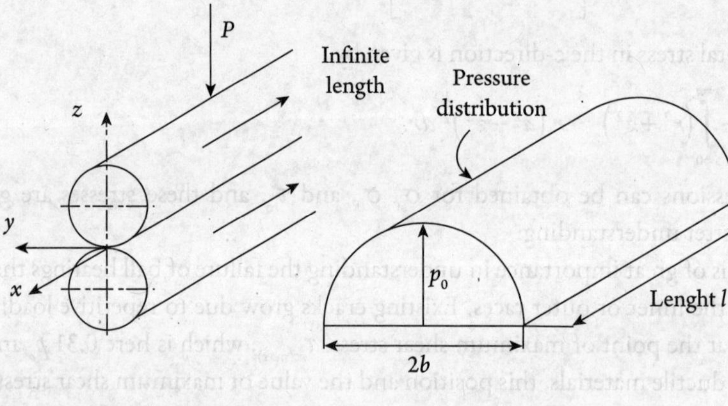

Figure 11.21 Contact between two long cylinders under a load and the pressure distribution at the contact

$$\frac{\left(\frac{\pi b p_o}{2}\right)}{2b} = \frac{\pi p_0}{4}.$$

Average pressure on the contact area is also

$$\frac{P}{2bl} = \frac{p'}{2b}.$$

Therefore, we may write

$$\frac{\pi p_o}{4} = \frac{p'}{2b},$$

and this gives the maximum pressure at the contact area

$$p_o = \frac{2p'}{\pi b}. \qquad (11.5.2.3)$$

Considering the local deformation and following the procedure set out in Section 11.4 and earlier in this chapter, we may find half the contact width as

$$b=\sqrt{\frac{4p'(K_1+K_2)R_1R_2}{R_1+R_2}}. \tag{11.5.2.4}$$

K_1 and K_2 are defined as $K_1=\dfrac{1-v_1^2}{\pi E_1}$ and $K_2=\dfrac{1-v_2^2}{\pi E_2}$. R_1 and R_2 are the radii of the cylinders.

From the values of p_o and b, sub-surface stresses may be calculated, and they are given in graphical form in Figure 11.22 for $v=0.3$.

Here in the figure, $\tau_{yz}=\dfrac{1}{2}\left(\sigma_z-\sigma_y\right)$ and $v=0.3$.

Here the maximum shear stress of 0.31 p_o occurs at 0.78b depth. This position and maximum shear stress at this position are of importance in the failure analysis of rolling element bearings. Comments similar to those made for Figure 11.20 apply here.

Finally, we briefly discuss the general case of contact between two curved bodies, given by equation (11.4.3.1). It is important to evaluate the semi-axes a and b. Timoshenko and Goodier (1970) have discussed this in detail. Here we briefly reproduce the results as follows:

$$a=M\sqrt[3]{\frac{3P\Delta}{4(A+B)}}\text{ and }b=N\sqrt[3]{\frac{3P\Delta}{4(A+B)}} \tag{11.5.2.5}$$

where the elastic constant

$$\Delta=\frac{1-v_1^2}{E_1}+\frac{1-v_2^2}{E_2}.$$

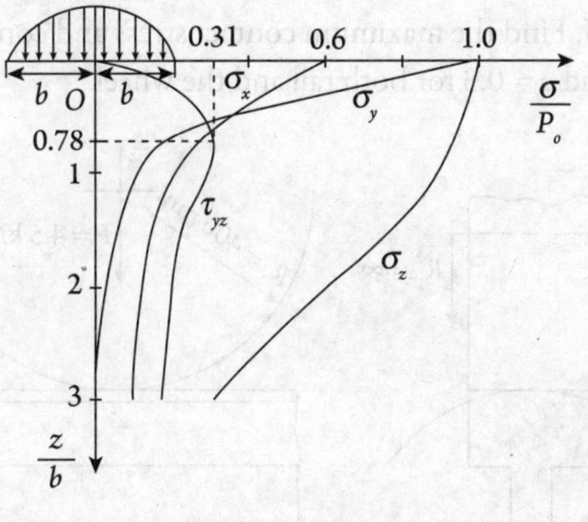

Figure 11.22 Typical sub-surface stress distribution at the area of contact between two cylinders with parallel axes under load P

Table 11.1 Values of M and N

$\cos^{-1}\left(\dfrac{B-A}{B+A}\right)$	30°	35°	40°	45°	50°	55°	60°	65°	70°	75°	80°	85°	90°
M	2.731	2.397	2.136	1.926	1.754	1.611	1.486	1.378	1.284	1.202	1.128	1.061	1.000
N	0.493	0.530	0.567	0.604	0.641	0.678	0.717	0.759	0.802	0.846	0.893	0.944	1.000

Here A, B, M, N, and ψ are the functions of the geometry. A, B, and ψ are defined in equation (11.2.2.3) and M and N are shown in Table 11.1.

If we know the pressure distribution at the contact, the stress at any point can be found. It can be shown that the maximum shear stress occurs along the z-axis, a small distance away from the contact surface. The ratio of b/a decides the position and magnitude of the maximum shear stresses.

Taking x- and y-axes as the directions of the semi-axes a and b of the ellipse of contact, the principal stresses at the center of the contact surface are

$$\sigma_x = -2\upsilon p_o - (1-2\upsilon)\, p_o \frac{b}{a+b}; \quad \sigma_y = -2\upsilon p_o - (1-2\upsilon)\, p_o \frac{a}{a+b}; \quad \sigma_z = -p_o.$$

At the ends of the axes of the ellipse $\sigma_x = -\sigma_y$ and $\tau_{xy} = 0$. The tensile stress in the radial direction is equal to the compressive stress in the circumferential direction. At these points, there exists a state of pure shear. More detailed analysis and the magnitudes of the shear stresses have been discussed by Timoshenko and Goodier (1970).

However, it is interesting to note that when $a = b$ in equation (11.5.2.5), the contact area reduces to a circle and the analysis is given in Section 11.5. If, however, $\dfrac{a}{b} \to \infty$, the contact area reduces to a narrow and long rectangle, and this is the contact between two cylinders of parallel axes.

Example 11.1

A steel railway wheel with a 400 mm radius rolls on a steel rail with a cross radius of 300 mm. The load on the wheel is 4.5 kN. Find the maximum contact stress and contact area. Take the material property as $E = 210$ GPa and $\upsilon = 0.3$ for both rail and the wheel.

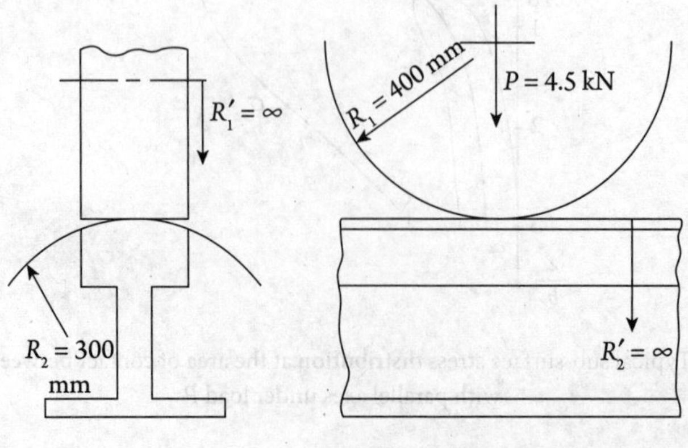

Solution:

Here the contact area is an ellipse of semimajor axes a and b, where using equation (11.5.2.5)

$$a = M\sqrt[3]{\frac{3P\Delta}{4(A+B)}} \quad \text{and} \quad b = N\sqrt[3]{\frac{3P\Delta}{4(A+B)}},$$

we have

$$\Delta = \frac{2(1-0.09)}{210\times10^9} = 8.7x10^{-12}\ \text{m}^2/\text{N}$$

$$A+B = \frac{1}{2}\left(\frac{1}{0.4}+\frac{1}{0.3}\right) = 2.92\,\text{m}^{-1}$$

$$B-A = \frac{1}{2}\left[\left(\frac{1}{0.4}\right)^2+\left(\frac{1}{0.3}\right)^2+2\left(\frac{1}{0.4}\times\frac{1}{0.3}\right)\cos\left(180^0\right)\right]^{\frac{1}{2}} = 0.408\ \text{m}^{-1}$$

$$\cos^{-1}\left(\frac{B-A}{B+A}\right) = \cos^{-1}(0.14) = 82^0.$$

From Table 11.1, with these values

$M = 1.101$; $N = 0.913$.

This gives

$$a = 1.101\sqrt[3]{\frac{3\times4500\times8.7\times10^{-12}}{4\times2.92}} = 0.2316\times10^{-2}\,\text{m}$$

$$b = 0.913\sqrt[3]{\frac{3\times4500\times8.7\times10^{-12}}{4\times2.92}} = 0.192\times10^{-2}\,\text{m}.$$

Maximum contact pressure $p_0 = \dfrac{3P}{2\pi ab} = 483.3$ MPa.

Contact area $= \pi ab = 14\times10^{-6}\ \text{m}^2$.

Example 11.2

A sphere of 100 mm diameter is pressed into the plane surface of a solid block, assuming that both the sphere and the block are made of steel of Young's modulus $E = 210$ GPa and Poisson's ratio $\upsilon = 0.3$, but the block is softer than the sphere and has a tensile yield stress of 200 MPa. Find the maximum pressing load that can be applied to the sphere before the yielding of the block sets in.

Solution:

The problem is shown in the figure below:

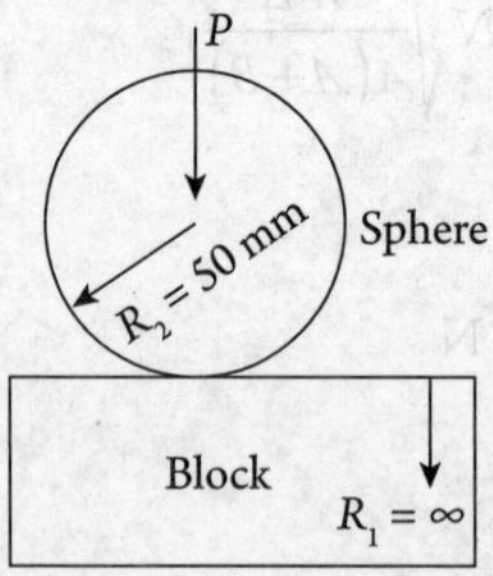

The situation was discussed in Section 11.5, and it was shown that the maximum contact pressure p_o is given in equation (11.5.4)

$$p_o = 0.388\sqrt[3]{\frac{PE^2}{R_2^2}} = 0.388\sqrt[3]{\frac{P\times(210\times10^9)^2}{(0.05)^2}} = 101\times10^6 P^{\frac{1}{3}}.$$

In Section 11.5, it was also shown that the maximum shear stress τ_{rz} is $0.31\,p_o$, and it occurs at 0.47 times the contact radius. For the yielding of the block, this means

$$0.31\,p_o = \frac{\sigma_{ty}}{2},$$

since the shear yield stress for ductile material is ½ (tensile yield stress).

Therefore, $0.31\times101\times10^6 P^{\frac{1}{3}} = \dfrac{200\times10^6}{2}$.

This gives the maximum pressing load $P = 32.6$ N.

Example 11.3

In a milling mill, two rolls are 500 mm diameter and 80 mm diameter and their axial length is 1 m. The maximum load between the rolls is 2 MN, and the rolls are made of a hardened steel having Young's modulus $E = 200$ GPa and Poisson's ratio $\upsilon = 0.3$. Find the maximum contact pressure p_o.

Solution:

For the given milling mill consisting of two rolls,

$R_1 = 0.25$ m, $R_2 = 0.04$, $l = 1$ m

$P = 2\times10^6$ N, $E = 200\times10^9$ Pa, $\nu = 0.3$.

The maximum contact pressure is given by equation (11.5.2.3) as

$$p_0 = \frac{2p'}{\pi b}.$$

In the above equation, load per unit length $p' = \frac{P}{l} = 2\times10^6$ N/m

and the contact width is given by equation (11.5.2.4) as $b = \sqrt{\frac{4p'(K_1+K_2)R_1R_2}{R_1+R_2}}$,

where $K_1 = K_2 = \frac{1-\nu^2}{\pi E} = 1.44\times10^{-12}$ m²/N.

Putting the numerical values, we get $b = 0.89\times10^{-3}$ m.

Now putting the values of p' and b in equation (11.5.2.3), we get the maximum contact pressure $p_0 = 1431.33$ MPa.

Example 11.4

A ball of diameter "d" and a cylindrical roller of the same diameter "d" and a length-to-diameter ratio $l/d = 1$ are loaded between two plates successively until the same maximum pressure p_o is reached in each case. Show that

$$P_{roller} = \frac{\sigma}{\pi} \frac{E}{(1-\upsilon^2)p_o} P_{ball},$$

where P_{ball} and P_{roller} represent the load on the ball and load on roller per unit length respectively.

Solution:

For contact of the ball with the plate, the contact radius is obtained from equation (11.5.1) as

$$a = \sqrt[3]{\frac{3}{2}\frac{(1-\nu^2)R}{E}P_{ball}}.$$

Hence, the contact pressure for the ball is obtained as

$$p_0 = \frac{3P_{ball}}{2\pi a^2} = \frac{3}{2\pi}\sqrt[3]{\frac{4P_{ball}E^2}{9(1-\nu^2)^2R^2}}.$$

Cubing both sides, we obtain the expression for P_{ball} as

$$P_{ball} = \frac{2\pi^3(1-\nu^2)^2R^2}{3E^2}p_0^3. \qquad \text{(i)}$$

For the contact of the roller with the plate, contact width is obtained from equation (11.5.2.4) as

$$b = \sqrt{\frac{8(1-\nu^2)R}{lE}P_{roller}}.$$

Substituting the above-presented expression of b in equation (11.5.2.3), we obtain the contact pressure for the roller as

$$p_0 = \frac{1}{\pi}\sqrt{\frac{P_{roller}E}{2(1-\nu^2)lR}}.$$

Squaring both sides, we obtain the expression for P_{roller} as

$$P_{roller} = \frac{2\pi^2\left(1-\nu^2\right)lR}{E} p_0^{\ 2}. \qquad \text{(ii)}$$

From equations (i) and (ii), noting $l/R = 2$, we get

$$P_{roller} = \frac{6}{\pi} \frac{E}{\left(1-\nu^2\right)} \frac{1}{p_0} P_{ball}.$$

Example 11.5

A 75 mm diameter ball made of hardened steel is held between two parallel plates and they are pressed together with a force of 900 N. The torque needed to rotate the ball about an axis perpendicular to the plate is 20 Nm. The ball is now replaced by a 100 mm diameter ball, and the applied pressing force is increased to 1 kN. Assuming that a similar frictional condition occurs in both cases and the stresses developed do not exceed the elastic limit, find the torque required to rotate the 100 mm diameter ball.

Solution:

First case

$$a = \sqrt[3]{\frac{3\pi}{4} \frac{P\left(K_1 + K_2\right)}{\frac{1}{R_1} + \frac{1}{R_2}}}.$$

Let $k_3 = \sqrt[3]{\left(K_1 + K_2\right)}$. Hence, the above equation becomes

$$a = \sqrt[3]{\frac{3\pi}{4} P R_1} k_3.$$

Putting $P = 900$ N and $R_1 = 0.0375$ m, we get

$$a = 4.3 k_3.$$

Now $p = \frac{p_0}{a}\sqrt{a^2 - r^2} = p_0\left(1 - \frac{r^2}{a^2}\right)^{1/2} = p_0\left(1 - \frac{1}{2}\frac{r^2}{a^2} \ldots\right).$

$$\text{Resulting torque} = \int_0^a p\, 2\pi r\, dr\, \mu r$$

$$= \int_0^a 2\pi\mu p_0\left(1 - \frac{1}{2}\frac{r^2}{a^2}\right) r^2\, dr$$

$$= 2\pi\mu p_0 \frac{7a^3}{30}$$

$$= \frac{7}{10}\mu a P \quad \left[\text{As } P = \frac{2}{3}\pi p_0 a^2\right]$$

$$= \frac{7}{10} \times 4.3 \times 900\, \mu k_3 = 20$$

$$\mu k_3 = 7.38 \times 10^{-3}.$$

Second case

$$a = 4.9 k_3.$$

$$\text{Resulting torque} = 2\pi\mu p_0 \frac{7}{30} a^3.$$

Substituting values, the resulting torque = 25.3 Nm.

Example 11.6

The dimensions of inner and outer races and the balls of a lightly loaded high-speed ball bearing are shown in the figure. To reduce slip and hence wear, some preloading is developed by increasing the ball diameter 1×10^{-2} mm larger than the clearance between the outer and inner races. If the races can be assumed to be rigidly supported so that the flexural deformation can be avoided and if, because of the large difference in curvatures of the balls and the races, the ball/race contact situation may be taken to be that of a ball and flat surface, determine the preload experienced by each ball. Find also the contact radius and the maximum Hertzian pressure for the preload condition for both the ball and races. $E = 200$ GPa and $\upsilon = 0.3$.

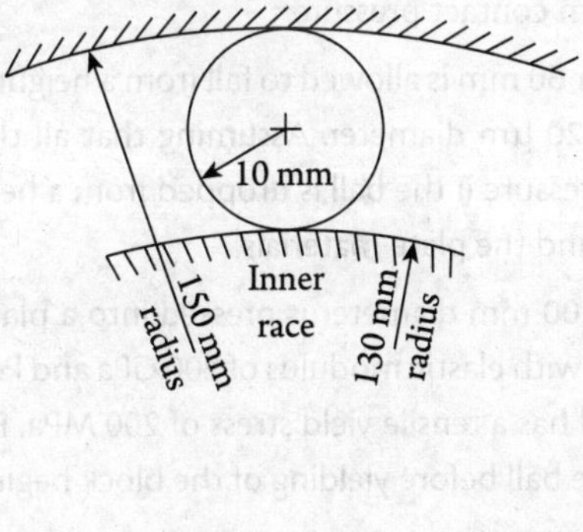

Solution:

Half the interference is equal to the approach on each side of the ball.

From equation (11.5.4), $\alpha = 1.23\sqrt[3]{\frac{P^2}{E^2 R_2}}$.

Putting $E = 200 \times 10^9$ Pa, $R_2 = 0.01$ m, and $\alpha = 5 \times 10^{-6}$ m.

$$P = 164 \text{ N}$$

$$a = 1.109\sqrt[3]{\frac{PR_2}{E}} = 0.22 \text{ mm}$$

$$p_0 = 1564.8 \text{ MPa}.$$

Exercises

1. Show by the stress function approach that a concentrated force acting vertically on the boundary of a semi-infinite solid produces a stress field that is purely radial compressive, directed toward the line of application of the load. State the essential boundary conditions of the problem.
2. Stresses in the *z*-direction in the cylindrical coordinate system at the boundary and sub-surface of a semi-infinite body due to a concentrated load *P* are given by

$$\sigma_z = -\frac{3P}{2\pi} z^3 \left(r^2 + z^2\right)^{-\frac{5}{2}}.$$

Find the total stress at the contact area of radius *a* between two curved bodies pressed under a force *P*. [Hints: Refer to equation (11.5.2.2).]
3. The intensity of pressure *p* over the surface of contact between two curved bodies is represented by the ordinates of a semi-ellipsoid constructed on the contact surface as follows:

$p = p_o\sqrt{1 - \frac{x^2}{a^2} - \frac{y^2}{b^2}}$, where p_o is the maximum pressure and *a* and *b* are the semi-axes of the contact ellipse.

Apply this representation to the case of two rollers of radii R_1 and R_2 made of the same material and having the same Poisson's ratio $\upsilon = 0.3$, pressed together by a load *P* per unit length of contact and derive expressions for the peak contact pressure and contact area.
4. Two cylindrical rollers, each of 100 mm diameter and 150 mm long, are mounted on parallel shafts and loaded by a force of 1 MN. Taking elastic modulus $E = 200$ GPa and Poisson's ratio $\upsilon = 0.3$, find the maximum contact pressure. [3053.715 MPa]
5. An elastic ball of diameter 60 mm is allowed to fall from a height *h* on a steel plate. This produces a contact impression of 20 μm diameter. Assuming that all the deformations are elastic, find the maximum contact pressure if the ball is dropped from a height of 3h. Take $E = 200$ GPa and $\upsilon = 0.3$ for both the ball and the plate materials. [29.06 MPa]
6. A steel spherical ball of 200 mm diameter is pressed into a plane surface of a solid block, both made of similar materials with elastic modulus of 200 GPa and Poisson's ratio of 0.3, but the block is softer than the ball and has a tensile yield stress of 200 MPa. Find the maximum pressing load that can be applied to the ball before yielding of the block begins. [143.66 N]
7. When a normal load *P* is applied to the surface of a semi-infinite body, the normal displacements of the surface are given by

$$\omega = \frac{P\left(1 - \nu^2\right)}{\pi E r},$$

where in a cylindrical coordinate system, *r* is the radial distance from the point of application of the load. Derive the expression for the maximum Hertzian stress "P_o" and the radius of contact "*a*" when a solid sphere is pressed into the plane surface of a solid block.

[Hints: Refer to Section 11.4.]

12

Energy Methods

Learning Objectives

After careful study of this chapter, students should be able to do the following:

LO1: Describe strain energy in different loading conditions.

LO2: Explain the principle of superposition and reciprocal relations.

LO3: Apply the first theorem of Castigliano.

LO4: Analyze the theorem of virtual work.

LO5: Apply the dummy load method.

LO6: Analyze the theorem of virtual work.

12.1 INTRODUCTION [LO1]

There are in general two approaches to solving equilibrium problems in solid mechanics: Eulerian and Lagrangian. The first approach deals with vectors such as force and moments, and considers the static equilibrium and compatibility equations to solve the problems. In the second approach, scalars such as work and energy are used, and here solutions to problems are based on the principle of conservation of energy. There are many situations where the second approach is more advantageous, and here some powerful methods, such as the method of virtual work, based on this approach, are used.

Eulerian and Lagrangian approaches to solving solid mechanics problems are much more involved. However, here we have chosen to describe these in a simplified manner, which is suitable as a prologue to the present discussion on energy methods.

In mechanics, energy is defined as the capacity to do work, and this may exist in different forms. We are concerned here with elastic strain energy, which is a form of potential energy stored in a body on which some work is done by externally applied forces. Here it is assumed that the material remains elastic when work has been done so that all the energy is recoverable and no permanent deformation occurs. This means that strain energy U = work done. If the load is applied gradually in straining, the material load–extension graph is as shown in Figure 12.1, and we may write $U = \frac{1}{2} P\delta$.

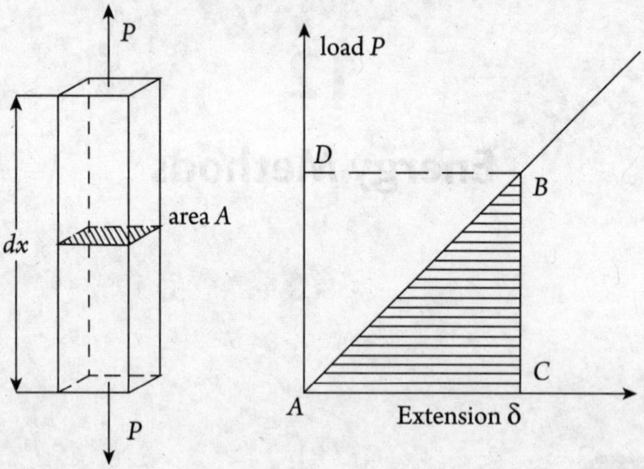

Figure 12.1 Load–extension graph of a uni-axially loaded elemental bar

The hatched portion of the load–extension graph represents the strain energy and the unhatched portion ABD represents the complementary energy that is utilized in some advanced energy methods of solution. We shall first consider the strain energy U per unit volume, which may also be written as

$$U = \frac{1}{2}\sigma \in = \frac{1}{2}\frac{\sigma^2}{E}. \qquad (12.1.1)$$

We may also define the strain energy in more detail by considering a pin-connected truss loaded at the pin end, as shown in Figure 12.2.

The work done W by the force P as it increases is given by

$$W = \int P d\delta.$$

We assume here that all the work done by W is transferred to the truss and produces deformation in the structure. Following the principle of conservation of energy, the total energy spent in deformation of the truss is conserved as a form of potential energy, known as strain energy in the structure. Therefore, we may write

$$\text{Strain Energy } U = \int P d\delta. \qquad (12.1.2)$$

Differentiating both sides of equation (12.1.2) with respect to δ, we have

$$P = \frac{dU}{d\delta}. \qquad (12.1.3)$$

Equations (12.1.2) and (12.1.3) are two important equations in solving deflection problems in solid mechanics.

Equation (12.1.1) defines the strain energy in terms of stress and strain for a uni-axially loaded member. However, if we consider all six stress elements in a three-dimensional (3D) element as shown in Figure 12.3, we may write a general expression for strain energy as

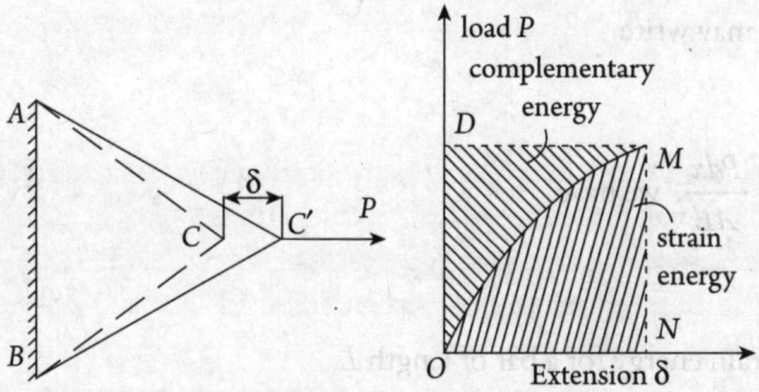

Figure 12.2 Load–extension graph of a loaded pin-connected truss

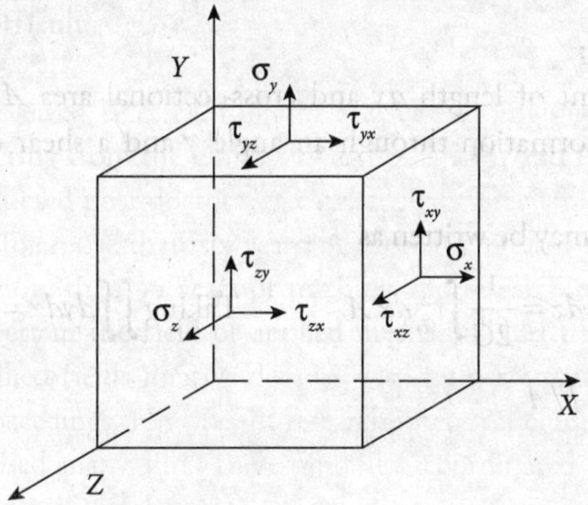

Figure 12.3 Stress elements on an infinitesimal body

$$U=\frac{1}{2}\iiint_{v}\left(\sigma_{x}\in_{x}+\sigma_{y}\in_{y}+\sigma_{z}\in_{z}+\tau_{xy}\gamma_{xy}+\tau_{yz}\gamma_{yz}+\tau_{zx}\gamma_{zx}\right)dxdydz. \tag{12.1.4}$$

We may now consider the strain energy in tension or compression, shear, bending, and torsion of a member individually.

<u>Strain energy due to tension or compression:</u>

Consider the small element of length dx and cross-sectional area A in Figure 12.1. As seen earlier in this case for axial loading in the x-direction

Strain Energy $U_t=\frac{1}{2}\iiint_{v}\sigma_x\in_x dxdydz=\frac{1}{2E}\int_{L}\sigma_x^2 dx\,A.$ Taking $\iint dydz=A.$

Putting $\sigma_x=\frac{P}{A}$ $\qquad U_t=\int_0^L\frac{P^2dx}{2AE}.$

This gives the strain energy for a bar of length L as

$$U_t=\frac{P^2L}{2AE}. \tag{12.1.5}$$

Alternatively, we may write

$$U_t = \frac{1}{2} P\delta$$

and substituting $\delta = \dfrac{Pdx}{AE}$, we have

$$U_t = \int \frac{P^2 dx}{2AE}.$$

Therefore, the strain energy for a bar of length L,

$$U_t = \int_0^L \frac{P^2 dx}{2AE} = \frac{P^2 L}{2AE}.$$

Strain Energy due to Shear:

Consider the small element of length dx and cross-sectional area A being subjected to shear force F. This causes a deformation through an angle γ and a shear deflection δ, as shown in Figure 12.4

Here the strain energy may be written as

$$U_s = \frac{1}{2}\iiint_v \tau_{xy}\gamma_{xy}\, dxdydz = \frac{1}{2G}\int_L \tau_{xy}^2\, dx\, A. \qquad \text{Taking} \iint dydz = A.$$

Now substituting $\tau_{xy} = F/A$

$$U_s = \int_L \frac{F^2 dx}{2AG}$$

and for a bar of length L

$$U_s = \frac{F^2 L}{2AG}.$$

Alternatively, we may write

$$U_s = \frac{1}{2} F\delta = \frac{1}{2} F\gamma\, dx$$

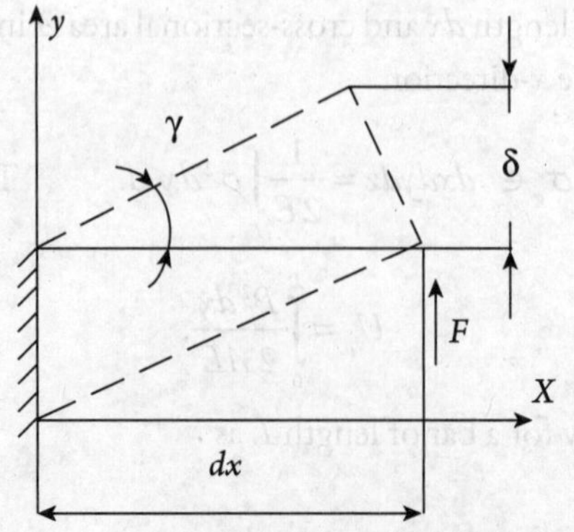

Figure 12.4 An element of length dx subjected to a shear F causing a deformation through an angle γ

and substituting $\gamma = \frac{\tau}{G} = \frac{F}{AG}$, we have

$$U_s = \int \frac{F^2}{2AG} dx,$$

the strain energy for a bar of length L

$$U_s = \int_0^L \frac{F^2}{2AG} dx = \frac{F^2 L}{2AG}.$$

Strain energy due to pure bending:

Consider the same small element of length dx and cross-sectional area A being subjected to a constant bending moment M that causes the element to bend into an arc of radius R and subtending an angle $d\theta$ at the center, as shown in Figure 12.5.

Here the strain energy may again be written as

$$U_B = \iiint_v \frac{\sigma_x^2}{2E} dxdydz.$$

But here $\sigma_x = \frac{My}{I}$,

where M is the applied moment, y is the distance from the neutral axis, and I is the second moment of area given by $\int_A y^2 dydz$.

We therefore have

$$U_B = \iiint_v \frac{M^2 y^2 dxdydz}{2I^2 E} = \int \frac{M^2 dx}{2EI}.$$

Alternatively,

Strain energy = work done in bending = $\int \frac{1}{2} M d\theta$.

Now, $dx = Rd\theta$ and $\frac{M}{I} = \frac{E}{R}$.

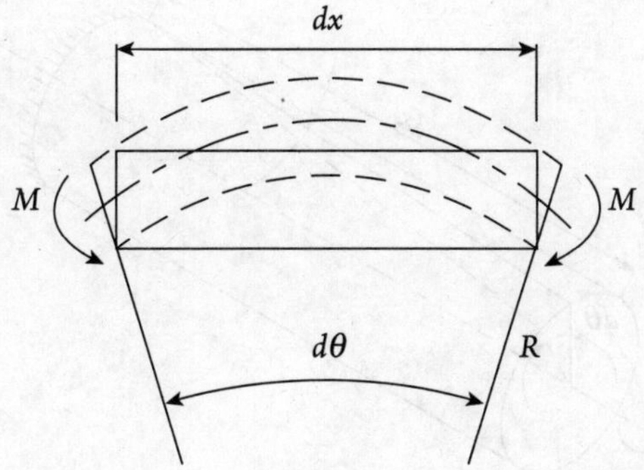

Figure 12.5 A small element of length of *dx* subjected to a bending moment *M*

This gives, strain energy $U_B = \int \frac{M^2 dx}{2EI}$.

For a constant bending moment on a bar of length L, this reduces to

$$U_B = \frac{M^2 L}{2EI}.$$

Strain energy due to torsion:

We consider here a circular member of radius r and length dx subjected to a torque T producing an angle of twist $d\theta$. This is shown in Figure 12.6.

Again following the general expression for strain energy, we may write here

$$U_T = \frac{1}{2}\iiint_v \frac{\tau_{xy}^2}{G} dxdydz.$$

Since $\frac{T}{J} = \frac{\tau}{r}$,

$$U_T = \iiint_v \frac{T^2 r^2}{2J^2 G} dxdydz.$$

We also have $dydz = dA$ and $\int_A r^2 dA = J$, and this gives strain energy in torsion

$$U_T = \int_L \frac{T^2 dx}{2GJ}.$$

For constant torque and for a bar of length L, we have

$$U_T = \frac{T^2 L}{2GJ}.$$

Alternatively, here strain energy in torsion $U_T = \int \frac{1}{2} T d\theta$.

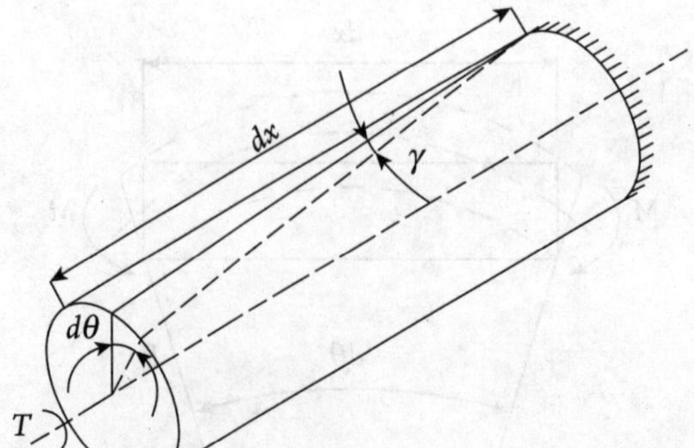

Figure 12.6 A circular member of length dx is subjected to a torque T producing an angle of twist $d\theta$

Since $\frac{T}{J}=\frac{Gd\theta}{dx}$, we may write

$$U_T=\int_0^L \frac{T^2dx}{2GJ}=\frac{T^2L}{2GJ}.$$

Strain energy of a three-dimensional system:

Referring to equation (12.1.4), the total strain energy for a 3D system can be given in terms of principal stresses σ_1, σ_2, and σ_3 as

$$U_T=\frac{1}{2E}\left\{\sigma_1^2+\sigma_2^2+\sigma_3^2-2\upsilon\left(\sigma_1\sigma_2+\sigma_2\sigma_3+\sigma_1\sigma_3\right)\right\} \text{ per unit volume}$$

Volumetric strain energy is given in terms of mean stress or hydrostatic stress σ_m, given by $\frac{1}{3}\left(\sigma_1+\sigma_2+\sigma_3\right)$ as

$$U_v=\frac{(1-2\upsilon)}{6E}\left[\left(\sigma_1+\sigma_2+\sigma_3\right)^2\right] \text{ per unit volume.}$$

Deviatoric strain energy is given as

$$U_D=\frac{1+\upsilon}{6E}\left[\left(\sigma_1-\sigma_2\right)^2+\left(\sigma_2-\sigma_3\right)^2+\left(\sigma_1-\sigma_3\right)^2\right] \text{ per unit volume.}$$

12.2 BASIC THEOREMS OF DEFLECTION OF ELASTIC BODIES [LO2]

Relation between the forces acting on an elastic body and the deformation or deflection is an important issue in solving solid mechanics problems for two main reasons: (a) The maximum permissible load that can be applied to a machine member or structure is often limited by their allowable deformation. (b) For statically indeterminate members, the equation for static equilibrium is often not sufficient and there an extra relation is needed. For many problems, this is supplied by the load–deflection relation.

Some basic principles and theorems are useful in the development of this load–deflection relation and in the following sections, we discuss them in detail.

12.2.1 Principle of Superposition

This principle has been discussed in earlier chapters, and this basically states that for elasticity problems governed by linear equations, solutions to problems of a system subjected to (say) two different sets of body forces and traction forces F_1, T_1 and F_2, T_2 would be the same if the system is subjected to the algebraic sum of the body and traction forces (F_1+F_2) and (T_1+T_2) respectively. This is a very useful principle in solving many complicated problems. It has also been discussed earlier that the total displacement at a point (say) 2 in Figure 12.7 in an elastic medium acted upon by several forces F_1, F_2, F_3, ... is given by

$$\delta_2=a_{21}F_1+a_{22}F_2+a_{23}F_3+\ldots$$

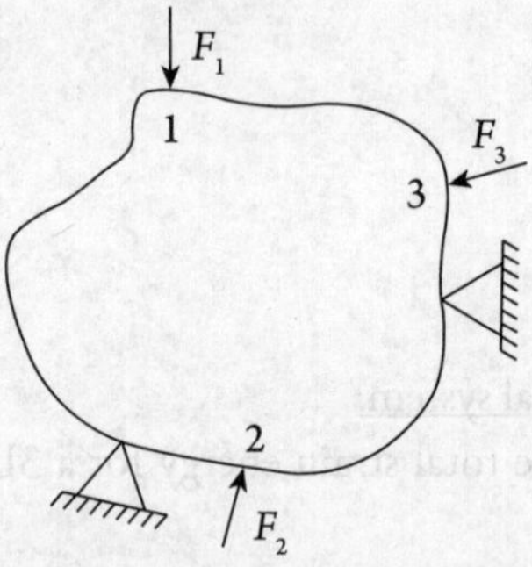

Figure 12.7 An elastic body subjected to multiple concentrated force

Indicating that deflections due to force at any point on the elastic body may be superposed. Here a_{21}, a_{23}, ... are the influence coefficients. The component of the total displacement at a point in the direction of the applied force at that point is called the corresponding displacement.

We may therefore write the total elastic strain energy stored in the body as

$$U=\frac{1}{2}\left(F_1\delta_1+F_2\delta_2+\ldots\right), \tag{12.2.1.1}$$

where δ_1, δ_2, δ_3.... are the corresponding displacements at points 1, 2, 3, ... etc.

It is important to note that the above expression is independent of the order of application of forces, provided that the forces are increased in constant proportion and gradually.

This principle is widely used in solid mechanics and one of the applications is in finding deflection, slope, bending moment, and shear force in a beam loaded by multiple concentrated loads. The results due to any one of the loads are considered to be independent of those due to other loads. Therefore, the values of deflection, slope, bending moment, and shear force at any section of the beam may be obtained by algebraically summing up the results due to all the concentrated loads. To demonstrate this, we shall consider a simple beam loaded by multiple concentrated loads, as shown in Figure 12.8.

The deflection y, slope θ, bending moment M, shear force V, and loading q are given by the following differential equations:

$\theta=\dfrac{dy}{dx}$: Slope of elastic curve

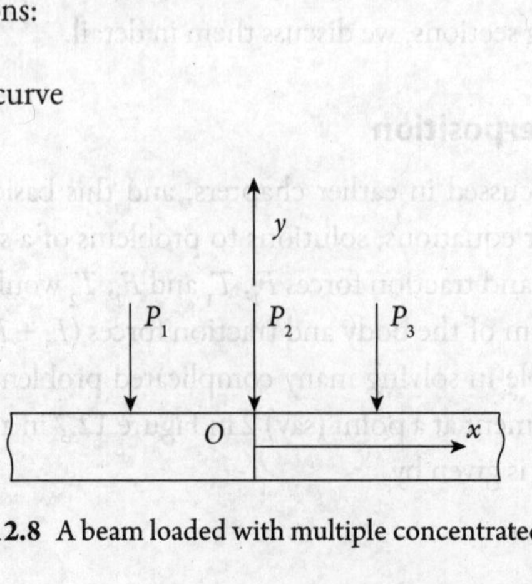

Figure 12.8 A beam loaded with multiple concentrated loads

$$M = EI\frac{d^2y}{dx^2}\text{: Bending moment}$$

$$V = EI\frac{d^3y}{dx^3}\text{: Shear force}$$

$$-q = EI\frac{d^4y}{dx^4}\text{: Beam loading} \tag{12.2.1.2}$$

Usually, downward loading is considered to be negative.

Considering the elasticity of the beam, we may write the last equation in (12.2.1.2) as

$$EI\frac{d^4y}{dx^4} = -Ky, \tag{12.2.1.3}$$

where K represents the spring constant per unit length of the elastic support. The solution to this equation for a long beam subjected to a concentrated force P may be given as

$$y = e^{-\beta x}\left(C_1\cos\beta x + C_2\sin\beta x\right). \tag{12.2.1.4}$$

Since the equation (12.2.1.3) is a fourth order homogeneous differential equation, the solution should ideally have four terms. But since the beam under consideration is assumed to be long, two of the terms are zero and the final solution is valid for positive values of x only. This restriction is not serious since, for long beams, we may consider symmetry with respect to the load.

It can be shown (see Seely and Smith, 1952) that $\beta = \sqrt[4]{\frac{K}{4EI}}$ and $C_1 = C_2 = C = P\beta/2K$, P being the concentrated load. This is based on the assumption that the slope is zero at $a = 0$ and using the equilibrium equation that states that the total integrated upward force on a beam subjected to a singe concentrated force P is equal to P.

This gives

$$y = \frac{P\beta}{2K}(A+B)$$

$$\theta = -\frac{P\beta^2}{K}B$$

$$M = \frac{P}{4\beta}(A-B) \tag{12.2.1.5}$$

$$V = -\frac{P}{2}A,$$

where $A = e^{-\beta x}\cos\beta x$ and $B = e^{-\beta x}\sin\beta x$.

Displacements for different loads P_1, P_3, etc. can be algebraically summed to obtain the displacements for the total loading using the principle of superposition. For an example, consider a long beam loaded at three points by forces P_1, P_2, and P_3, as shown in Figure 12.9. Approximate deflection curves due to the forces P_1, P_2, and P_3 using the first of the equations (12.2.1.5) are shown as y_1, y_2, and y_3, respectively. Using the principle of superposition, the total deflection due to the three forces was calculated using again the first of the equations (12.2.1.5) and algebraically

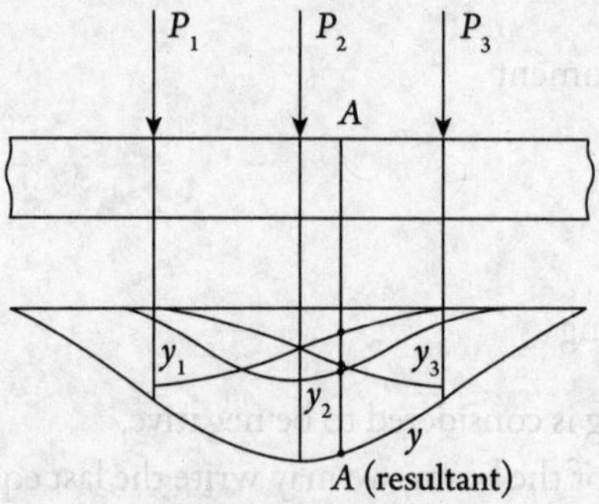

Figure 12.9 Deflection of a long beam loaded with multiple concentrated loads

summing up the results of P_1, P_2, and P_3 for different values of x. Deflection of a point (say) A may be found from the resultant elastic curve shown as y in Figure 12.9. Readers may refer to Seely and Smith (1952) for further details.

12.2.2 Reciprocal Relations

This is another important theorem on displacement. It simply indicates that the proportionality constant or the influence coefficient for the deformation at point '1' due to force F_2 is the same as that for the deformation at point '2' due to force F_1, as shown in Figure 12.10.

This essentially means

$$a_{12} = a_{21}, \text{ i.e., } a_{ij} = a_{ji}. \tag{12.2.2.1}$$

To prove this, we consider the following arguments. When only F_1 acts, the corresponding displacement is δ_1, and the strain energy is

$$U_1 = \frac{1}{2}F_1\delta_1 = \frac{1}{2}F_1\left(F_1a_{11}\right) = \frac{1}{2}a_{11}F_1^2.$$

If now F_2 also acts at point "2", then the additional strain energy is

$$U_2 = \frac{1}{2}F_2\left(a_{22}F_2\right) + F_1\left(a_{12}F_2\right).$$

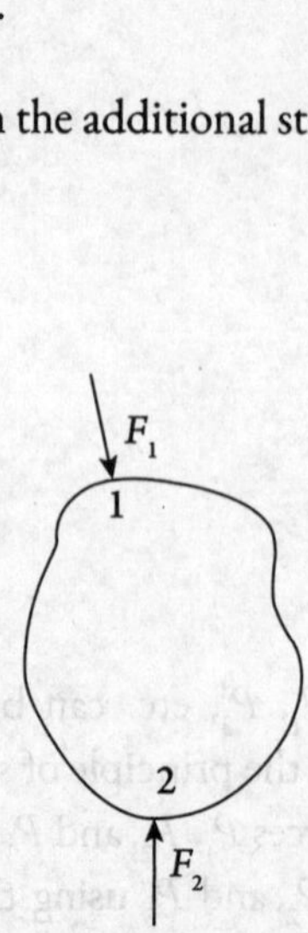

Figure 12.10 An elastic body subjected to force F_1 at point "1" and force F_2 at point "2"

Therefore, the total strain energy due to both F_1 and F_2 is

$$U = U_1 + U_2 = \frac{1}{2}a_{11}F_1^2 + \frac{1}{2}a_{22}F_2^2 + a_{12}F_1F_2. \quad (12.2.2.2)$$

If now the order of application of forces is changed and F_2 is applied first and then F_1, the total strain energy is given by

$$U' = \frac{1}{2}a_{22}F_2^2 + F_2\left(a_{21}F_1\right) + \frac{1}{2}a_{11}F_1^2,$$

since the total strain energy is independent of the order in which the forces are applied

$U = U'$.

And this means $a_{12} = a_{21}$.

This leads us to an interesting conclusion. We may write the total strain energy of an elastic body subjected to a number of forces at different points as

$$U = \frac{1}{2}F_1\delta_1 + \frac{1}{2}F_2\delta_2 + \ldots + \frac{1}{2}F_n\delta_n.$$

Substituting the expression for $\delta_1, \delta_2, \delta_3, \ldots\ldots \delta_n$, we have

$$U = \frac{1}{2}F_1\left(a_{11}F_1 + a_{12}F_2 + \ldots + a_{1n}F_n\right) + \frac{1}{2}F_2\left(a_{22}F_2 + a_{21}F_1 + a_{23}F_3 + \ldots + a_{2n}F_n\right) + \ldots$$

This gives a simplified expression as

$$U = \frac{1}{2}\left(a_{11}F_1^2 + a_{22}F_2^2 + \ldots + a_{nn}F_n^2\right) + \left(a_{12}F_1F_2 + a_{13}F_1F_3 + \ldots + a_{1n}F_1F_n\right)$$
$$+ \left(a_{23}F_2F_3 + a_{24}F_2F_4 + \ldots + a_{2n}F_2F_n\right)\ldots \quad (12.2.2.3)$$

This equation helps in extending the reciprocal theorem to more useful theorems.

12.2.3 Maxwell–Betti Theorem

The theorem first stated by Enrico Betti in 1872 considers systems of forces, say, F_i ($i = 1, 2, 3, \ldots. n$) and F_i' ($i = 1, 2, 3, \ldots. n$), acting on an elastic medium with corresponding deflections δ_i ($i = 1, 2, 3, \ldots. n$) and δ_i' ($i = 1, 2, 3, \ldots. n$) respectively for the force systems. The theorem states that the work done by the first system of force F_i through the displacements caused by the second system force F_i' equals the work done by the second system of force F_i' through the displacements caused by the first system of force F_i.

The theorem has wide application in both structural and mechanical engineering. Let us consider this in more detail with forces and displacements on an elastic beam. Consider a simply supported beam subjected to two systems of forces F_i and F_i' as shown in Figure 12.11.

The work done by F_i system of forces through the displacements caused by F_i' system of forces is given by

$$W_1 = F_1\delta_1' + F_2\delta_2' + F_3\delta_3' \quad (12.2.3.1)$$

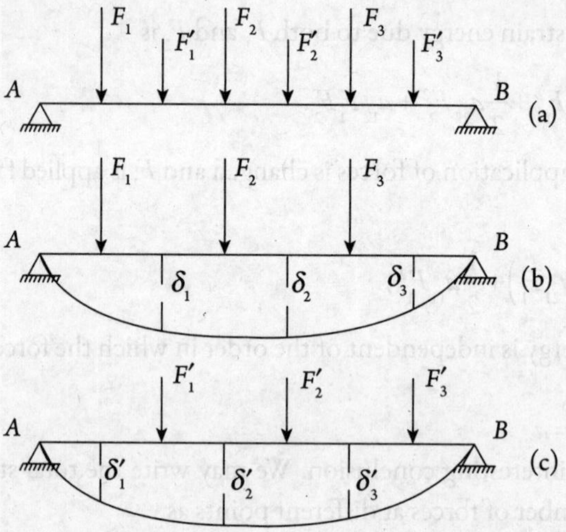

Figure 12.11 (a) A simply supported beam subjected to two systems of forces F_1, F_2, F_3 and F_1', F_2', F_3' (b) The beam subjected to F_i system of forces and the corresponding elastic curve (c) The beam subjected to F_i' system of forces and the corresponding elastic curve

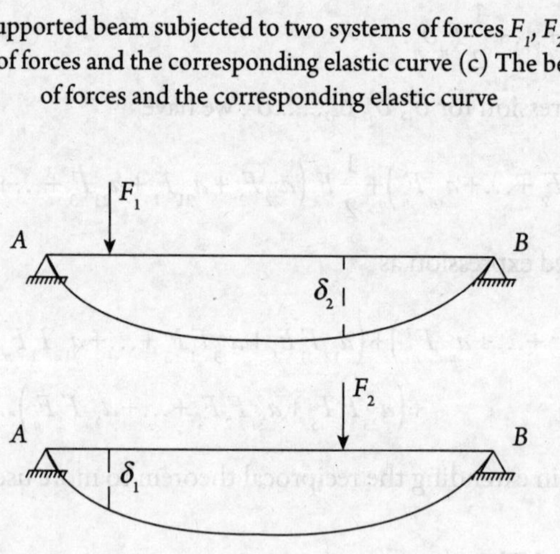

Figure 12.12 A simply supported beam loaded with two loads F_1 and F_2 with deflections δ_1 and δ_2 respectively

and the work done by F_i' system of forces through the displacement caused by F_i system of forces is given by

$$W_2 = F_1'\delta_1 + F_2'\delta_2 + F_3'\delta_3. \tag{12.2.3.2}$$

According to Betti's theorem, $W_1 = W_2$. (12.2.3.3)

Maxwell's reciprocal theorem, proposed in 1864, is a special case of Betti's theorem. It states that the deformation at a point "1" on a linearly elastic medium due to load at point "2" equals the deformation at point "2" due to the same load at point "1". This is explained in Figure 12.12.

According to Betti's theorem, $F_1\delta_2 = F_2\delta_1$. (12.2.3.4)

If we consider a special case where $F_1 = F_2$, then the two deflections are the same,

$$\delta_1 = \delta_2. \tag{12.2.3.5}$$

Maxwell's theorem is valid only if the two loads are the same. This is a very useful theorem in structural analysis.

The theorem may be proved using the results of reciprocal relations in Section 12.2.2.

An important extension to the above theorem is Maxwell–Betti–Rayleigh theorem, which may be written in a slightly different way. Here we consider two different systems of forces F_i $(i = 1, 2, ...n)$ and F_i' $(i = 1, 2, ...n)$ acting at the same points and in the same directions as shown in Figure 12.13. Let the deflections due to F_i system of forces be δ_i $(i = 1, 2, ...n)$ and those due to F_i' system of forces be δ_i' $(i = 1, 2, ...n)$. We now state the theorem as follows:

Total work done by the first system of forces F_i through the deflection due to the second system of forces F_i' is equal to the work done by the second system of forces F_i' through the deflection due to the first system of forces F_i. This means

$$F_1\delta_1' + F_2\delta_2' + \ldots F_n\delta_n' = F_1'\delta_1 + F_2'\delta_2 + \ldots F_n'\delta_n.$$

To prove this, consider the right-hand side (RHS) first

$$RHS = F_1'\delta_1 + F_2'\delta_2 + \ldots + F_n'\delta_n$$

$$= F_1'\left(a_{11}F_1 + a_{12}F_2 + a_{13}F_3 + \ldots\right) + F_2'\left(a_{22}F_2 + a_{21}F_1 + a_{23}F_3 + \ldots\right) + \ldots + F_n'\left(a_{n1}F_1 + a_{n2}F_2 + \ldots\right)$$

$$= \left(a_{11}F_1F_1' + a_{22}F_2F_2' + \ldots + a_{nn}F_nF_n'\right) + a_{12}\left(F_1'F_2 + F_2'F_1\right) + a_{13}\left(F_1'F_3 + F_3'F_1\right) + \ldots + a_{1n}\left(F_1'F_n + F_n'F_1\right).$$

And then we consider the left-hand side (LHS)

$$LHS = F_1\left(a_{11}F_1' + a_{12}F_2' + \ldots\right) + F_2\left(a_{21}F_1' + a_{22}F_2' + a_{23}F_3' \ldots\right)$$

$$= \left(a_{11}F_1F_1' + a_{22}F_2F_2' + \ldots + a_{nn}F_nF_n'\right) + a_{12}\left(F_1'F_2 + F_2'F_1\right) + \ldots + a_{1n}\left(F_1'F_n + F_n'F_1\right).$$

This proves the theorem.

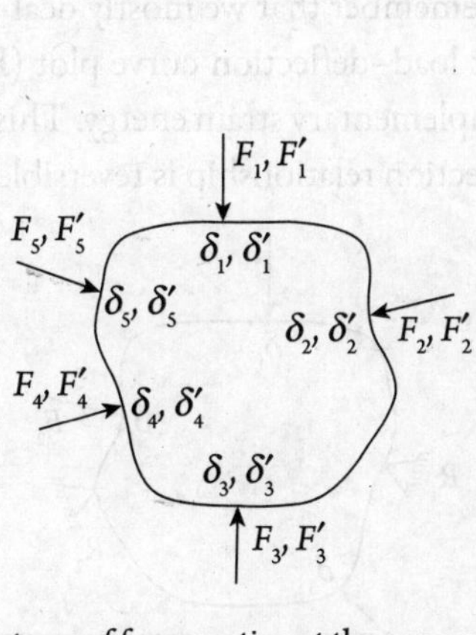

Figure 12.13 Two different systems of forces acting at the same points and in the same directions on an elastic body

12.3 FIRST THEOREM OF CASTIGLIANO [LO3]

The theorem states that if external forces act on an elastic body that is subjected to small deflections, then the deflection in the direction of any one of the forces at the point of application of the force is given by the partial derivative of the total internal strain energy in the member with respect to the force. This may be given with reference to Figure 12.14 as

$$\delta_1 = \frac{\delta U}{\delta F_1}; \delta_2 = \frac{\delta U}{\delta F_2}, \text{etc.}$$

Here U represents the total strain energy of the system. The theorem may be simply proved using the expression of the strain energy derived in the reciprocal theorem (equation 12.2.2.3). There, it was shown

$$U = \frac{1}{2}\left(a_{11}F_1^2 + a_{22}F_2^2 + \ldots + a_{nn}F_n^2\right) + \left(a_{12}F_1F_2 + a_{13}F_1F_3 + \ldots\right) + \left(a_{23}F_2F_3 + a_{24}F_2F_4 + \ldots\right) + \ldots$$

Differentiating this with respect to F_1, we have

$$\frac{\delta U}{\delta F_1} = a_{11}F_1 + a_{12}F_2 + a_{13}F_3 + \ldots$$

And referring to Section 12.2.1, this is nothing but the deflection at point "1" in the direction of F_1. This gives $\frac{\delta U}{\delta F_1} = \delta_1$. Similarly, we can show $\frac{\delta U}{\delta F_2} = \delta_2$, $\frac{\delta U}{\delta F_3} = \delta_3$. In the above expressions, the term "force" represents not only a concentrated normal force but also shear force, bending moment, and torsion. Similarly, the term "displacement" includes both linear and angular displacements. In view of this, it is important to be able to express the total strain energy in terms of the force, moment, or torque. The success of the theorem depends on how well we can express the total strain energy in terms of the force at the point of application for which we need to find the deflection. Methods of expressing the strain energy in tension or compression, shear, bending, and torsion of a member are given in Section 12.1. More details would follow while solving problems. However, it is important here to remember that we mostly deal with linearly elastic materials and therefore, strain energy in a linear load–deflection curve plot (Figure 12.1) is relevant. Here the elastic strain energy equals the complementary strain energy. This also implies that the force system is conservative, and the force–deflection relationship is reversible.

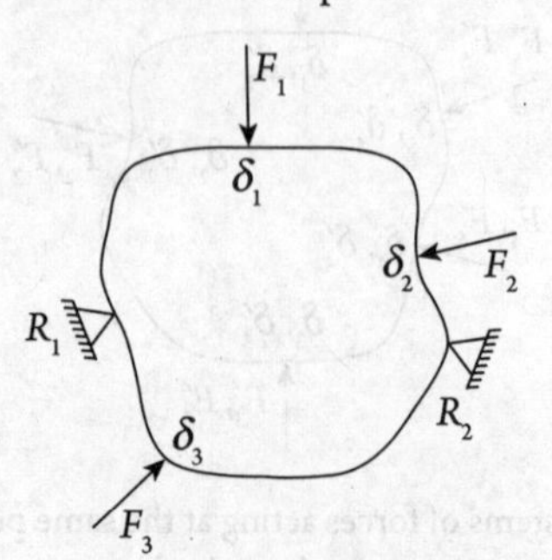

Figure 12.14 Forces and deflections on an elastic body

12.4 THEOREM OF VIRTUAL WORK [LO4]

This theorem essentially considers a small hypothetical displacement on an elastic system in equilibrium and the corresponding small force needed to maintain the equilibrium and to satisfy the conservation of energy criterion. To illustrate this briefly, consider a mechanical system consisting of a member or a structure and certain other bodies capable of exerting forces F_1, F_2, etc., on the member or the structure. We also assume that the system is isolated such that no energy is transferred into or out of the system as changes in energy within the system take place. Now, the energy in the system may be considered to be made up of two components: firstly, the strain energy U in the strained member itself, and secondly, the energy E available in the rest of the system associated with the external loads F_1, F_2, etc., to the member or structure. Therefore, the principle of conservation of energy of such an isolated system may be expressed by the following equation.

$$U+E=\text{constant} \tag{12.4.1}$$

If more energy is used in deforming the members, then that energy must be taken from E, and we may write

$$(U+\Delta U)+(E-\Delta E)=\text{constant},$$

such that $\Delta U=\Delta E$. It is also important to remember that the process is reversible. For stable equilibrium, we may write

$$\frac{\partial}{\partial\delta}(U+E)=0, \tag{12.4.2}$$

since this indicates the condition of minimum energy. Principle of virtual displacement or work is in fact the statement of equation (12.4.1) in terms of small changes in U and E.

The theorem may also be stated as: If a virtual displacement is given to an elastic medium, initially in equilibrium, then the work done by all the external forces acting on the body is equal to the increase in stored elastic energy or strain energy. We may, in a way, state that the internal virtual work equals the external virtual work.

To illustrate this further, we again consider an elastic body subjected to several concentrated forces F_1, F_2, F_3 with corresponding displacements, say, δ_1, δ_2, δ_3... at the point of application of forces as shown in Figure 12.14. Now, let one of these displacements, say δ_i, be increased by a small amount $\Delta\delta_i$, such that the displacements at the points of applications of other forces remain unaltered and the increase in δ_i takes place in a constrained manner. This means that all the other displacements remain constant. However, to do this, an additional force ΔF is needed to maintain the condition. A hypothetical displacement of this kind is known as *virtual displacement*. When this virtual displacement is applied, no other force except F_i is doing the work and the work done is

$$\Delta U=\left(F_i+\Delta F_i\right)\Delta\delta_i=F_i\Delta\delta_i+\Delta F_i\Delta\delta_i.$$

The second term on the RHS, $\Delta F_i\Delta\delta_i$, is small and can be neglected.

This gives

$$\frac{\Delta U}{\Delta\delta_i}=F_i$$

and this may be written as

$$\frac{\partial U}{\partial \delta_i} = F_i. \tag{12.4.3}$$

This is the theorem of virtual work. In this case, the strain energy must be expressed in terms of displacements, whereas in Castigliano's theorem, the strain energy must be expressed in terms of force.

It is also important to note that in deriving the theorem of virtual work, it was not assumed that the material was linearly elastic, i.e., Hookean. Thus, the theorem may be used for both linear and nonlinear elastic bodies. For Castigliano's theorem, however, linear elasticity was assumed and the theorem is strictly applicable to Hookean solids.

12.5 DUMMY LOAD METHOD [LO5]

Castigliano's theorem is very useful in finding the deflection or rotation at the point of application of a concentrated force or moment. For example, we consider a beam loaded as shown in Figure 12.15.

To find the deflection at point "*B*", we need to find the total strain energy *U* in the beam and differentiate it with respect to *P*.

Here,

$$U = \int_0^L \frac{M_x^2}{2EI}\,dx \text{ and } M_x = Px$$

and this gives

$$U = \int_0^L \frac{P^2 x^2}{2EI}\,dx = \frac{P^2}{2EI}\frac{L^3}{3} = \frac{P^2 L^3}{6EI}.$$

This gives the deflection at point *B*

$$\delta_B = \frac{\partial U}{\partial P} = \frac{PL^3}{3EI}.$$

However, if we like to find deflection at point "*C*", where no force is acting, then we need to apply a fictitious force, say P_1, and redo the above process and finally let $P_1 = 0$ in the final expression. We may therefore write the steps to be followed in the dummy load method as follows:

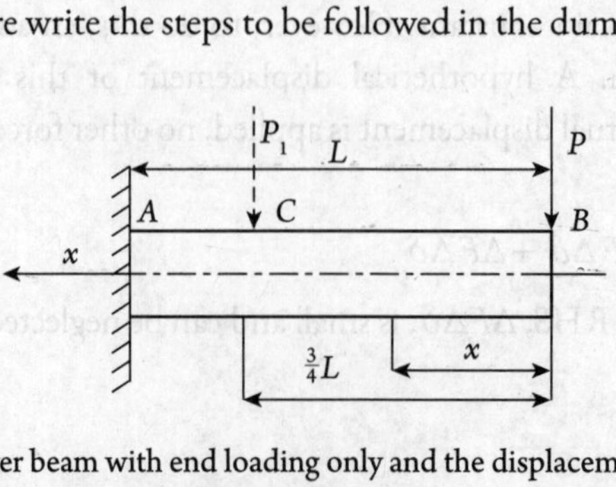

Figure 12.15 A cantilever beam with end loading only and the displacement is desired at point *C*, where no force acts

1. Apply a fictitious load P_1 at the point and in the direction in which the displacement is desired. If rotation is needed at a point, we need to apply a fictitious moment M_1.
2. We then write an expression for the total strain energy of the system consisting of components for axial force, shear force, moment, and torsion, as discussed in Section 12.1.
3. The expression for the total strain energy U is partially derived with respect to the fictitious load to find the deflection at the point of application of the force, and if rotation is needed, then the total strain energy is partially derived with respect to the fictitious moment.
4. In the final expression for deflection or rotation, we set $P_1 = 0$ or $M_1 = 0$. This method will be clarified in more detail while solving problems.

12.6 MENABREA–CASTIGLIANO'S THEOREM OF LEAST WORK [LO6]

L. F. Menabrea, an Italian scientist, studied linear elastic redundant systems during 1858–1873 and proposed theorems on the principle of least work. His proposal may be roughly stated as: "The elastic energy of a body in perfect elastic equilibrium is a minimum with respect to any possible system of stress variation compatible with the equations of static continua in addition to the boundary conditions".

This is a version of the principle of least work. However, Menabrea made several revisions to this statement later on. These statements were the basis for theorems of redundant frames proposed by Castigliano in 1873. Castigliano modified the theorem proposed by Menabrea, and this is now recognized as a general tool for solving many linear elasticity problems.

The principle of least work may now be stated as: In an elastic body or system, there may arise several stress distributions that satisfy equilibrium but not necessarily compatibility equations. The only compatible stress distribution has the least elastic energy in the system. The principle of least work is also stated in relation to redundant reaction components of a statically indeterminate structure. It states: The redundant reaction components of a statically indeterminate system are such that they make the internal work (strain energy) a minimum.

To demonstrate the principle, we consider a statically indeterminate system as shown in Figure 12.16. In this case, we may take reaction R as redundant. We may now write the bending moments and strain energies for two sections and then sum up the strain energies to give the total strain energy of the system as $U(P, R)$. Using Castigliano's theorem, we may write

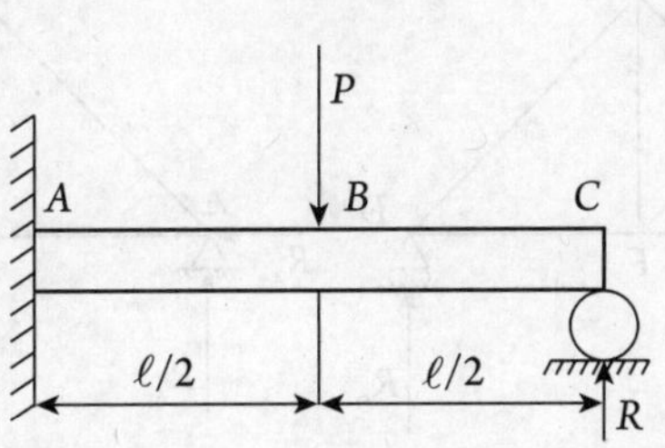

Figure 12.16 A cantilever beam with end support and a single mid-span load

$$\frac{\partial U(P,R)}{\partial R} = \delta_c.$$

But in this configuration, δ_c needs to be zero. Thus, we may write

$$\frac{\partial U(P,R)}{\partial R} = 0. \tag{12.6.1}$$

We note here that the beam is in equilibrium for any value of P and R, and therefore, all the stress distributions satisfy the equilibrium equation, but they may not be compatible. The only stress distribution that gives $\delta_c = 0$ is geometrically compatible and this follows the theorem of least work.

We may now apply this principle to a truss or structure as shown in Figure 12.17, and state the principle as follows:

Forces in redundant structures are such that the total elastic strain energy is the minimum.

This means for every internal redundant force or moment and external redundant reactions or moments, say $F_1, F_2, F_3, \ldots$ or $M_1, M_2, M_3, \ldots$ in the structures, we have

$$\frac{\partial U}{\partial F_i} = 0; \frac{\partial U}{\partial M_i} = 0, \text{ etc.,} \tag{12.6.2}$$

where U is the total strain energy.

Before we go further on the truss redundancy analysis, we shall briefly discuss the degree of indeterminacy. There may be two types of indeterminacy: static and kinematic.

In planar statics problems, there are only three equations of equilibrium:

$$\sum F_x = 0; \sum F_y = 0; \sum M_z = 0. \tag{12.6.3}$$

In a cantilever beam, for example, with end loading, there are three unknowns, and with three equations, the system is a determinate one. However, if we add a prop, as shown in Figure 12.16, there are four unknowns and only three equations. The problem is now indeterminate, and the degree of indeterminacy is one.

The degree of indeterminacy may here be defined as the excess of unknowns over the equations.

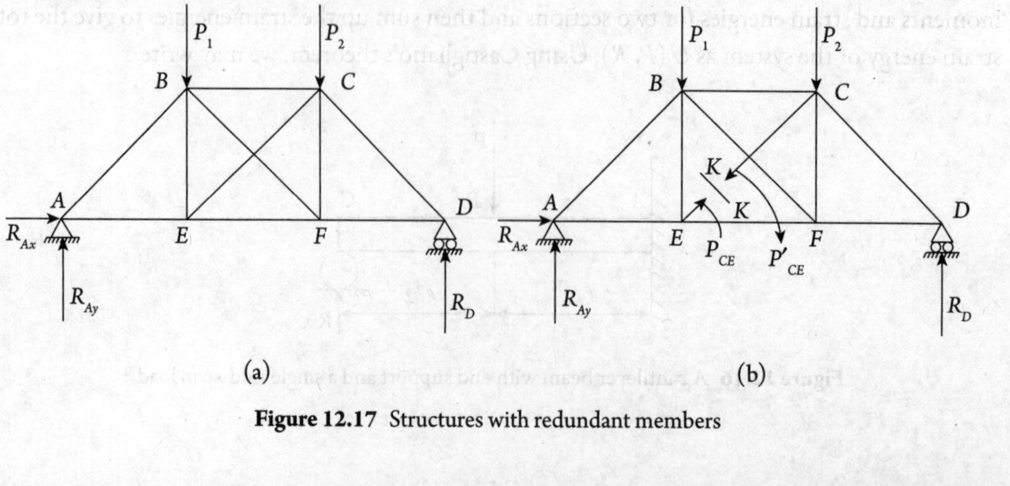

Figure 12.17 Structures with redundant members

In kinematic indeterminacy, the degree is given by the number of unknown displacements. Take, for example, a tripod with three legs fixed on the ground but the common top may have three displacements *u*, *v*, and *w* in the *x*, *y*, and *z* directions. If the three displacements are known, then the forces in the members can be found.

Now we consider the degree of redundancy in a truss, for example, the one shown in Figure 12.17. A truss is considered to be statically determinate if all the support reactions and member forces can be determined using only the equations of static equilibrium (12.6.3). For a planer framework, the degree of redundancy *n* may be given as

$$n = m + r - 2j, \tag{12.6.4}$$

where *m* is the number of members, *r* is the number of reactions, and *j* is the number of joints.

The truss in Figure 12.17 has 10 members, 3 reactions, and 6 joints that gives the degree of redundancy to be one, according to equation (12.6.4). The member EC (or BF) is redundant in the truss, and if this member is removed, then the truss becomes determinate. The only redundant member EC is subjected to internal force $P_{CE} = P'_{CE}$; two equal and opposite forces at a section *K–K* passing through EC. Since the displacements due to P_{CE} and P'_{CE} are collinear and are of equal magnitude, we may consider the displacement at the section *K–K* in EC to be zero, and we may write

$$\frac{\partial U}{\partial P_{CE}} = 0.$$

This equation would apply equally well to any other redundant member. This then allows us to find the internal force in the redundant member. The method is illustrated more in solved examples.

Example 12.1

Find the maximum deflection of the loaded beam as shown below:

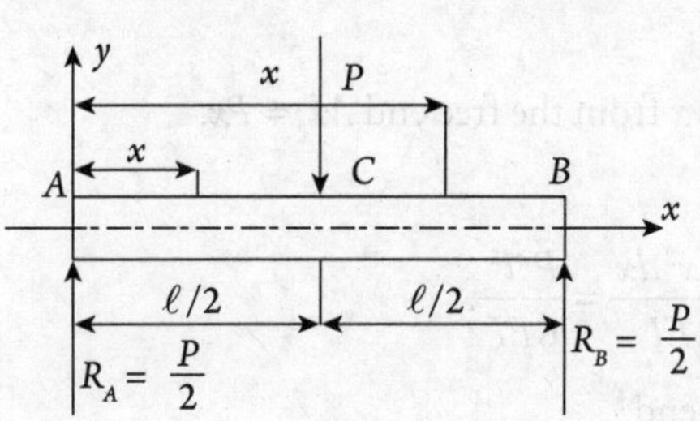

Solution:

Section AC

$$M_x = \frac{Px}{2}$$

$$U_1 = \int_0^{\frac{l}{2}} \frac{M_x^2}{2EI} dx = \int_0^{\frac{l}{2}} \frac{P^2 x^2}{8EI} dx = \frac{P^2 l^3}{192EI}.$$

Section CB

$$M_x = \frac{P}{2}(l-x)$$

$$U_2 = \int_{\frac{l}{2}}^{l} \frac{P^2}{8EI}(l-x)^2\,dx = \int_{\frac{l}{2}}^{l} \frac{P^2}{8EI}(l^2-2lx+x^2)dx = \frac{P^2 l^3}{192EI}.$$

Total strain energy

$$U = U_1 + U_2 = \frac{P^2 l^3}{96EI}.$$

Maximum deflection at C

$$\delta_c = \frac{\partial U}{\partial P} = \frac{Pl^3}{48EI}.$$

Example 12.2

(a) Find the deflection and slope at point A of the end loaded cantilever beam as shown in the Figure.

(b) Find also the deflection at point A if load P is shifted by distance l_1 away along the beam.

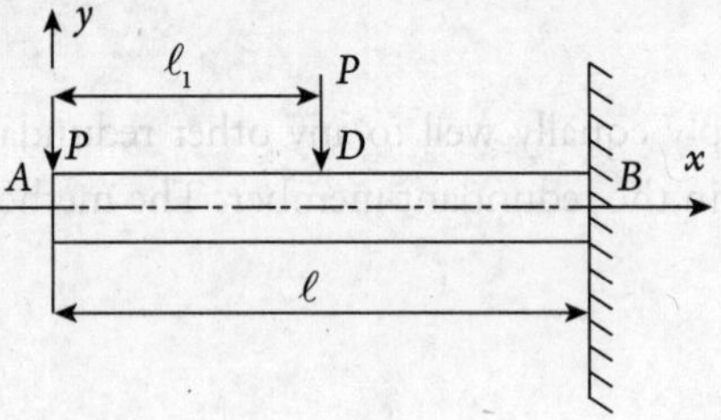

Solution:

(a) Moment at a distance x from the free end $M_x = Px$.

Strain energy

$$U = \int_0^l \frac{M_x^2 dx}{2EI} = \int_0^l \frac{P^2 x^2 dx}{2EI} = \frac{P^2 l^3}{6EI}.$$

Deflection at the free end

$$\delta_c = \frac{\partial U}{\partial P} = \frac{\partial}{\partial P}\left(\frac{P^2 l^3}{6EI}\right) = \frac{Pl^3}{3EI}.$$

An extension of Castigliano's theorem is that the change in slope at any section of a loaded member is the partial derivative of the total strain energy with respect to the couple moment applied at that section, and this is given as $\theta = \partial U / \partial C$, where θ is the slope and C the applied couple.

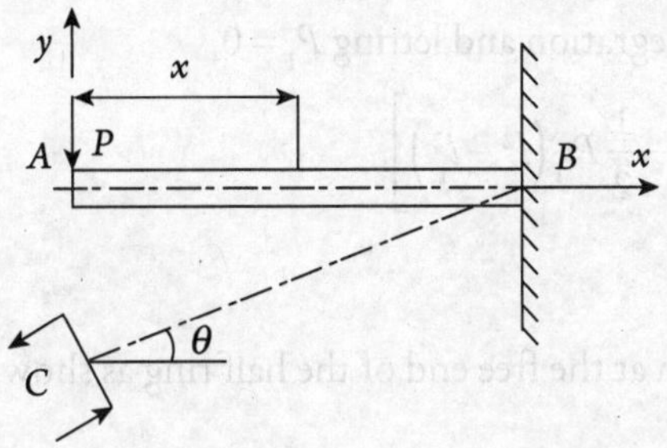

Since there is no couple acting at section A, we apply a fictitious couple C at A. Therefore, the moment at a distance x from the free end in $M_x = Px + C$.

Since $\theta_A = \dfrac{\partial}{\partial C}\left[\int \dfrac{M_x^{\,2} dx}{2EI}\right]$, we may also write

$$\theta_A = \int_0^l \frac{M_x}{EI}\frac{\partial M_x}{\partial C}dx.$$

This gives

$$\theta = \int_0^l \frac{1}{EI}(P\cdot x + C)dx, \text{ since } \frac{\partial M_x}{\partial C} = 1$$

$$= \frac{Pl^2}{2EI} + Cl.$$

Here we let $C = 0$, and this gives $\theta = \dfrac{Pl^2}{2EI}$.

(b) To find the vertical deflection at point A, we apply a vertical fictitious load P_1 at "A".

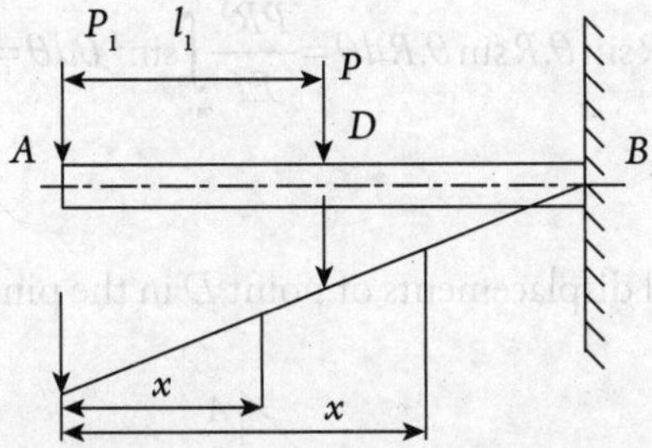

In segment AD, $M_{1x} = P_1 x$. Therefore, $\partial M_{1x} / \partial P_1 = x$.

In segment DB, $M_{2x} = P_1 x + P(x - l_1)$. Therefore, $\partial M_{2x} / \partial P_1 = x$.

If we now write the deflection equation as

$$\delta_A = \int_0^{l_1} \frac{M_{1x}}{EI}\frac{\partial M_{1x}}{\partial P_1}dx + \int_{l_1}^{l} \frac{M_{2x}}{EI}\frac{\partial M_{2x}}{\partial P_1}dx$$

and this gives, after integration and letting $P_1 = 0$,

$$\delta_A = \frac{1}{EI}\left[\frac{1}{3}P\left(l^3 - l_1^3\right) - \frac{1}{2}Pl_1\left(l^2 - l_1^2\right)\right].$$

Example 12.3

Find the horizontal deflection at the free end of the half ring as shown below.

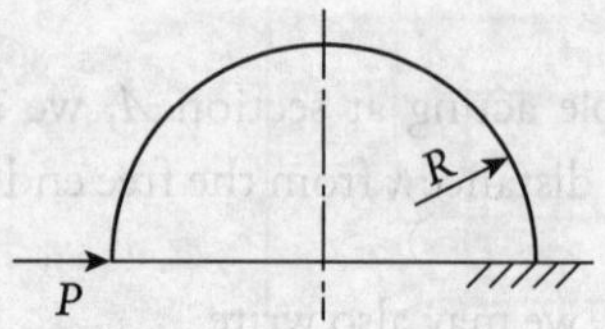

Solution:

Consider a small section of included angle $d\theta$ at angle θ from the horizontal.

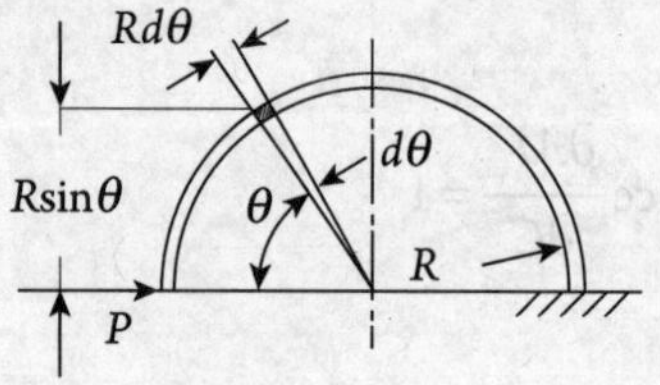

Bending moment at the section, $M = PR\, sin\theta$

$\frac{\partial M}{\partial P} = R sin\theta$ and $dx = Rd\theta$

We write deflection δ as

$$\delta = \int \frac{M}{EI}\frac{\partial M}{\partial P}dx = \frac{1}{EI}\int_0^{\pi} PR\sin\theta.R\sin\theta.Rd\theta = \frac{PR^3}{EI}\int_0^{\pi}\sin^2\theta d\theta = \frac{\pi PR^3}{2EI}.$$

Example 12.4

Find the horizontal and vertical displacements of point D in the pin-connected truss shown in the figure.

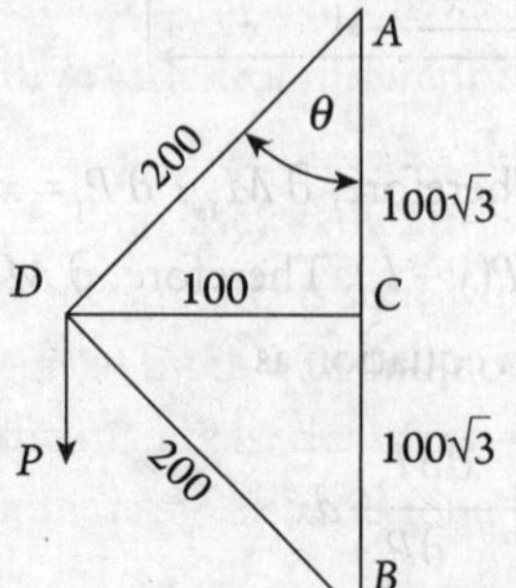

Solution:

Let the vertical and horizontal displacements of point D be δ_1 and δ_2 respectively, as shown in Figures (a) and (b) below:

Extension of AD due to $\delta_1 = \delta_1 \cos\theta$

Extension of AD due to $\delta_2 = \delta_2 \sin\theta$

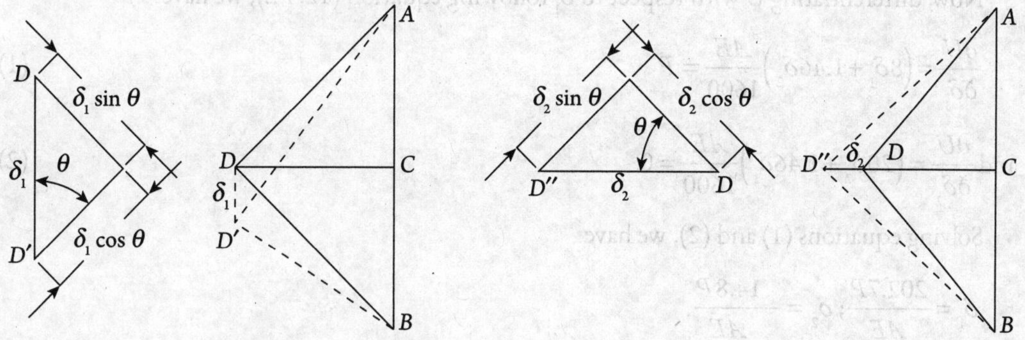

(a) (b)

Extension of BD due to $\delta_2 = \delta_2 \sin\theta$

Contraction of BD due to $\delta_1 = \delta_1 \sin\theta$

Extension of $CD = \delta_2$

This gives:

Total extension of $AD = \delta_1 \cos\theta + \delta_2 \sin\theta = \delta_{AD}$

Total extension of $BD = \delta_2 \sin\theta - \delta_1 \sin\theta = \delta_{BD}$

Total extension of $CD = \delta_2 = \delta_{CD}$

From the geometric configuration of the truss:

$$\cos\theta = \frac{\sqrt{3}}{2};\ \sin\theta = \frac{1}{2}.$$

This gives

$$\delta_{AD} = \frac{1}{2}\left(\delta_2 + \sqrt{3}\delta_1\right)$$

$$\delta_{BD} = \frac{1}{2}\left(\delta_2 - \delta_1\right)$$

$$\delta_{CD} = \delta_2$$

Since $\delta = \dfrac{Pl}{AE}$ and strain energy $U = \dfrac{1}{2}P\delta = \dfrac{AE\delta^2}{2l}$,

where A is the cross-sectional area of the member and l the length and E the elastic modulus of the material.

Total strain energy:

$$U=\frac{AE}{2}\left[\frac{\left(\delta_2+\sqrt{3}\delta_1\right)^2}{800}+\frac{\left(\delta_2-\delta_1\right)^2}{800}+\frac{\delta_2^2}{100}\right]=\left(10\delta_2^2+4\delta_1^2+1.46\delta_1\delta_2\right)\frac{AE}{1600}.$$

Now, differentiating U with respect to δ_1 following equation (12.4.2), we have

$$\frac{\partial U}{\partial \delta_1}=\left(8\delta_1+1.46\delta_2\right)\frac{AE}{1600}=P \quad (1)$$

and $$\frac{\partial U}{\partial \delta_2}=\left(20\delta_2+1.46\delta_1\right)\frac{AE}{1600}=0. \quad (2)$$

Solving equations (1) and (2), we have

$$\delta_1=\frac{202.7P}{AE};\ \delta_2=-\frac{14.8P}{AE}.$$

Example 12.5

Determine the reactions of the statically indeterminate beam, shown below, by the method of least work. Consider that the elastic modulus E and the second moment of inertia I are constant for the beam.

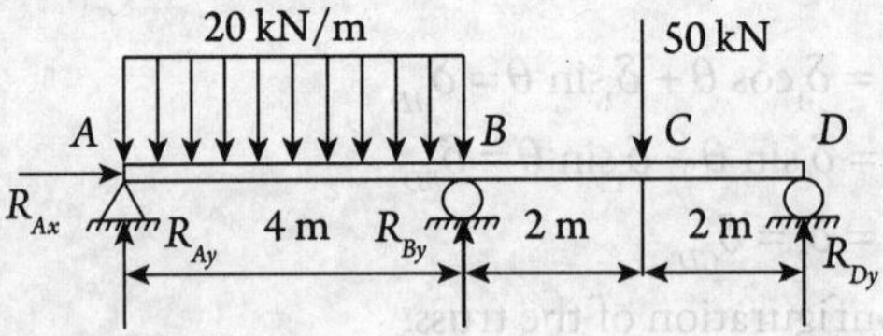

Solution:

The beam is supported by four reactions. Therefore, the degree of indeterminacy of the beam is one since there are only three equations of equilibrium. Taking R_{By} as the redundant reaction, we may proceed as follows:

Here we can write the condition for redundancy as

$$\frac{\partial U}{\partial R_{By}}=\int\frac{M}{EI}\frac{\partial M}{\partial R_{By}}dx=0.$$

Now, applying the force and moment equilibrium equation, we need to find reaction forces at A and D in terms of R_{By}. Accordingly, we obtain

$$R_{Ax}=0$$

$$R_{Ay}=72.5-0.5\times R_{By}$$

$$R_{Dy}=57.5-0.5\times R_{By}.$$

Considering the three segments separately, as shown in the figure below, we obtain the moments as mentioned in the Table.

Segment	*Origin*	*Limits (m)*	*M*	$\frac{\partial M}{\partial R_{By}}$
AB	*A*	0 to 4	$\left(72.5-0.5\times R_{By}\right)x-10x^2$	$-0.5x$
DC	*D*	0 to 2	$\left(57.5-0.5\times R_{By}\right)x$	$-0.5x$
CB	*D*	2 to 4	$7.5x-0.5\times R_{By}\,x+100$	$-0.5x$

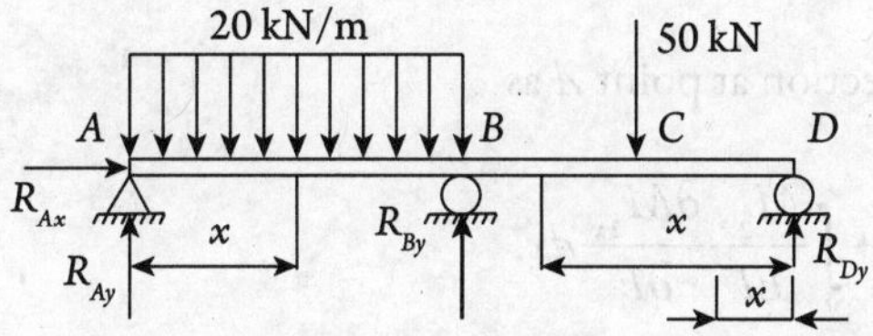

Combining the effects of all three segments, we have

$$\frac{1}{EI}\left[\left\{\int_0^4\left(72.5x-0.5R_{By}\,x-10x^2\right)\times\left(-0.5x\right)dx\right\}+\left\{\int_2^4\left(7.5x-0.5R_{By}\,x+100\right)\times\left(-0.5x\right)dx\right\}\right.$$

$$\left.+\left\{\int_0^2\left(57.5x-0.5R_{By}\,x\right)\times\left(-0.5x\right)dx\right\}\right]=0.$$

This yields $R_{By}=84.375$ kN.

Substituting this, $R_{Ay}=30.3125$ kN and $R_{Dy}=15.3125$ kN.

Example 12.6

A cantilever beam is loaded as shown in the figure below:

Find the deflection at the free end in the direction of the load, using Castigliano's theorem.

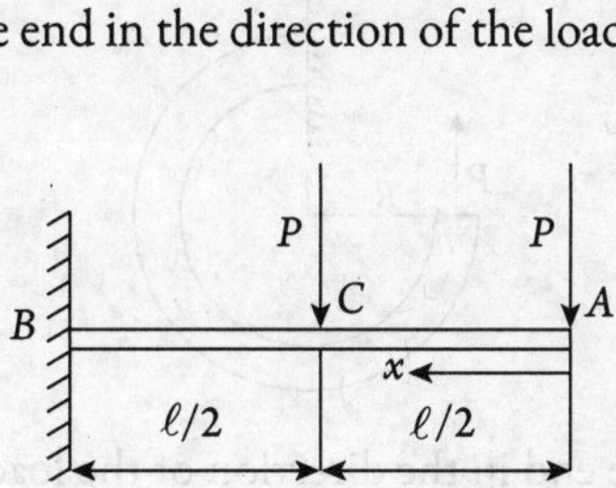

Solution:

To find the deflection at the free end A using Castigliano's theorem, we denote the vertical load P at A by F. So, we have $F = P$.

In section AC

$$M_{1x} = Fx$$

$$\frac{\partial M_{1x}}{\partial F} = x.$$

In section CB

$$M_{2x} = Fx + P\left(x - \frac{l}{2}\right)$$

$$\frac{\partial M_{2x}}{\partial F} = x.$$

We now write the deflection at point A as

$$\delta_A = \int_0^{\frac{l}{2}} \frac{M_{1x}}{EI}\frac{\partial M_{1x}}{\partial F}dx + \int_{\frac{l}{2}}^{l} \frac{M_{2x}}{EI}\frac{\partial M_{2x}}{\partial F}dx.$$

Substituting the moment expressions, we get

$$\delta_A = \frac{1}{EI}\int_0^{\frac{l}{2}} Fx^2 dx + \frac{1}{EI}\int_{\frac{l}{2}}^{l}\left(Fx^2 + Px^2 - \frac{Pl}{2}x\right)dx.$$

After integration and putting $F = P$, we get

$$\delta_A = \frac{7Pl^3}{16EI}.$$

Example 12.7

A curved beam of radius of curvature R is fixed at one end and is loaded, as shown in the figure below:

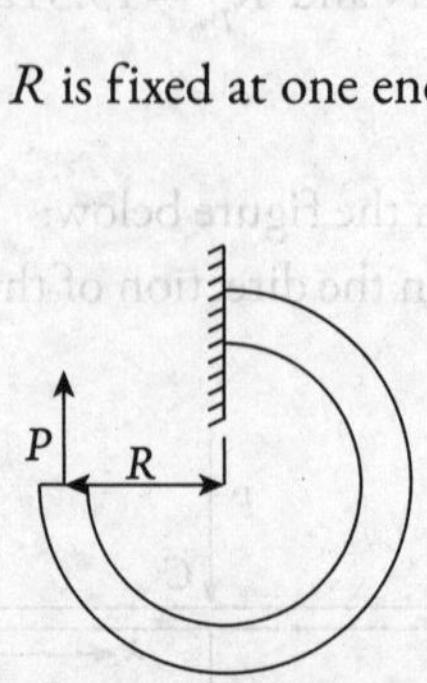

Find the deflection at the free end in the direction of the loading using Castigliano's theorem and considering strain energy due to bending alone.

Solution:

Let us consider a small section of included angle $d\theta$ at angle θ from the horizontal line shown in the Figure passing through the free end.

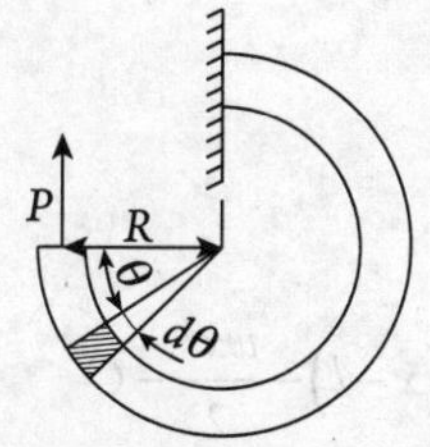

Bending moment at the section is obtained as

$$M = PR(1-\cos\theta).$$

Partial derivative of the bending moment with respect to P is given by

$$\frac{\partial M}{\partial P} = R(1-\cos\theta).$$

Deflection of the curved beam at the free end in the direction of load P is given by

$$\delta = \int \frac{M}{EI}\frac{\partial M}{\partial P}dx = \frac{1}{EI}\int_0^{\frac{3\pi}{2}} PR(1-\cos\theta)R(1-\cos\theta)Rd\theta$$

$$= \frac{PR^3}{EI}\int_0^{\frac{3\pi}{2}} (1-\cos\theta)^2\, d\theta = 9\frac{PR^3}{EI}.$$

Example 12.8

Find the change in slope of the section at "B" in the beam loaded as shown in the figure below. Consider a constant value of the product EI.

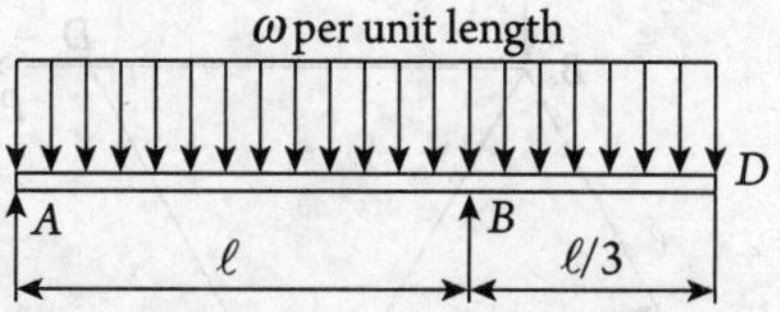

Solution:

Let us consider a fictitious anticlockwise couple C at section B.

The reaction forces at A and B are obtained from equilibrium conditions, as given by

$$R_A = \frac{C}{l} + \frac{4}{9}wl$$

$$R_B = \frac{8}{9}wl - \frac{C}{l}$$

In section AB

$$M_{1x} = \left(\frac{C}{l} + \frac{4}{9}wl\right)x - \frac{wx^2}{2}$$

$$\frac{\partial M_{1x}}{\partial C} = \frac{x}{l}$$

In section BD

$$M_{2x} = \left(\frac{C}{l} + \frac{4}{9}wl\right)x + \left(\frac{8}{9}wl - \frac{C}{l}\right)(x-l) - \frac{wx^2}{2} - C$$

$$\frac{\partial M_{2x}}{\partial C} = \frac{x}{l} - \frac{x-l}{l} - 1 = 0$$

Slope at section B is given by

$$\theta_B = \int_0^l \frac{M_{1x}}{EI}\frac{\partial M_{1x}}{\partial C}dx + \int_l^{\frac{4l}{3}} \frac{M_{2x}}{EI}\frac{\partial M_{2x}}{\partial C}dx$$

$$= \frac{1}{EI}\int_0^l \left\{\left(\frac{C}{l} + \frac{4}{9}wl\right)x - \frac{wx^2}{2}\right\}\frac{x}{l}dx + 0$$

Evaluating the integration and putting $C = 0$, we get

$$\theta_B = \frac{5}{216}\frac{wl^3}{EI}.$$

Example 12.9

Find the horizontal and vertical deflections at point D in the pin-connected truss shown in the figure below:

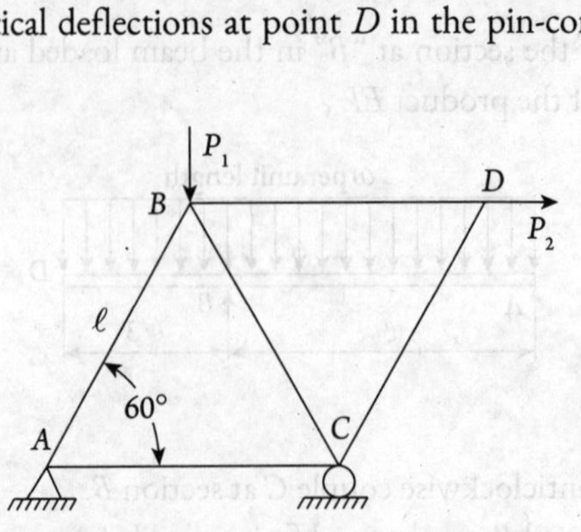

Assume all the members have the same length l, cross-sectional area A, and the same elastic modulus E.

Solution:

Let us consider a vertical fictitious force P_v at point D.

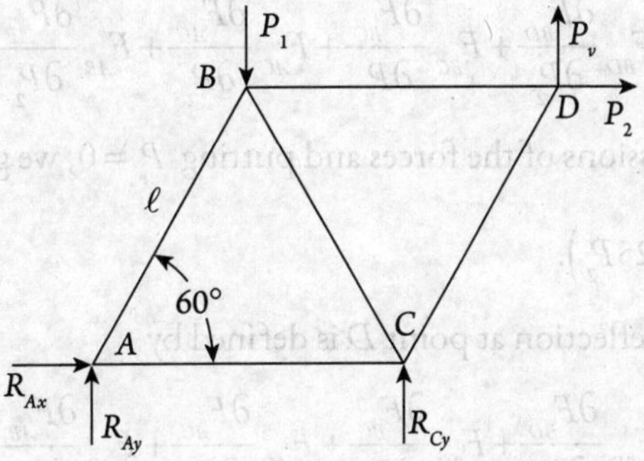

From equilibrium conditions, the reaction forces at supports A and C are obtained as

$$R_{Ax} = -P_2$$

$$R_{Ay} = \frac{P_v}{2} - \frac{\sqrt{3}}{2}P_2 + \frac{P_1}{2}$$

$$R_{Cy} = \frac{P_1}{2} - \frac{3P_v}{2} + \frac{\sqrt{3}}{2}P_2$$

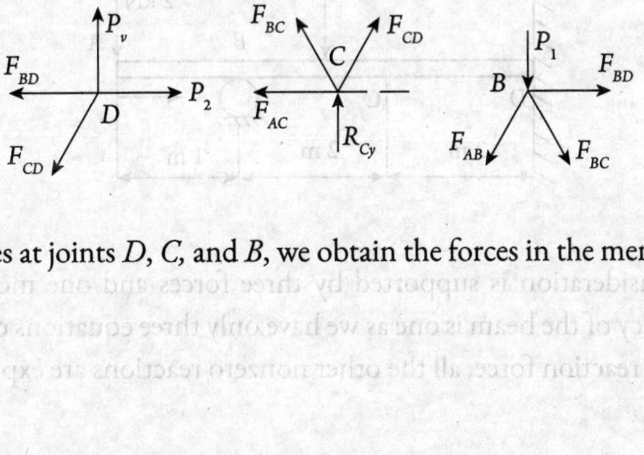

From force analyses at joints D, C, and B, we obtain the forces in the members as given by

$$F_{CD} = \frac{2P_v}{\sqrt{3}}$$

$$F_{BD} = P_2 - \frac{P_v}{\sqrt{3}}$$

$$F_{BC} = -\frac{P_1}{\sqrt{3}} - P_2 + \frac{P_v}{\sqrt{3}}$$

$$F_{AC} = \frac{P_1}{2\sqrt{3}} + \frac{P_2}{2} + \frac{P_v}{2\sqrt{3}}$$

$$F_{AB} = -\frac{P_1}{\sqrt{3}} + P_2 - \frac{P_v}{\sqrt{3}}.$$

The horizontal deflection at point D is obtained by the superposition of component displacements

$$\delta_h = \frac{l}{EA}\left[F_{CD}\frac{\partial F_{CD}}{\partial P_2} + F_{BD}\frac{\partial F_{BD}}{\partial P_2} + F_{BC}\frac{\partial F_{BC}}{\partial P_2} + F_{AC}\frac{\partial F_{AC}}{\partial P_2} + F_{AB}\frac{\partial F_{AB}}{\partial P_2}\right].$$

Substituting the expressions of the forces and putting $P_v = 0$, we get

$$\delta_h = \frac{l}{EA}\left(0.144P_1 + 3.25P_2\right).$$

Similarly, the vertical deflection at point D is defined by

$$\delta_v = \frac{l}{EA}\left[F_{CD}\frac{\partial F_{CD}}{\partial P_v} + F_{BD}\frac{\partial F_{BD}}{\partial P_v} + F_{BC}\frac{\partial F_{BC}}{\partial P_v} + F_{AC}\frac{\partial F_{AC}}{\partial P_v} + F_{AB}\frac{\partial F_{AB}}{\partial P_v}\right].$$

Substituting the expressions of the forces and putting $P_v = 0$, we get

$$\delta_v = \frac{l}{EA}\left(0.083P_1 - 1.587P_2\right).$$

Example 12.10

Find the reactions in the beam shown in the figure below using the least work method. Consider a constant value of the product EI for the beam.

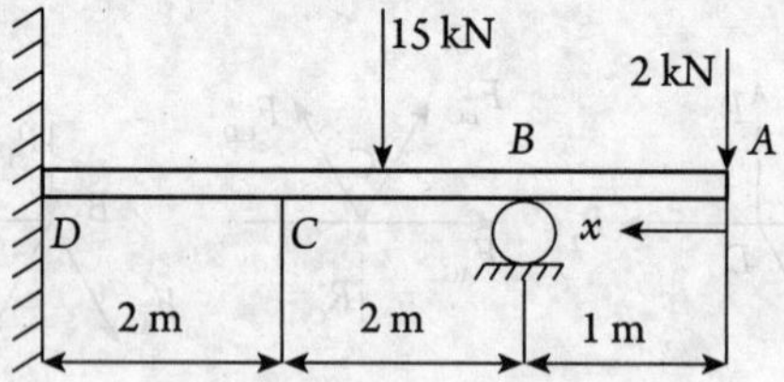

Solution:

The beam under consideration is supported by three forces and one moment. Therefore, the degree of indeterminacy of the beam is one as we have only three equations of equilibrium. Taking R_{By} as the redundant reaction force, all the other nonzero reactions are expressed in terms of R_{By} as given below.

$$R_{Dy} = 17 - R_{By}$$

$$M_D = 40 - 4R_{By}.$$

We can write the condition of redundancy as

$$\frac{\partial U}{\partial R_{By}} = \int \frac{M}{EI}\frac{\partial M}{\partial R_{By}}dx = 0.$$

In section AB

$$M_{1x} = -2x$$

$$\frac{\partial M_{1x}}{\partial R_{By}} = 0.$$

In section BC

$$M_{2x} = R_{By}(x-1) - 2x$$

$$\frac{\partial M_{2x}}{\partial R_{By}} = x-1.$$

In section CD

$$M_{3x} = R_{By}(x-1) - 2x - 15(x-3)$$

$$\frac{\partial M_{3x}}{\partial R_{By}} = x-1.$$

Combining the effects of all three segments of the beam

$$\frac{1}{EI}\left[\int_0^1 0\,dx + \int_1^3 \{R_{By}(x-1)-2x\}(x-1)dx + \int_3^5 \{R_{By}(x-1)-2x-15(x-3)\}(x-1)dx\right] = 0.$$

After performing the integrations and computing, we get $R_{By} = 7.44$ kN.

Putting the value of R_{By}, we get the other reactions as $R_{Dy} = 9.56$ kN and $M_D = 10.24$ kNm.

Exercises

1. (a) Derive formulae for strain energies resulting from tension or compression, bending, and torsion.

   (b) Calculate the strain energy stored in a bar of circular cross-section of diameter 100 mm and 500 mm long, (i) subjected to a tensile load of 10 kN, (ii) a uniform bending moment of 5 kNm, and (iii) a torque of 10 kNm.
   Take E and G for the material to be 200 GPa and 80 GPa.

   [Ans.: 0.016 Nm, 6.369 Nm, 31.847 Nm]

2. (a) Discuss the reciprocal theorem on displacements in detail.

   (b) State Castigliano's first theorem and show how the reciprocal theorem can be used to prove Castigliano's theorem.

3. (a) Discuss the dummy load method for finding displacement or rotation at a point where no concentrated force or moment acts.

   (b) Derive the equation for the slope at the free end of a cantilever beam of length l, carrying a uniformly distributed load of w per unit length. Assume that the product EI is a constant.

   [$\theta = \frac{1}{EI}\int_0^l \left(M_1 - \frac{wx^2}{2}\right)dx$; M_1 is a fictitious moment.]

4. Determine the reactions R_{cy} and R_{cx} at point C and also the displacement at point B in the member ABC loaded, as shown in Figure. Take E = 200 GPa, R = 100 mm.

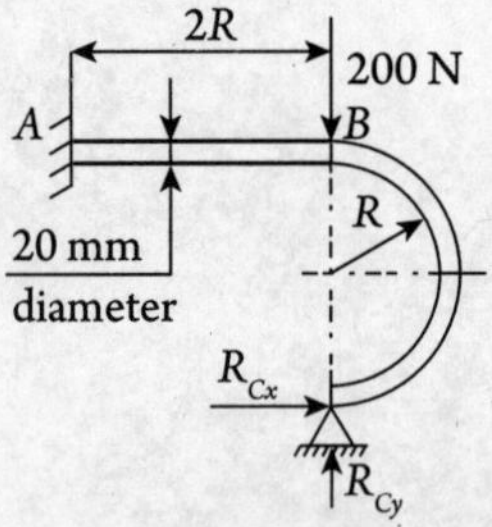

[Ans.: 103.85 N, 46.6 N, 44 μm]

5. A steel ball of 2 kg is dropped from a height of 1 m over the edge of a steel cantilever beam, as shown in the figure. Find the deflection of the beam at the edge where the ball strikes. Consider the strain energy due to both bending and shear. Take E = 200 GPa and G = 80 GPa.

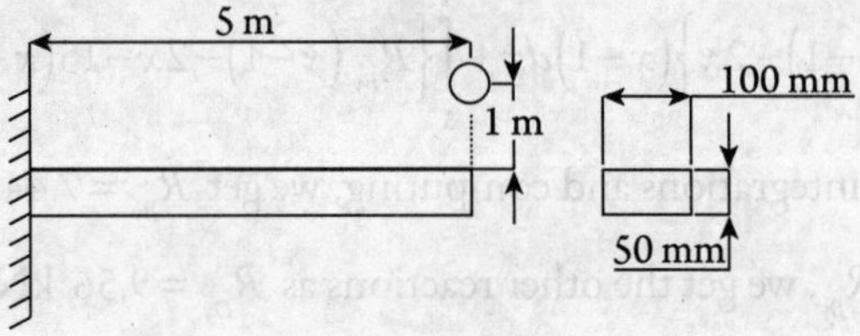

[Ans.: 0.089 m]

6. The cantilever beam, as shown below, is supported by a spring that is compressed by 3 mm. Find the spring force R at the free end of the beam.

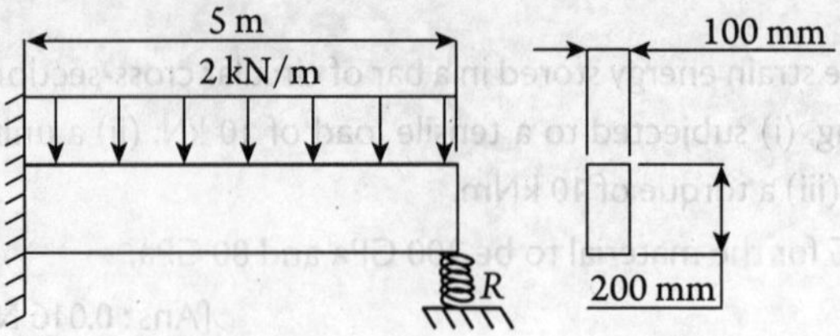

[Ans.: 4.71 kN]

Selected Bibliography

Ameen, M. (2005). *Computational Elasticity*. New Delhi: Narosa Publishing House.

Archer, R. R, N. H. Cook, S. H. Cradnall, N. C. Dahl, T. J. Lardner, F. A. Mclintock, E. Rabinowicz, and G. S. Reichenbach (1978). *An Introduction to the Mechanics of Solids*, 2nd edition. New York: McGraw-Hill.

Boresi, A. P., and K. P. Chong (2000). *Elasticity in Engineering Mechanics*, 2nd edition. New York: Wiley Interscience.

Boresi, A. P., and R. J. Schmidt (2014). *Advanced Mechanics of Materials*, 6th edition. New Jersey: Wiley.

Boussinesq, J. (1885). *Application dès potentiels a l'étude de l'équilibre et du mouvement des solides élastiques*. Paris: Gauthier-Villars.

Castigliano, A., and E. S. Andrews (1966). *The Theory of Equilibrium of Elastic Systems and Its Applications*. New York: Dover Publication Inc.

Den Hartog, J. P. (1987). *Advanced Strength of Materials*. New York: Dover Publications Inc.

Dugas, R. (1988). *A History of Mechanics*. New York: Dover Publication Inc.

Durelli, A. J., E. P. Phillips, and C. J. Tsao (1958). *Introduction to the Theoretical and Experimental Analysis of Stress and Strain*. New York: McGraw-Hill.

Fung, Y. C. (1965). *Foundation of Solid Mechanics*. New Jersey: Prentice Hall, Englewood Cliffs.

Gere, J. (2001). *Mechanics of Materials*, 5th edition. Pacific Grove CA: Brooks/Cole.

Green, A. E., and W. Zerna (1968). *Theoretical Elasticity*. Oxford: Oxford University Press.

Griffith, A. A., and G. I. Taylor (1917). The use of soap films in solving torsion problems. *Proceedings of the Institution of Mechanical Engineers* 93(1):755–809. doi:10.1243/PIME_PROC_1917_093_015_02.

Hearn, E. J. (1997). *Mechanics of Materials*, 2 vols, 3rd edition. Oxford, UK: Butterworth–Heinemann.

Hertz, H. (1881). On the contact of elastic solids. *J. Reineangew Math*. 92: 56.

——— (1896). *Miscellaneous Papers*. New York: Macmillan.

Inglis, C. E. (1913). Stresses in a plate due to the presence of cracks or sharp corners. *Transaction of the Institute of Naval Architects (British)*. Vol. LX(Pt. I): p.219.

Jacobson, L. S. (1925). Torsional stress concentration in shafts of circular cross-section and variable diameter. *Trans. Am. Soc. Mech. Eng*. 47: 619.

Jeffreys, H. (1957). *Cartesian Tensor*. London: Cambridge University Press.

Johnson, K. L. (1985). *Contact Mechanics*. Cambridge: Cambridge University Press.

Johnson, K. L. (1982). One hundred years of Hertz contact. *Proceedings of the Institution of Mechanical Engineers* 196(1): 363–378. doi:10.1243/PIME_PROC_1982_196_039_02.

Kazimi, S. M. A. (1974). *Solid Mechanics*. New Delhi: Tata McGraw-Hill.

Kreyszig, E. (1999). *Advanced Engineering Mathematics*, 8th edition. New York: John Wiley.

Lamb, H. (1902). On Boussinesq's problem. *Proc. London Math. Soc.* 34: 276.

Langhaar, H. L. (1989). *Energy Methods in Applied Mechanics*. Malabar, FL: Krieger.

Love, A. E. H. (1929). The stress produced in a semi-infinite solid by pressure on part of the boundary. *Phil. Trans. Roy. Soc. (London), Ser. A*. 228: 377–420.

Love. A. E.H. (1963). *Mathematical Theory of Elasticity*. New York: Dover Publications.

Malvern L. E. (1969). *Introduction to the Mechanics of Continuous Medium*. New Jersey: Prentice Hall.

Mindlin, R. D. (1949). Compliance of elastic bodies in contact. *J. Appl. Mech.* 16(3): 259.

Morley, A. (1935). *Strength of Materials*, 8th edition. London: Longmans Green.

Muki, R. (1960). Asymmetric problems of the theory of elasticity for a semi-infinite solid and a thick plate. In: *Progress in Solid Mechanics*, ed. I. N. Sneddon and R. Hill. North Holland, Amsterdam. pp. 399–439.

Nadai, A. (1950). *Theory of Flow and Fracture of Solids*, vol. 1. New York: McGraw-Hill.

Oliveira, A. R. E. (2000). A history of ancient mechanics: From Aristotle to Pappus. In: *International Symposium on History of Machines and Mechanisms Proceedings,* HMM 2000, ed. M. Ceccarelli. Dordrecht: Springer. https://doi.org/10.1007/978-94-015-9554-4_29.

Popov, E. P. (1965). *Mechanics of Materials*. New Delhi: Prentice Hall of India.

Prandtl, L. (1903). Zur torsion von prismatischenstäben. *Phys. Z.* 4: 758–770.

Sadd, M. H. (2021). *Elasticity: Theory, Applications and Numerics*, London: Academic Press.

Saint-Venant. (1856). *Mémoiresur la torsion des prismes, avec des considérationssurleur flexion ainsiquesurl'équilibreintérieur des solidesélastiques en général, et des formulespratiques pour le calcul de leurrésistance à divers efforts s'exerçantsimultanément* [Imprimerienationale], Paris.

Seely, F. B., and Smith, J. O. (1952). *Advanced Mechanics of Materials*, 2nd edition. New York: John Wiley & Sons Inc.

Shiegley, J. E. (1977). *Mechanical Engineering Design*, 3rd edition. New York: McGraw Hill.

Smith, J. O., and C. K. Liu (1953). Stresses due to tangential and normal loads on an elastic solid with application to some contact stress problems. *J. Appl. Mech.* 20(2): 157–166. doi: https://doi.org/10.1115/1.4010643.

Sneddon, I. N., and R. Hill (eds.) (1960). *Progress in Solid Mechanics*. Amsterdam: North-Holland Publish Co.

Southwell, R. V. (1941). *An Introduction to the Theory of Elasticity for Engineers and Physicists*. Oxford: Oxford University Press.

Sternberg, E. (1954). On Saint Venant's principle. *Quar. Jour. Appl. Math.* 11: 393–402.

Stodola, A. (1981). *Steam and Gas Turbine*. New York: Peter Smith Publisher Inc.

Timoshenko, S. P. (1941). *Strength of Materials*, 2 vols. Princeton, New Jersey: D. Van Nostrand Co. Inc.

Timoshenko, S. P. and J. H. Goodier (1970). *Theory of Elasticity*. New York: McGraw-Hill.

Toupin, R. A. (1945). Saint Venant's principle. *Arch. Rat. Mech. Anal.* 18: 83–96.

Von Misses, R. (1945). On Saint Venant's principle. *Bull. Amer. Math.* 51: 555–562.

Wang, C. T. (1953). *Applied Elasticity*. New York: McGraw-Hill.

Index